X Meeting on Recent Advances in the Physics of Fluids and their Applications 2008

Journal of Physics: Conference Series Volume 166

Santa Fe, Argentina
19-21 November 2008

ISBN: 978-1-61738-317-5

Printed from e-media with permission by:

Curran Associates, Inc.
57 Morehouse Lane
Red Hook, NY 12571
www.proceedings.com

Some format issues inherent in the e-media version may also appear in this print version.

Printed by Curran Associates, Inc. (2010)

Institute of Physics
Dirac House, Temple Back
Bristol BS1 6BE UK

Phone: 44 1 17 929 7481
Fax: 44 1 17 920 0979

techtracking@iop.org

X Meeting on Recent Advances in the Physics of Fluids and their Applications 2008

Journal of Physics: Conference Series Volume 166

Santa Fe, Argentina
19-21 November 2008

ISBN: 978-1-61738-317-5

IOP Institute of Physics

Accepted papers received: 2 April 2009
Published online: 1 June 2009

PREFACE

PAPERS

The X Meeting on Recent Advances in Physics of Fluids and Related Applications (Fluids 2008) was held in Santa Fe, Argentina, on 19–21 November 2008. It belongs to a series of meetings that started in 1989 and has continued – except for just one occasion – every other year. Thus, the first meeting took place in the city of Tandil in 1989 followed by three events in the city of La Plata (1991-93-95), Tunuyán (Mendoza) in 1997, Paraná (Entre Rios) in 1999, Buenos Aires in 2001, Tandil in 2003 and Mendoza in 2006.

These meetings gather together most of the people working in Fluid Mechanics and related problems in Argentina. The objective of the meetings is to provide a forum to facilitate the interactions between participants in a friendly academic atmosphere. This goal is achieved by means of lectures and technical presentations on different subjects and from different points of view, the only constraint being the current academic/technical interest. Applications usually deal with problems of local interest.

In the present meeting a variety of lecture topics were presented, among them we might mention Capillary Hydrodynamics, Wetting, Density Currents, Instabilities, Elastic-Dynamics, Flows in Porous Media, Sediment Transport, Plasma Dynamics, etc. In particular, we would like to highlight the specially invited lectures given by Dr Ramon Cerro (Chemical and Material Engineering Department University of Alabama in Huntsville, USA), Dr David Quéré (Physique et Mécanique des Milieux Hétérogènes ESPCI, FRANCE), Dr Marcelo García (College of Engineering University of Illinois at Urbana-Champaign) and Dr Víctor Calo (Earth and Environmental Science and Engineering, King Abdullah University of Science and Technology (KAUST) and Institute for Computational Engineering and Sciences (ICES), University of Texas at Austin). In addition, we had 18 invited talks and more than fifty contributions that were presented in poster sessions.

On behalf of both the Honorary Committee and the Local Committee, we would like to thank the Institutions that sponsored the Meeting, namely: the National Scientific Research Council of Argentina (CONICET) and the National Agency for Scientific and Technology Promotion (ANPCyT), who supported the meeting with grants-in-aid; and the Universidad Nacional del Litoral, who provided all the logistic support needed to achieve a successful and enjoyable Conference. All the papers constituting this volume of *Journal of Physics: Conference Series* have been reviewed by experts in the field; though they are just a limited number of the works presented at the Meeting the volume provides a scientific record of the topics discussed. We hope that the readers will enjoy the reading.

Fernando Adolfo Saita and María Delia Giavedoni

Flexible fiber transport by a fluid flow in fractures with smooth and rough walls.

M V D'Angelo[1,2], H Auradou[2], G Picard[3], M Poitzsch[3] and J P Hulin[2]

[1] Grupo de Medios Porosos, Facultad de Ingeniería, Universidad de Buenos Aires, Paseo Colón 850 C1063 ACV, Buenos Aires, Argentina, CONICET

[2] Univ Pierre et Marie Curie-Paris 6, Univ Paris-Sud, CNRS, Lab FAST, Bat 502, Campus Univ, Orsay, F-91405, France

[3] Schlumberger-Doll Research, 1 Hampshire St, Cambridge, MA 02139, USA.

E-mail: vdangelo@fi.uba.ar

E-mail: auradou@fast.u-psud.fr

Abstract. The transport of flexible fibers by a flowing fluid has been studied experimentally in transparent model fractures. Both finite length segments ($20\,\mathrm{mm} \leq l \leq 150\,\mathrm{mm}$) and continuous fibers penetrating freely into the model were used; their motion is monitored by means of a digital camera and of an image thresholding technique and is induced by the flow of water or of a polymer solution at a mean velocity U ($50 \leq U \leq 400\,\mathrm{mm.s^{-1}}$). In a model with plane smooth parallel walls, the influence of the friction with the walls is small: fiber segments reach quickly a constant velocity as their distance x_m to the inlet increases, the velocity of continuous fibers increases slower with distance before reaching a constant value. The second model fracture has two complementary rough self-affine walls with a relative lateral shift; it displays, in addition, a gradient of the aperture in the direction transverse to the mean flow. For this model, the transport of the fibers by flowing water is only possible in the region of largest aperture and is of a "stop and go" type at low velocities. If water is replaced by the shear thinning polymer solution, the fibers move faster and continuously in high aperture regions; fiber transport also becomes possible in narrower regions.

1. Introduction

Understanding the dynamics of fiber transport by flowing fluids is relevant to many fields of physics, biology and engineering. Examples include paper pulp [1], fiber-reinforced composites [2], the dynamics and rheology of biological polymers [3] and the motility of microscopic organisms [4, 5]. Recently, in-situ measurements on natural water flows by means of long optical fibers have been suggested [6]. Yet, relatively little work has been devoted to flow channels with rough walls such as fractures of natural rocks or of materials of major industrial interest in civil, environmental and petroleum engineering. In that case, the interactions of the fibers with the walls are particularly strong and may stop their motion and/or lead to a blockage of the flow channels. In addition to these applications, such processes raise important fundamental questions on the motion of flexible solid bodies in complex flow fields, particularly regarding the coupling between their mechanical deformations and the flow.

The present paper reports laboratory experiments performed on a model of a single fracture and investigating the transport of a single fiber by a fluid flow. Two experimental fracture models have been considered: the first one has two plane parallel flat walls and the roughness of the walls of the second one reproduces that of natural fractures. In this second model, the roughness and the relative position of the walls creates long range correlations in the aperture field; also, the interaction of the fiber with the walls becomes very strong and may block the motion of the fiber. Several factors may influence this blockage: it would seem that, to ease the motion in the fracture, the fiber would need to be very thin and flexible in order to pass through local constrictions and to remain in the central part of the gap where the velocity is highest. However, additional factors may prevent the fiber from following the streamlines and lead to its trapping. One needs in particular to take into account tension forces (*i.e.* the mechanical cohesion of the fiber along its length), elastic forces (non-zero stiffness of the fibers), and, also, the rheology of the fluids which influence the hydrodynamic forces on the fibers. The objective of the experiments reported here is to analyze the transport (and/or the trapping) of a fiber in a smooth and a rough model fracture and their dependence on these different parameters.

2. Experimental set-up and procedure

The fiber used in this work is a commercial Polyester fine thread of radius varying from 110 to 170μm and of the type used for needlework. The thread is made of two filaments twisted together. The density of the fiber is $\rho = 1.8 \pm 0.1\,\mathrm{g.cm}^{-3}$ and its bending modulus is of the order of 10^3 Pa [7].

Both finite length and continuous fibers have been used. In the first case, fiber segments of length $20\,\mathrm{mm} \leq l \leq 150\,\mathrm{mm}$ are cut out of longer samples. The continuous fibers are longer than the fracture and are dragged into the fracture from its upper side by hydrodynamic forces and by their own weight during the experiments.

The model fractures are obtained by milling with a computer controlled machine two plexiglas plates which are then clamped together with, between them, a gap saturated by the flowing fluid. A detailed description of the procedure is given in ref. [8]. Two fracture models were machined. The first one, referred to in the following as the smooth walled fracture, has two plane walls at a fixed distance $a(x, y) = a_0 = 0.65\,\mathrm{mm}$. The second model has rough self-affine walls and is referred as the rough fracture in the rest of the paper. There is a small angle between the mean planes of these rough walls: as a result the mean distance between them is constant in the direction x of the mean flow $\vec{U}$ but varies linearly in the transverse direction y. The minimum aperture (averaged along x) is $a_x(y = 0) = 0.8\,\mathrm{mm}$ while the maximum one is reached on the opposite side with $a_x(y = w) = 1.1\,\mathrm{mm}$ where $w = 90\ mm$ is the width of the cell. The average of the aperture over the full model surface is $\overline{a}_{x,y} = 0.91\,\mathrm{mm}$. This wedge-like shape of the fracture reproduces aperture gradients often encountered near the edge of natural fractures.

The model rough fracture surfaces used in this work have statistical properties similar to those of real rocks, *i.e.* they are self-affine with and the amplitude of the roughness is of the order of 25 mm [9, 10]. The two walls have complementary geometries and can be brought in contact over their full area. In the model, both a normal relative displacement equal to $a_x(y)$ and a lateral shear displacement $\delta = 0.65\,\mathrm{mm}$ normal to $\vec{U}$ (*i.e.* $\parallel y$) have been introduced with respect to the contact position. The shear displacement introduces a random disorder of the aperture field with a standard deviation $\sigma_a = 0.127\,\mathrm{mm}$.

The models are held vertically and saturated with a fluid; a vertical downward flow is established by sucking this fluid at the bottom side through a leak-tight injector part distributing evenly the flow by means of a gear pump. In order to establish a continuous permanent flow, the pumped fluid is reinjected into a rectangular bath attached onto the upper side of the model and connected to the fracture gap through a 5 mm wide slot. The mean flow velocity is $50 \leq U \leq 400\,\mathrm{mm.s}^{-1}$. For water, this corresponds to a range of Reynolds numbers $R_e = U\overline{a}\rho/\mu$: $40 \leq Re \leq 320$ for which substantial inertial effects are expected [11, 12], particularly for rough fractures. The fibers are injected into the model through the free surface of the bath: in order to ease their insertion, the top part of the fracture gap has a Y-shaped section with an aperture increasing upward and acting like a funnel for the fiber.

The back of the models is illuminated by a light panel and a digital camera allows one to record images of the moving fiber at a rate of 30 frames per second. Each image has 1024 × 768 pixels; it corresponds to a field of view of vertical length 150 mm with its upper boundary located at 110 mm below the upper side of the model and of width corresponding to that of the fracture. Finally, the location of the fiber is determined for each picture by a binary thresholding technique. Quantitatively, the velocity of the fibers is characterized by the time variation of the distance x_m of their center of mass (for segments) or x_t of their tip (for continuous fibers) to the inlet of the model. Unless otherwise mentioned, dx_m/dt or dx_t/dt are referred to as the fiber velocity V_f.

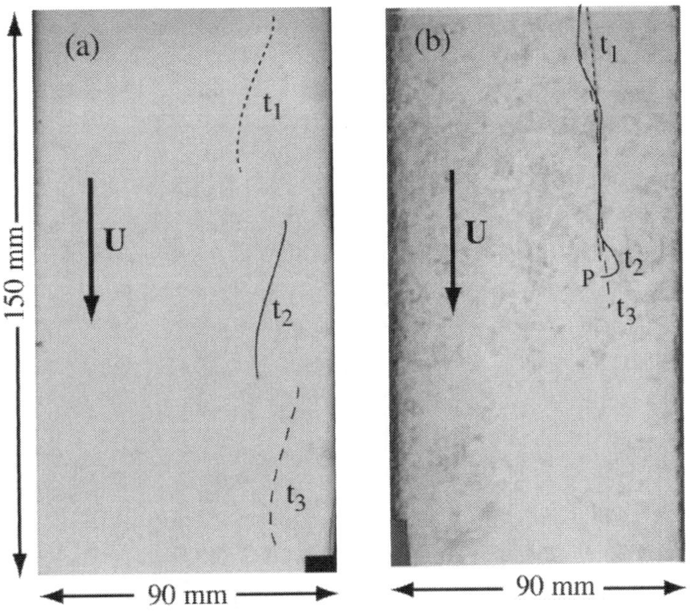

Figure 1. Experimental pictures of the motion of fibers in two model fractures at successive times $t_1 < t_2 < t_3$. (a): 50 mm long fiber segment moving in the fracture with flat parallel walls. (b) Transient pinning event at a point P for a continuous fiber in the fracture with rough walls.

The Newtonian fluid used in the experiments is high purity water (Millipore - Milli-Q grade) with a density close to $\rho = 1\,\mathrm{g.cm}^{-3}$ and a dynamic viscosity of 10^{-3} Pa.s. The shear thinning fluid is a solution of concentration $C_p = 1000$ ppm of high molecular weight Scleroglucan (Sanofi Bioindustries) in water. The variation of its effective viscosity μ with the shear rate $\dot{\gamma}$ (see Figure 2 of ref. [13]) is well adjusted by the Carreau function:

$$\mu = \frac{1}{(1 + (\frac{\dot{\gamma}}{\dot{\gamma}_0})^2)^{\frac{1-n}{2}}}(\mu_0 - \mu_\infty) + \mu_\infty. \tag{1}$$

In Eq. (1), the parameter $\dot{\gamma}_0$ corresponds to the transition between a "Newtonian plateau" domain at low shear rates ($\dot{\gamma} < \dot{\gamma}_0$) in which $\mu = \mu_0$ and a domain in which μ decreases with $\dot{\gamma}$ following the power law $\mu \propto \dot{\gamma}^{(n-1)}$ ($\dot{\gamma} > \dot{\gamma}_0$). For the present solution, rheological measurements give $n = 0.26$, $\dot{\gamma}_0 = 26 \pm 4 \times 10^{-3}\,\mathrm{s}^{-1}$ and μ_∞ is taken equal to the viscosity of the solvent, *i.e. water* with $\mu_\infty = 10^{-3}$ Pa.s. The corresponding transition value of $\dot{\gamma}$ may be estimated by approximating the rheological curve at high shear rates by a truncated power law, leading to $\dot{\gamma}_\infty \sim 2300\,\mathrm{s}^{-1}$ [14].

The maximum value of the shear rate is reached at the walls and is equal to $\dot{\gamma}_w = 6U/a$ for a Poiseuille Newtonian flow (the actual value will be higher because of the shear thinning properties of the fluid at intermediate shear rates). In the present experiments for which $50 \leq U \leq 400\,\mathrm{mm.s}^{-1}$, this leads to the rough estimation $300 \leq \dot{\gamma}_w \leq 2400\mathrm{s}^{-1}$: therefore, at the walls, $\dot{\gamma}_w$ is much larger than $\dot{\gamma}_o$ and may become of the order of $\dot{\gamma}_\infty$. Since, in the center of the pipe, $\dot{\gamma}$ always cancel out, we conclude that, in the upper range of velocities U of the present experiments, both limiting effective viscosities μ_∞ and μ_o are reached respectively at the walls and in the center of the pipe with a "shear thinning" region in between.

3. Experimental results

3.1. Fiber transport by water in the fracture with smooth walls

A first series of experiment was realized in the fracture with flat walls, considered as a reference case. In a first step, the influence of the length and the flow velocity on the transport of fiber segments has been studied. Fibers with four different lengths $20 \leq l \leq 150\,\mathrm{mm}$ have been used and 20 measurements of the fiber velocity V_f were performed for each value of l at velocities $50 \leq U \leq 400\,\mathrm{mm.s}^{-1}$. Only experiments in which the angle of the fibers with respect to $\vec{U}$ is less than 10° are retained for the analysis. In all these cases, the fibers have a constant velocity inside the field of view and do not get deformed during their motion (see Fig. 1a). Figure 2

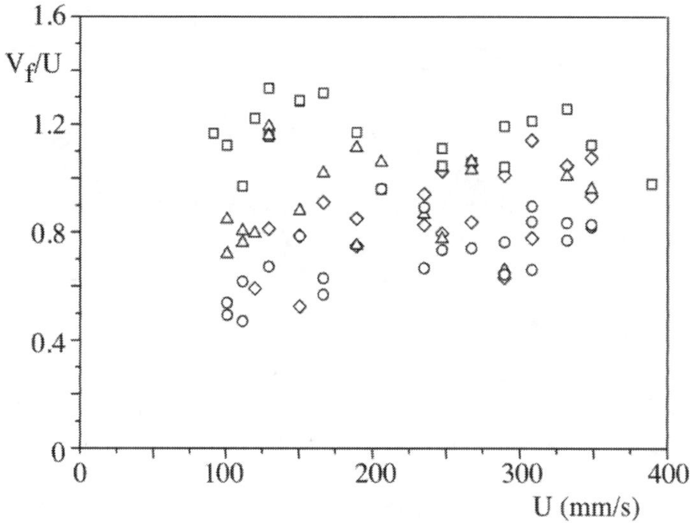

Figure 2. Variation of the normalized velocity V_f/U of the center of mass of fiber segments as a function of the mean fluid velocity U for different lengths l of the fibers. □: $l = 20\,\mathrm{mm}$, △: $l = 50\,\mathrm{mm}$, ◇: $l = 100\,\mathrm{mm}$, ○: $l = 150\,\mathrm{mm}$.

displays the variation of the normalized fiber velocity V_f/U as a function of U for different values of the length l. The dispersion of these experimental values may reflect a variability of the initial injection and, also, the natural curvature of the fiber. For a given length l, the distribution of the values of V_f/U becomes narrower at high velocities; also, there is no clear trend for the variation of V_f/U with U, particularly if one takes into account the broad range of values of the U (almost a decade). At most, V_f/U increases slightly with U for the longer fibers and decreases for the shorter ones: this latter trend is also visible for $l = 50\,\mathrm{mm}$ in Fig. 3 (open symbols).

At the highest velocities, V_f/U is of the order of 1 ($0.9 \leq V_f/U \leq 1.2$): this implies that the relative velocity of the fluid and the fiber is small and that the fiber remains in the center part of the gap.

Overall, V_f/U also decreases for longer fibers, particularly at low velocities (nearly a factor of 2 in this case): long fibers will more often than shorter ones be curved or, even, display meanders. They are then more likely to be in the lower velocity zones close to the walls over, at least, a part of their length. Friction with the walls may also increase and recirculation inside meanders may dissipate energy.

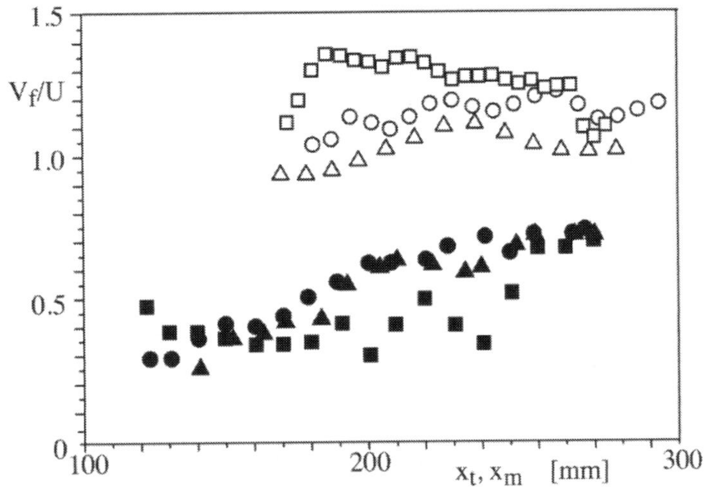

Figure 3. Full symbols: variation of the normalized velocities V_f/U of the tip of the continuous fibers as a function of x_t : (■) $U = 110\,\mathrm{mm.s^{-1}}$, (•) $U = 190\,\mathrm{mm.s^{-1}}$, (▲) $U = 290\,\mathrm{mm.s^{-1}}$. Open symbols: variation of the normalized velocity V_f/U of the center of mass of fiber segments ($l = 50\,\mathrm{mm}$) as a function of x_m: (□) $U = 110\,\mathrm{mm.s^{-1}}$, (∘) $U = 190\,\mathrm{mm.s^{-1}}$, (△) $U = 290\,\mathrm{mm.s^{-1}}$.

A second series of experiments has been performed using a continuous long fiber sucked into the model by hydrodynamic forces at the inlet side with no external applied force. The variations of the corresponding values of $V_f/U = (1/U)dx_t/dt$ with the distance x_t are compared in Figure 3 to the corresponding values of V_f/U for a 50mm long segment plotted as a function of x_m. While the velocity of the segments is globally independent of distance x_m (open symbols), the normalized velocity of the continuous fiber clearly increases with the distance x_t from the inlet (full symbols) before leveling off for $v_t \sim 250\,\mathrm{mm}$. The variations are identical for the two highest mean velocities. This variation may be interpreted as reflecting the larger local aperture both in the upper part of the model (of height equal to 52 mm) which has an Y-shaped section and in the upper bath. This higher aperture reduces the local velocity and, therefore, the drag force on the fiber: the velocity dx_t/dt corresponds then to an average of the mean velocities over the full distance from the free surface down to the tip. This explains why the normalized velocity V_f/U of the tip of the continuous fiber is at most 0.7, *i.e.* significantly lower than that of short segments. This effect should be independent on the velocity provided the fiber retains the same shape and remains at the same distance from the walls in the gap.

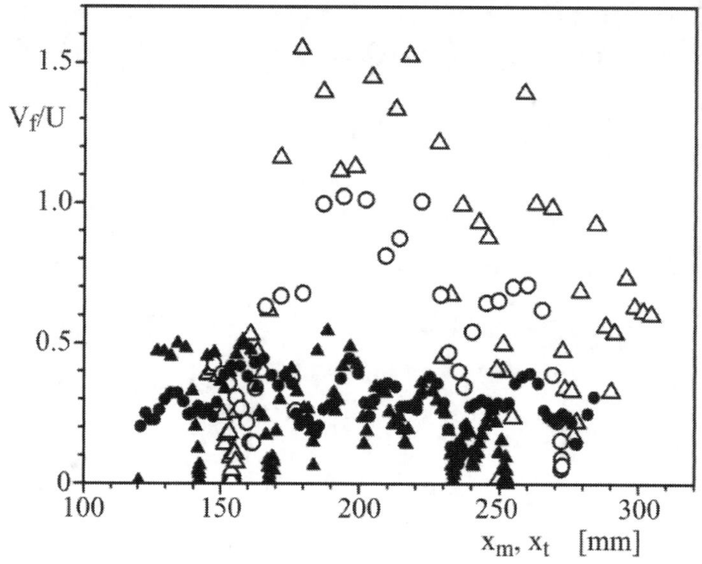

Figure 4. Variation of the normalized fiber velocity V_f/U in a water flow as a function of the distance x_m (fiber segments) or x_t (continuous fibers) from the inlet of the rough model. Continuous fibers: (•) $U = 165\,\mathrm{mm.s^{-1}}$, (▲) $U = 220\,\mathrm{mm.s^{-1}}$. Fiber segments ($l = 50\,\mathrm{mm}$): (○) $U = 165\,\mathrm{mm.s^{-1}}$, (△) $U = 220\,\mathrm{mm.s^{-1}}$.

3.2. Fiber transport by water in the rough fracture

As mentioned above, the rough model has a wedge-like section transverse to the flow and we observed that the fiber could penetrate deeply into the fracture only on the most open third of the width. Figure 4 displays the normalized velocity of continuous fiber and of a fiber segment ($l = 50\,\mathrm{mm}$) as a function of distance along their path for two different mean velocities U; the initial location of the fiber at the inlet is the same for all 4 experiments.

Compared to the smooth flat fracture, a striking feature is the very large fluctuations of V_f/U with the distance. Also, the mean velocity is significantly larger for the segment of fiber than for the continuous one. Moreover, the mean velocity of the continuous fiber does not increase steadily with distance as it does for the smooth fracture. Several pinning events were observed for both types of fibers, particularly at lower mean velocities U: they remained stopped during some time before getting released with a sharp jump of the velocity (see Fig. 1b). For different U values, a given fiber slows down at the same locations but, at the higher U's, it does not stop before accelerating again.

Figure 5 displays the variations of the geometry of a continuous fiber during a pinning event. While the tip of the fiber remains motionless, it displays first a global curvature and then two meanders with their curvatures in opposite directions. The local curvatures increase with time which results in an accumulation of bending elastic energy. When this energy becomes large enough, depinning occurs and the fiber pursues its progression in the model fracture.

3.3. Fiber transport by a polymer solution in the rough fracture

The influence of the fluid rheology on these results has been studied by performing experiments with the same continuous fibers as above but in a flow of the polymer solution with the rheological properties discussed in Sec. 2. Since its effective viscosity will be larger than that of water, except near the walls, one may expect that the hydrodynamic forces will be significantly enhanced.

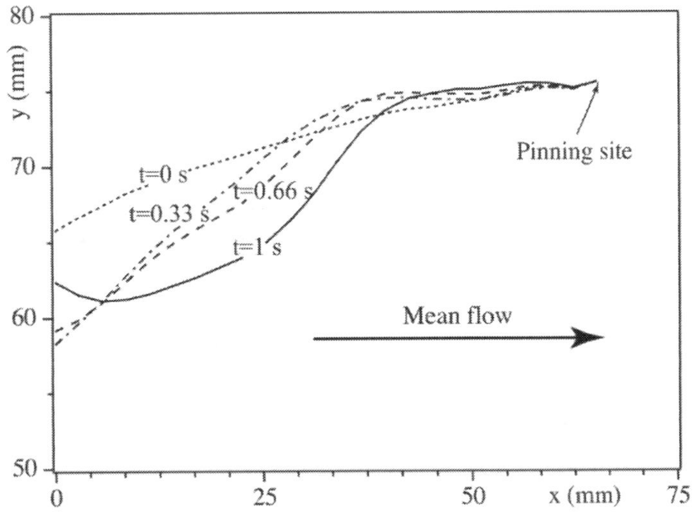

Figure 5. Time variation of the shape of a fiber during a pinning event in a water flow with $U = 230\,\mathrm{mm.s^{-1}}$. Time elapsed after pinning: $t = 0\,\mathrm{s}$ (dots), $t = 0.33\,\mathrm{s}$ (dash-dots), $t = 0.66\,\mathrm{s}$ (dashes), $t = 1\,\mathrm{s}$ (continuous line).

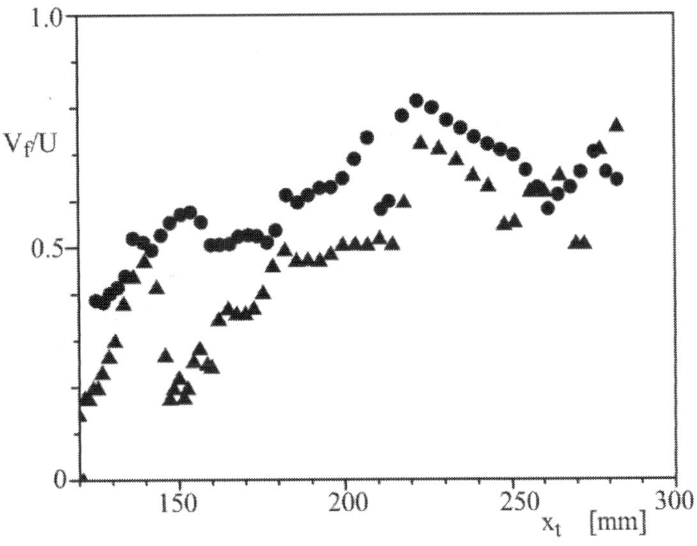

Figure 6. Variation of the normalized fiber velocity V_f/U of the tip of a continuous fiber in a water-polymer flow as a function of the distance x_t of the tip from the inlet of the rough model: (•) $U = 165\,\mathrm{mm.s^{-1}}$, (▲) $U = 220\,\mathrm{mm.s^{-1}}$.

At a similar velocity (Fig. 6), the ratio V_f/U is larger for the shear thinning solution than for water, except at short distances. The amplitude of the velocity fluctuations is also smaller but the minima of the velocity are still mostly located at the same distances x_t as the pinning sites observed for water flows at low velocities U. The motion of the fiber is therefore still influenced by the local constrictions encountered along its path, although much more weakly.

Also, the global increase of the fiber velocity of these continuous fibers with the distance x_t which had been discussed in the case of a water flow and of smooth walls is again observed here: this suggests that using the polymer solution instead of water reduces significantly the relative influence of the friction forces between the fibers and the rough walls.

4. Conclusions

In the present work, the transport of fibers in two model transparent fractures with (a) smooth parallel walls and (b) rough walls has been compared. The transport of both finite length fibers and continuous ones has displayed very different characteristics and a strong dependence on the rheology of the fluids.

For the model with smooth walls, the influence of the interaction of the fibers with the walls is small, even for a low viscosity fluid like water. The hydrodynamic forces applied to the fibers are transmitted all along their length and the fibers remain in the vicinity of the central part of the gap of the fracture during their motion (Their velocity is then of the order of the fluid velocity). For the rough model, the friction of the fibers with the walls is much larger and reduces strongly their mobility, particularly for a water flow: however, in this case, the transport of the fibers along the fracture still remains possible as long as the aperture is about twice the fiber diameter or larger. Also the propagation of the hydrodynamic forces along the fiber is more difficult so that the influence of the length of the fibers on their velocity is reduced, particularly for continuous fibers. An important result of the present work is the enhancement of the mobility of the fibers when water is replaced by a water-polymer solution with a high viscosity at low shear rates. Further studies will allow to evaluate the feasibility of fiber transport through narrow constrictions when such fluids are used.

An important feature of fiber transport in the rough fracture, particularly at low velocities and in regions of narrow aperture, is the occurrence of pinning resulting in an intermittent, stop and go-like, progression of the tip of the fiber. In the pinning events, the deformability of the fibers is a key factor: for relatively rigid fibers and moderate pinning forces, the fiber becomes increasingly deformed by the hydrodynamic forces until the total force on the fiber is large enough to induce depinning. New studies will be focused on the dynamics of the deformation of the fiber during the pinning stage.

Finally, it will be necessary to extend these results to the transport of fibers in natural fracture networks. A first issue is the influence of the orientation of a channelization of the fracture with respect to the direction of the motion of the fiber (and of the mean flow). For that purpose, experiments with channels perpendicular to the mean flow will, in particular, be needed [15]. Another important step towards the application of these results to fracture networks will be the transport of fibers at an intersection between fractures.

Acknowledgments

We are indebted to R. Pidoux for the experimental setup. We are grateful to B. Semin, E.J. Hinch and J. Koplik for their enlightening comments. This work was supported by the Schlumberger Foundation in Paris and by Schlumberger-Doll Research in Boston through a fellowship and a research grant. We also benefited from funding by CNRS and ANDRA through the EHDRA (European Hot Dry Rock Association) and the GdR 2990-HTHS.

References

[1] J. Stockie and S. Green, *J. of Comput. Phys.* **147**, 147-165 (1998).
[2] K. Yasude, N. Mori and K. Nakamura, *Int. J. Eng. Sci.* **40**, 1037-1052 (2002).
[3] M. Lagomarsino, I. Pagonabarraga and C. Lowe, *Phys.Rev. Lett.* **94**, 148104 (2005).
[4] C. P. Lowe, *Philos Trans R Soc Lond B Biol Sci.* **358**, 1543-1550 (2003).
[5] E. M. Purcell, *Am. J. Phys.* **45**, 3-11 (1977).

[6] J. Selker, M. Van de Giesen, N. Westhoff, W. Luxemburg and M. B. Parlange *Geophys. Res. Lett.* **33**, L24401 (2006).
[7] C. Chu and Z. Kizil, *Surg. Gynecol. Obstet.* **168**, 233-238 (1989).
[8] A. Boschan, H. Auradou, I. Ippolito, R. Chertcoff and J.P. Hulin, *Water Resour. Res.* **43**, W03438 (2007).
[9] E. Bouchaud, G. Lapasset and J. Planès, *Europhys. Lett.* **13**, 73-79 (1990).
[10] C. Poon, R. Sayles and T. Jones, *J. Phys. D: Appl. Phys.* **25**, 1269-1275 (1992).
[11] E. Skjetne, A. Hansen, and J. S. Gudmundsson, *J. Fluid Mech.* **383**, 1-28 (1999).
[12] W. Zimmerman, A. Al-Yaarubi, C. Pain and C. Grattoni, *Int. J. Rock Mech. Min. Sci.* **41**, 384-384 (2004).
[13] V. D'Angelo, H. Auradou, C. Allain, J-P Hulin, *Phys. Fluids* **19**, 033103 (2007).
[14] S. Gabbanelli, G. Drazer and J. Koplik, *Phys. Rev. E.* **72**, 046312 (2005).
[15] H. Auradou, G. Drazer, A. Boschan, J.P. Hulin, *Geothermics* **35**, 576-588 (2006).

Journal of Physics: Conference Series **166** (2009) 012002 doi:10.1088/1742-6596/166/1/012002

Flow of wet natural pozzolana

M E Médici[1], O A Benegas[1], F Aguirre[1], M R Baudino[1], R O Uñac[2], A M Vidales[2] and I Ippolito[3]

1 Departamento de Minería, Universidad Nacional de San Luis, 5700 San Luis, Argentina.
2 INFAP-CONICET, Departamento de Física, Universidad Nacional de San Luis, 5700 San Luis, Argentina.
3 GMP, CONICET y Facultad de Ingeniería, Universidad de Buenos Aires, 1063 Buenos Aires, Argentina

E-mail: avidales@unsl.edu.ar

Abstract. We present experimental results on the flow and stability conditions for natural pozzolana, a natural volcanic sand widely used in concrete production. We measured different angles involved in equilibrium conditions for sand piles and relate them to the flux parameters necessary to produce a silo evacuation. We vary some of the geometrical parameters in the silo to inspect the different flux responses of the system. Results are showed as a function of humidity present in the system. In this way, we related critical angles with flux conditions through a silo under different geometric setups and different humidity degrees, thus setting up a basic phase diagram for flux.

1. Introduction

A fundamental understanding of the behavior of real granular materials is crucial in many industrial applications. In order to address many processing and manipulating problems, researchers take advantage of experiments at the laboratory scale as well as of numerical simulations. A large number of studies have been done in this direction over the last decades, especially in the field of dry granular matter. In contrast, wet granular materials have received little attention and researchers very often try to avoid or minimize humidity both in experiments and models. However, several important applications involve, unavoidably, mechanical properties of wet granular media. Some examples include food processing, pharmaceutics, mine industry, geology and construction. Due to this strong relationship with applications, most of the research is aimed at providing empirical solutions to practical issues of wet granular systems [1-3].

Feng and collaborators performed interesting experiments in the field of packing of wet particles when capillary forces are dominant [4-7]. They measured porosity vs. moisture content in packings of spheres of different sizes, and derived a relationship between porosity and moisture content for both mono- and polydisperse packings of spheres.

They showed that the packing fraction (porosity) is strongly affected by the presence of moisture and observed three different regimes: a wetting regime where porosity increases with moisture content; a filling regime where porosity does not change with the addition of liquid; and a sediment regime in which porosity decreases with moisture content.

The interstitial liquid impacts on different physical properties of granular media, such us the angle of repose, which is the characteristic angle exhibited by a granular slope after an avalanche. Experiments reported by Hornbaker *et al.* [8] demonstrated that even a nanometer scale coating of liquid on millimeter-size grains can result in large changes of the angle of repose. In addition to the capillary force, ''wet'' particles experience a viscous force resisting motion, which can be derived from lubrication theory [9,10].

Among many examples, wet sand is one of the model systems for studying the behavior of moist grains. Our model system here will be a sand-like with substantial impact on cement production: the pozzolana.

Pozzolana, also known as pozzolanic ash, is a fine, sand-like volcanic ash (grain size less than 0.2mm), which owes its name to the city of Pozzuoli, Italy, where it was originally discovered. Finely ground and mixed with lime it creates a hydraulic cement and can be used to make a strong mortar that will also set under water. It transformed the possibilities for making concrete structures. These properties have encouraged its ubiquitous use in industrial processes, resulting in many manipulation problems.

It is common industrial practice to store raw granular material in piles before being used. This implies several issues that may be complicated, such as dead volumes due to repose angle; lost of material if it is dry; wetting of the material if it is stored outside; volume expansion due to moisture, and many more. On the other hand, a wet granular material is likely to get jammed when it is placed inside a chute to feed some mixing process. Thus, it is crucial to study the possible humidity dependence of the parameters involved in these manipulation stages.

2. Experimental methods and results

2.1. Density of a packing: wet dependence

The compaction capacity of a granular system is crucial for its flowing properties. Therefore, we first determined the dependence of the apparent density of the material on its humidity degree. We used an experimental technique similar to that in [4]. First, we pour a known amount of water in a container with Pozzolana sand. We then turn the container approximately 20 times to obtain a homogeneous mixing between the grains and the liquid. The mixture is then carefully poured through a funnel into a calibrated test tube. The resultant packing is weighted before and after drying it in a stove at 80 °C during two or three days. The water content W is determined as the percentage weight ratio between water and dry particles. We performed ten measurements (removing each time the used grains) under identical conditions in order to average the properties of the different packings (we work here with "loose" packings). The approximate W values investigated were: 0, 0.64, 0.68 and 1%. The behavior found for the apparent density is shown in figure 1.

We can see that the apparent density decreases with the percent of water present in the sample, from its initial value corresponding to a dry sample (approximately 0.74).

Due to the chemical composition of pozzolana we do not expect the grains to expand due to water absorption. Therefore, we can conclude that the expansion observed under the presence of humidity must be due to the presence of liquid bridges between sand grains.

The results presented in figure 1 can be also expressed in terms of the porosity of the system. The results obtained in this way show a good agreement with those found in pioneering studies, such as those reported in [4] for $W < 2$ %, or in the work by Yu *et al.* [11] and Zou *et al.* [7]. This discussion will be presented elsewhere.

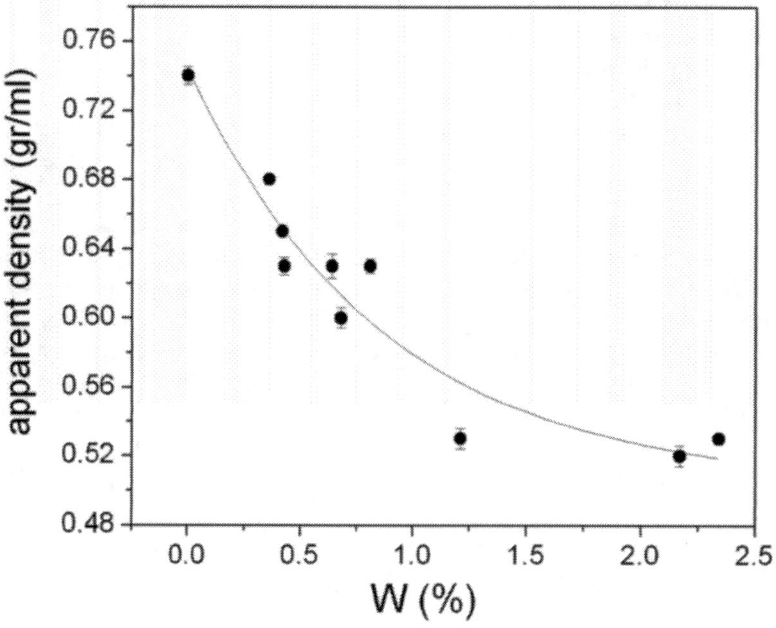

Figure 1. Apparent density as a function of the percentage of liquid content. The line is only to guide the eye.

2.2. Stability angles for pozzolanic piles

2.2.1. *Experimental set up.* Our objective in this part is to measure different critical angles related to the storing stage. We performed different kind of experiments based on the observed fact that the behavior of the banks of wet sand strongly depends on the history of formation of the pile. Thus, the measured angles depend on many details of the experimental set-up and the preparation procedure.

The experimental set-up consists of a glass box (25 × 45 × 30 cm). One of the vertical walls of the box can be removed and serves as a sliding gate (See figure 2). The height of the packing was chosen to be large enough so that the bottom wall has no significant influence on the stability of the pile. The box is attached to a rotation axis and a goniometer allows us to measure the angles with a precision of 0.5°. The granular packings are prepared by pouring the grains and filling the box up to the level of the lateral wall, as sketched in figure 2(b). All the measurements were performed with the same pozzolanic sample but at different wetting degrees. Humidity was controlled by drying the whole sample before a new experiment at a given humidity degree was started. Once the sample is completely dry, water was added in order to obtain the desired water-weight fraction. Note that this percentage is what we call humidity degree or liquid content. The added water is sprinkled over the grains and mixed in order to get a homogeneous sample.

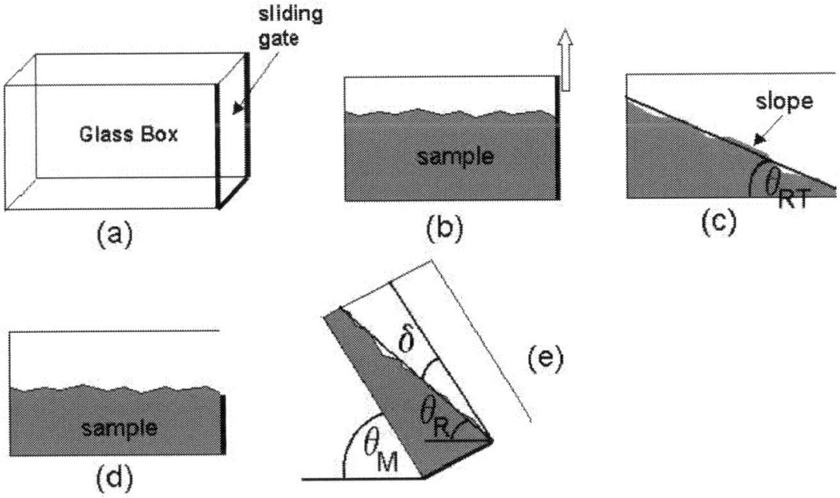

Figure 2. Set-up used for measuring critical angles. (a) Glass box used for the determination of all angles; (b) Front view of the box with a sample inside it; the right wall is being pulled upwards. (c) After the avalanche, the system relaxes to an angle θ_{RT}. (d) Front view of the box with upper half of the right wall removed (bold line). This set up is used to determine θ_R and θ_M. (e) After continuously tilting the wall, an avalanche occurs and rotation is stopped. Angles θ_R and θ_M are measured as indicated.

2.2.2. *Results.* The different critical angles as well as the experimental procedure used to measure them are described below.

Repose angle produced by sliding gate (θ_{RT}): a sample of the order of 0.1 m3 of volume is placed inside the glass box sketched in figure 2(a). The container is carefully set horizontal. The sliding gate is then quickly moved upwards and part of the grains flow out of the system spontaneously (figure 2(b)). A straight line that follows the surface of the grains left inside the box is then plotted on the glass wall (figure 2(c)). The box is then tilted until this line becomes horizontal which allows reading directly the value of the angle of repose θ_{RT} on the goniometer. We repeated similar measurements of for different filling heights of the glass box and no significant differences were found at the laboratory scale we were working with.

Repose angle by avalanche (θ_R) and maximum angle of stability (θ_M): The granular packing is prepared by completely filling up the box with grains, as depicted in figure 2(d). In this way, the height of the packing is determined by the height of the sliding wall. The box is initially horizontal and then carefully tilted at a constant rotation rate until an avalanche occurs (figure 2(e)). We define "avalanche" as the first sudden, simultaneous motion of several grain layers. The rotation is then stopped and the tilt angle of the container with respect to the horizontal is recorded. We also measure the slope of the surface after the avalanche stops and a new equilibrium has been reached (figure 2(e)). When the avalanche is triggered, the displaced material is always free to leave the container. Therefore, the slope of the surface after the avalanche corresponds to the angle of repose θ_R, and the tilt angle of the container is the maximum angle of stability θ_M.

Repose angle in conical pile (θ_C): this angle is the one related to the formation of a pile by pouring material from above at a constant, slow rate. The volume of the pile was of the order of 0.05 m^3. We measured the inclination of the conical surface at different positions over the surface, using an inclinometer, as sketched in figure 3. The average over these measurements determines θ_C. The values obtained in this way were compared with those measured over bigger heaps made at the industrial plant. No important scale effects were found on the results. This encourages further measurements at a laboratory scale.

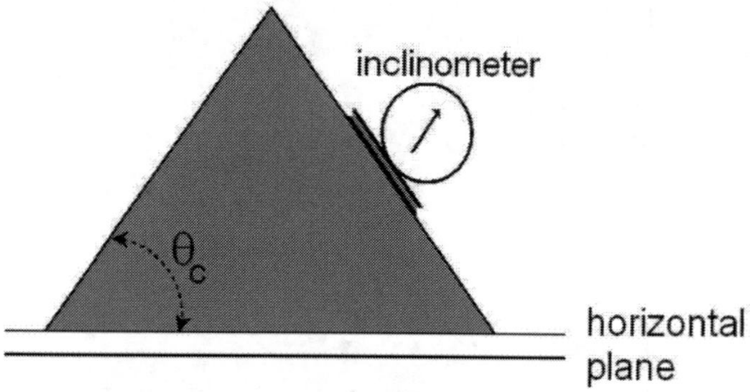

Figure 3. Sketch of a conical pile of pozzolana. An inclinometer was placed at different positions over the surface. θ_C and a position used for the inclinometer are indicated.

Figure 4 shows the variation of the characteristic angles as a function of the humidity degree. It is clear that the general trend of the stability angles is to increase with liquid content. The solid curves shown in the figure are guides to observe the tendencies. In experiments performed with dry material we observed that the grains were highly sliding and unstable. The measured angles in dry materials are systematically smaller than the corresponding ones for wet materials.

An interesting aspect is that the pozzolana system has "memory" with respect to the form in which the initial packing is prepared. Evidence of this is the large difference found between the values of the different characteristic angles θ_{RT}, θ_C and θ_R. This means that the chains of forces formed by grains in the initial packing (before inducing any type of avalanche) tend to maintain their network structure. These chains are reinforced by the presence of capillary forces due to inter-grain liquid in the sample. In addition, a significant increase in volume is observed as the humidity is increased. In figure 4 we see that θ_{RT} is systematically larger than the other angles, whereas the angle of rest in conical piles is always the smallest of the characteristic angles.

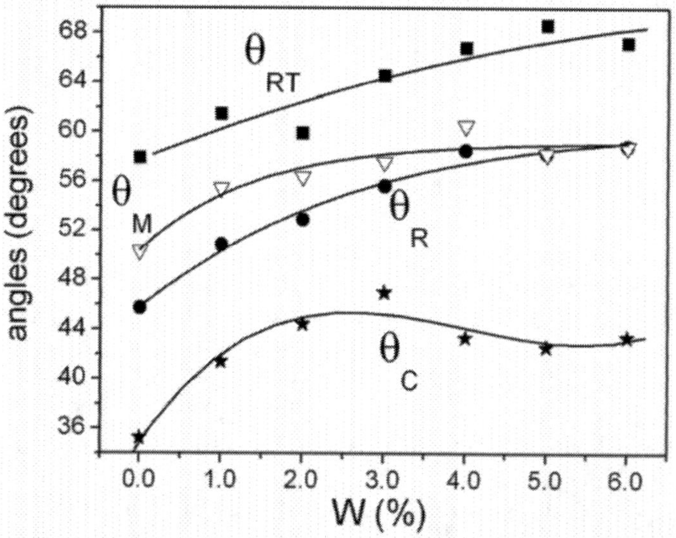

Figure 4. Stability angles for pozzolana sand as a function of wetting degree.

The behavior found for the characteristic angles is related to the method in which each avalanche is generated and, consequently, it depends on the effect of the chains of forces in the piling up. For that reason, they have different values. In the measurements of θ_{RT}, pozzolana is initially horizontal in the container, and the avalanche takes place due to the sudden rise of one of the sidewalls. Therefore, the effect on the internal chains is partial, and they are able to provide greater stability to the piling up. On the other hand, the measurement of θ_R involves a continuous rotation of the entire sample until an avalanche occurs. Therefore, the probability of breaking internal chains of forces is larger than in the previous case. Finally, the instability caused by the top flow of the material during the formation of a conical pile explains the smaller values of θ_C with respect to the previous angles.

Another important observation is that, as humidity degree increases, the evacuation of the material during the avalanche occurs in "blocks" of grains and superficial movements (so called "precursory" avalanches) are not observed. As humidity increases further, the values for θ_R and θ_M are practically the same, that is, the free surface is parallel to the base of the glass box. In this stage, the system behaves as a soft solid and the avalanche happens as a sliding block. This behavior (called Rankin effect) indicates that the characterization of the system as a granular material is no longer valid. For that reason, humidity values greater than 6% were not explored. Finally, it was observed that θ_R is the angle of stability that better characterizes the piles and it is the one showing the smaller fluctuations. In terms of its behavior as a function of humidity, this angle seems to reach a saturation value at which the solid-like features appear.

2.3. Flow in a silo

In order to determine the flow capacity of the material and relate it with stability conditions, we performed flow experiments in a cylindrical silo made of a zinc sheet, with diameter 33cm and height 100cm.

The exit consisted of a funnel that can be changed to vary the size of the outlet diameter. The diameter of the different outlets used in the experiments was: 11, 9, 7, 6, 5, 4, and 3 cm. The interchangeable funnels have the same height, h=21cm. Therefore, the angle α of the walls of the funnel can also be varied and is directly related to the outlet size of the hopper, i.e., to change the size the outlet means in our case to change the angle α of the cone of the funnel. A sketch of the silo and the interchangeable hoppers is shown in figure 5.

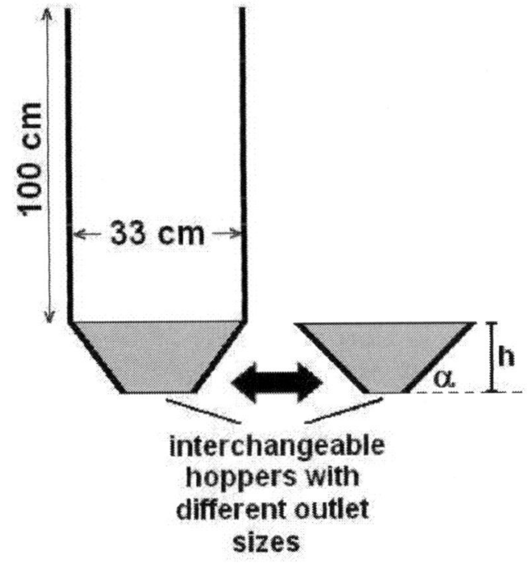

Figure 5. Sketch of the laboratory scaled silo used for the flux experiments. Interchangeable funnels were used with different outlet sizes. The height h is always the same and equal to 21cm and, thus, the angle α can be varied.

Journal of Physics: Conference Series **166** (2009) 012002 doi:10.1088/1742-6596/166/1/012002

To have a better idea about the relationship between outlet sizes and angles of inclination of the funnels we present their corresponding values in the following table:

Table 1. Correspondence between angles and outlet sizes

Outlet Sizes (cm)	Angle α (°)
11	74.7
9	69.7
7	65.1
6	62.9
5	60.8
4	58.7
3	56.8

The silo, with its outlet closed, was filled with a weighted mass of pozzolana (ranging between 12 and 13 kilograms, approximately). The wetting degree of the material, W, is varied in order to study its effect on the flowing behavior. After the silo is full we open the outlet and the material unloads. The total time taken to vacate the silo is recorded. The average flow rate is then calculated through the ratio between the mass and the evacuation time. For each combination of W and outlet size, we repeated the experiment 10 times in order to average the results. It is important to mention here that the manipulation of the material induces the evaporation of some of the water added to it. For this reason we took several small samples during the different experiments in order to check the humidity values. Although we found low variation in these measurements, the wetting values W reported here are always an average over several samples.

During the discharge, the flow could eventually be interrupted. This jamming effect is due to the presence of arches created by grains during the flow process. This feature is enhanced by the presence of capillary forces in wet sands. When flow disruption occurred, we immediately tap the lateral walls of the silo to promote flow again. The procedure was so that the time employed during the tapping was small compared with the time for complete evacuation thus assuring an almost permanent flow in all the operation. The total time needed to complete the discharge (including taps) is computed as the evacuation time.

We classified the different flow features during discharge into two main sets: those presenting *jamming*, even if it only occurred once in all the ten experiments; and those *without jamming* where the flow was always continuous over the ten repetitions, i.e., no clogging was observed in the silo or at the outlet.

The inspected percentages of humidity, W, did not exceed 1%. Beyond this value, the evacuation of the material becomes impossible, even using the largest outlet of 11cm.

In figure 6 we show the diagram of different flow regimes, taking into account the wetting degree and the size of the outlet during discharge. As we pointed out before, two main zones can be distinguished in this figure, i.e., *jamming* and *no-jamming*. The lines on the figure separate these zones. Note, for example, that the point corresponding to an outlet angle $\alpha = 62.9°$ and $W = 0.8\%$ (represented by a plus symbol) belongs to the jamming phase, while the one at $\alpha = 58.7°$ and $W = 0\%$, represented by a square symbol, belongs to the no-jamming phase.

As the lines indicate, for angles larger than 60°, the jamming phase starts if the wetting percentage is above 0.35%. In fact, the left triangle drawn on figure 6, just below the line, represents a border case. During the experiments used to calculate this point, the flux was interrupted half the time by clogging. All points above the solid line have funnel angles greater than 60° or have humidity rate lower than 0.35%, so that they do not offer jamming problems for the flow through the silo.

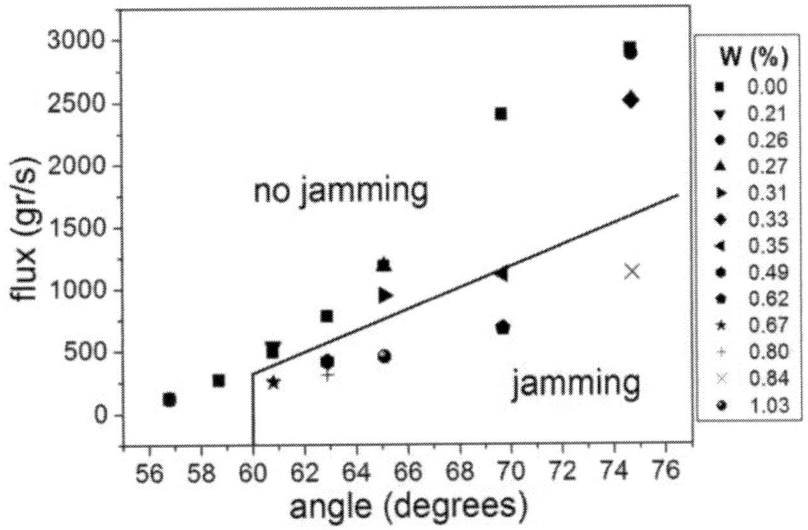

Figure 6. Phase diagram for the flux of pozzolana down a pilot silo. Jamming - no jamming zones are indicated. The angle indicated in the horizontal axis corresponds to the inclination of the walls of the exit funnel with respect to the horizontal.

It is useful to relate the results of figure 6 with those of figure 4 above. The angle θ_{RT} is the most relevant angle for the problem of flow down a hopper. This is the case because the experiments start from a static condition (full silo with its outlet closed) and move to a dynamic one triggered by a sudden perturbation (outlet opened). As explained in previous sections, the measurement of θ_{RT} is performed by suddenly removing the lateral wall of the container where the pozzolana was in a static condition. If we look at the values of θ_{RT} corresponding to a wetting degree between 0 to 1%, they are around 60°. This is in good agreement with our observations in flow experiments. Moreover, for W = 0%, figure 4 shows that θ_{RT} is close to 57°, which explains why all funnel outlets work fine for flow of dry pozzolana.

Finally, we observe in figure 6 that the flow rate increases non-linearly as the size of outlet in the funnel increases. This effect can be observed more clearly for W = 0% (squares), where we present the results for all the available outlet sizes. This may be related to the case where flow takes place through a channel formed within stagnant material and a funnel flow pattern will be developed. In that case, some material moves and the rest remains stationary during discharge from the silo. Funnel flow occurs when the sloping hopper walls of a silo are not steep enough and of the friction is not low enough for the material to flow along them [12]. Under these conditions, particles slide on themselves rather than on the hopper walls, and an internal flow channel appears. In this case, the flow rate depends on the hopper angle, on the size of particles and on the size of the outlet in a non-linearly way. The presence of humidity (typically > 10 %) does not meet the above pattern flow because the caking can occur when particles are stagnant.

3. Conclusions

The results presented here indicate that the stability angles as well as the flow of pozzolana have a strong dependence on humidity degree.

It was observed that dry material is highly sliding and unstable. All the measured angles are systematically smaller than the corresponding ones for wet material.

Journal of Physics: Conference Series **166** (2009) 012002 doi:10.1088/1742-6596/166/1/012002

We also observed that the way in which a pile is prepared has a significant effect on the angles of stability. This "memory" effect is observed in the different values obtained for θ_{RT}, θ_R, θ_M, θ_C. In all cases, stability increases with wetting degree W.

On the other hand, the measurement of either of the stability angles, allows predicting the wetting degree of a given sample, knowing *a priori* a typical curve of behavior like anyone of those shown in figure 4. This is an important point for industrial manipulations in which there might be stages where humidity is present and hoppers must be designed adequately.

For values greater than 6% of humidity, the system behaves like a solid, making difficult its characterization by means of the stability angles.

Flow experiments conducted on the silo, using hoppers of variable outlet diameter, allowed us to establish the conditions of flow and clogging for pozzolana at different wetting degrees. The stability conditions (slope angle) of the material were related to the conditions for flow. Thus, we determined a simple phase diagram jamming/no-jamming that contributes to understand the necessary conditions for manipulating a wet granular material like pozzolana.

Finally, it is important to remark that given a particular real scenario, one has to choose which is the appropriate stability angle to characterize and predict the behavior of the system.

The results presented here are of direct application to larger scales and contribute to the basic knowledge of a very common system in cement industry, among others.

References

[1] Herminghaus S 2005 *Adv. in Phys.* **54** 221–244
[2] Mitari N and Nori F 2006 *Adv. in Phys.* **55** 1–47
[3] Fournier Z, Geromichalos D, Herminghaus S, Kohonen M M, Mugele F, Scheel M, Schulz M, Schulz B, Schier Ch, Seemann R and Skudelny A 2005 *J. Phys.: Condens. Matter* **17** S477–02
[4] Feng C L and Yu A B 1998 *Powder Technol.* **99** 22–28
[5] Feng C L and Yu A B 2000 *J. Colloid Interface Sci.* **231** 136–142
[6] Zou R P, Feng C L and Yu A B 2001 *J. Am. Ceram. Soc.* **84** 504–508
[7] Zou R P, Zu J Q, Feng C L, Yu A B, Johnston S and Standish N 2003 *Powder Technol.* **130** 77–83
[8] Hornbaker D J, Albert R, Albert I, Barabási A-L and Schiffer P 1997 *Nature* **387** 765–765
[9] Adams M J and Edmonson B 1987 *Tribology in Particulate Technology* (Adam Hilger Publishers, Philidelphia USA)
[10] Adams M J and Perchard V 1985 *Inst. Chem. Eng. Symp.* **91** 147–160
[11] Yu A B, Bridgwater J, Burbidge A 1997 *Powder Technol.* **92** 185–194
[12] Nedderman R M 1992 *Statics and Kinematics of Granular Materials* (Department of Chemical Engineering, University of Camdridge, Cambridge Univ. Press)

Journal of Physics: Conference Series **166** (2009) 012003 doi:10.1088/1742-6596/166/1/012003

Segregation in a Galton Board

J G Benito[1], **A M Vidales**[1] **and I Ippolito**[2]
1 INFAP-CONICET, Departamento de Física, Universidad Nacional de San Luis, 5700 San Luis, Argentina.
2 GMP, CONICET y Facultad de Ingeniería, Universidad de Buenos Aires, 1063 Buenos Aires, Argentina

E-mails: jbenito@unsl.edu.ar, avidales@unsl.edu.ar, iippoli@fi.uba.ar

Abstract. This work deals with a numerical study of the problem of separation of particles with different elastic properties. The separation procedure uses a Galton Board which consist in a bidimensional system of obstacles arranged in a tiangular lattice. Disks of equal diameters but different elastic properties are launched from the top of the device. The Galton Board is commonly used for mixing particles, but here, we intend to find special conditions under which one can use it as a segregating device. We introduce a mixture of particles and generate, through simulations, different conditions to favor the segregation process based on the different elastic coefficients of the particles. We inspect which is the best configuration of size, density of obstacles and wall separation to favor the separations of particles. Our results prove that the Galton Board can be used as a segregation device under certain conditions.

1. Introduction

The problem of mixing and segregation in granular materials has been the subject of renewed and intense interest in recent years because of its great importance to the industry [1-3].

The bi-dimensional Galton Board (BGB) is a system of small pins forming a regular triangular lattice. A typical sketch of a board is shown in figure 1. There are plenty of examples in the literature of much reserch done using this set up to the interesting dynamics present in it. Those examples range from charged particles scattering to dispersive flow of glass beads [4-7].

Usually, the BGB is arranged in a vertical position and experiments consist in launching small disks (compared to the spacing between pins of the lattice) from the top row of the array and collecting them at the bottom part of the system by a set of collectors.

The implementation of the BGB like a particle mixer has been developed and successfully proved. Bruno *et al.* [8-11] studied experimentally and numerically the dynamics of gravity driven particles going through an ordered array of obstacles. Collisions with obstacles make particles to diffuse, and the diffusive regime is attained after a few number of collisions. Due to this process, particles are forced to mix. They also analyzed the feasibility and efficiency of the mixture of particles through the BGB. They developed an analytical method to predict the relative composition of a mixture as a function of the radii ratio of the particles in the mixture.

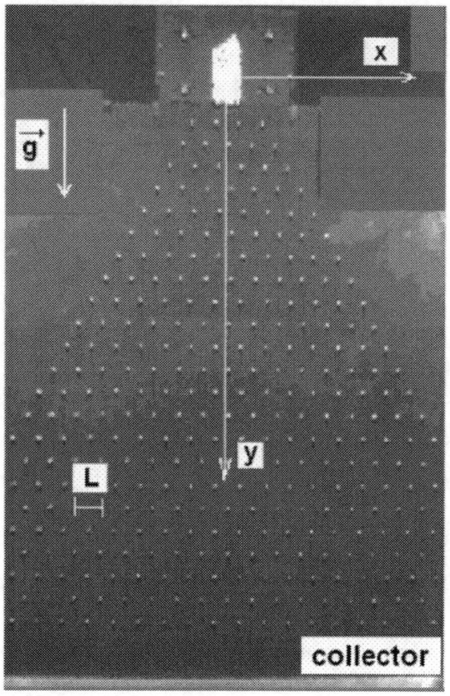

Figure 1. A typical bi-dimensional Galton Board device. At the top, there is a container with particles ready to be released. The spacing between pins, L, is indicated as well as the coordinates and gravity direction.

Recently, Benito *et al.* [12-14] have studied and characterized the behavior of the exit distributions of same species of particles falling down through a BGB.

They found a change in the behavior of these distributions as the spacing W between BGB walls is reduced. They observed a crossover in the behavior of the exit distributions changing from a Gaussian-like to a uniform-like shape. They determined a simple parabolic relationship that holds when searching for the necessary separation between lateral walls in order to obtain a uniform exit distribution of percolating particles. As a second stage, they have studied experimentally and numerically the influence of lateral walls on the mixing capability of a BGB, finding that this separation is crucial in a mixing process, and obtaining, for small separations W, a homogeneous mixture.

All above inspires the following question: can this particle mixer be used in a reversal way?. In other words, can it be used to segregate particles originally mixed?. For this reason, the present paper will focus on the problem of particle separation through the BGB.

2. Numerical model

Simulations were driven to reproduce the geometrical features of BGB device used by Bruno *et al.* [8-11] and Benito *et al.* [12-14]. We define sites (representing the pins of the board) arranged in a triangular lattice and keeping experimental distances.

Disks of the same size like those of the experiment are launched from the center of the top of the table. Trajectories followed by the disks were calculated in this way: a disk falls down by gravity until it encounters a pin in its trajectory. Then, and depending on the relative positions of the centres, the small disk will roll over the pin to the right (the x-coordinate of its centre is to the right of that of the pin) or to the left (the x-coordinate of its centre is to the left of that of the pin). At this stage, we have to decide which will be the new x-position of the particle, i.e., we have to introduce the effect of bouncing without calculating any force (we are not interested here in the dynamics of the problem).

Thus, we just choose a random number uniformly distributed between 0 and 1 (ran(0, 1)) and put the particle at the corresponding x-position given by:

$$x_p = x_{pin} + aR_{pin} + aR_p + a\delta d\, ran(0,1) \qquad (1)$$

Here, x_{pin} is the x-coordinate of the centre of the pin; R_{pin} is the radius of the pin (the same for all of them in the lattice), R_p is the particle radius, d is the distance between two neighboring pins minus two times their radius, δ is a factor to account for the hardness of the disks and, finally, a is a factor equal to +1(−1) if the x-coordinate of the centre of the particle is to the right (left) of that of the pin. Figure 2 depicts the details.

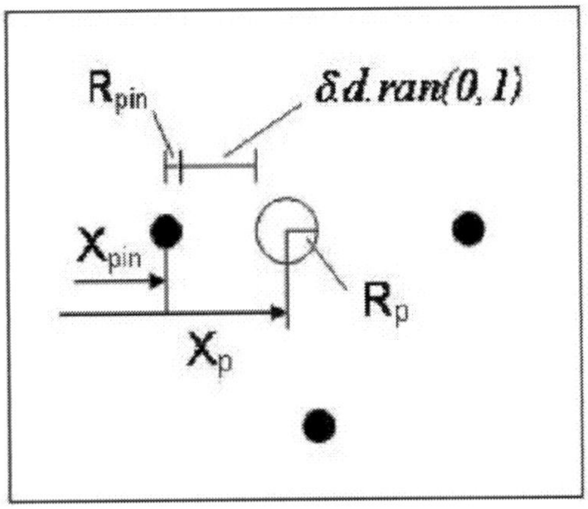

Figure 2. Schematic representation showing how we get the new x-position of the disk in simulations. this case relates to a = +1 in Equation 1.

We use parameters δ = 0.55, 0.6 and 1 simulating different hardness of the disks. This selection is totally arbitrary and these parameters will just represent some real bouncing parameters: the smaller the value of δ, the smaller the hardness of the percolating particle and thus, the smaller the lateral displacement.

After a new position is attained by the particle, a new calculation for the trajectory of the particle is performed, following the procedure explained above, and so on.

After the particle left the last row of the BGB, its final x-position is recorded to build the exit distribution histogram. In simulations, we have the advantage to know exactly the final position of each particle and thus one can classify these data to build a histogram as desired. Walls are consider sufficiently far appart to avoid colisions to a minimum.

Under these conditions, we begin a typical run by dropping through the BGB a binary mixture composed of 10^5 disks of each species. At this stage, it is worth mentioning that the effect of throwing disks one at a time in simulations is not relevant for the final results obtained. In fact, this has been demonstrated experimentally by Bruno *et al.* [8-11].

In a typical BGB experiment, while the discs are falling, they collide inelastically with the obstacles and other falling discs, losing energy and changing their local velocity direction. Bruno *et al.* [8-11] analyzed the influence of the collective interactions, i.e., the influence of other falling particles on a tracer disc. They studied the velocity distribution of tracer particle and compared it to that obtained when the discs are released one by one and they found no significant influence. Even more, they performed numerical simulations of the spatial frequency of disc–disc and disc–obstacle collisions. They have seen that the spatial frequency of disc–obstacle collisions remains constant in average during all the chute and disc–disc collisions occur mainly close to the entrance where the particles are clustered together. This is the region of high disc–disc interaction density corresponding to the first four or five rows of obstacles. Afterwards, disc–disc collisions became less probable when the particles have sufficiently spread due to the presence of obstacles. Thus, the system will behave as a single particle. For this reason, the results presented and discussed in simulations will refer to the one particle-at-a-time problem.

3. Segregation mechanism

The mechanism of separation through BGB is based on the fact that we use particles with different elastic properties.

At the end of a typical experiment, the final exit distribution shows a good degree of mixture in the central collectors, but in the lateral ones we only find particles belonging to the set of higher parameter δ. This is due to the fact that they can difuse far appart, compared to those with lower hardness, thus exploring more remote regions of the table.

Using this property, we extract the disks on the sides of the device not mixed with other species. The remaining mixture is re-injected into a new BGB, thus beginning a new cycle. This reinjection can be performed in two ways.

3.1. Design with funnel

In this design, the remaining mixture is redirected toward the middle of the new BGB simulating a funnel dynamics. Disks fall again through the BGB and we obtain the new final *x*-position of each reinserted disk. As explained above, those particles located in laterals collectors belonging to one species and not mixed with the other, will be extracted, while the remaining mixture will be redirected again by means of a funnel to a new BGB, beginning another new cycle. This process is repeated *n* times.

For simplicity we define a new parameter Δ as the ratio between the parameters δ of the disks composing the binary mixture:

$$\Delta = \frac{\delta_1}{\delta_2} \tag{2}$$

Here, the bouncing parameter δ_2 is always the higher value of the two parameters. In our experiences, δ_2 is always equal to one.

Two different mixtures were tried. One with parameter Δ = 0.55, and the other with Δ = 0.6. Figure 3 shows the results obtained. We have plotted the cumulative number of particles separated from the original mixture per number of cycles *n*.

In both cases, we note that after a certain number of cycles $n = N$, the species of disks initially mixed are completely separated.

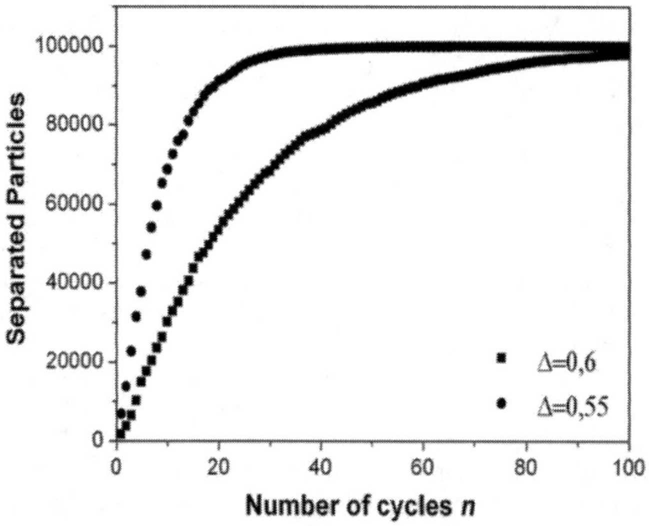

Figure 3. Number of accumulative disks separated with the funnel design vs. number of cycles for mixtures with Δ=*0.6* and *0.55*.

The number of cycles required for this depends on the parameter Δ of the mixture. The lower the parameter, the lower the number of cycles N.

For the mixture with $\Delta = 0.55$, the number of cycles N needed to completely segregate the particles is equal 35, while for the case $\Delta = 0.6$, it is approximately equal to 100.

3.2. Design without funnel

In this design, the reinjection of the remaining mixture comming from the first BGB is performed by retaining the final x-position of each disk for the reinjection to the next BGB. In other words, the initial position of each disk reinserted in a new cycle is equal to the final position obtained in the previous one. No funnel effect is present here.

In figure 4 we compare the results obtained with both designs in the case of a binary mixture with Δ =0.6 and composed of 10^5 disks of each species.

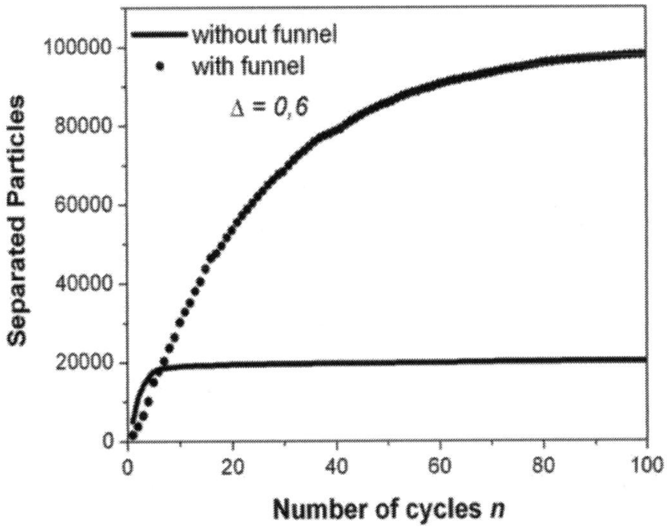

Figure 4. Accumulative number of separated disks vs. number of cycles for designs with and without funnel and Δ=0.6

We note that, after a few cycles, the design without funnel is unable to continue separating particles meaning that the algorithm fails to find disks with higher δ not mixed with the other species in the lateral collectors. This is due to border effects, as indeed it was recently proved by Benito *et al.* [14]. In that work, it was proved that, both experimental and numerically, the existence of lateral walls in a BGB is crucial to obtain a homogeneous mixture of disks. They performed experiments with particles with two different bouncing properties studying the exit distributions and the mixing indexes as the separation between lateral walls was varied. It was found that the mixing region changes as walls become closer. It goes from a normal distribution to a uniform one which is better achieved when particles with similar hardness properties are mixed. As separation between walls decreases, the mixing is more efficiently fulfilled and more homogeneous. Thus, in our present design without funnel, the disks that are reinjected into a new BGB (retaining their previous final x-coordinates values), can explore further regions and notice after a few cycles the presence of laterals walls. Under these conditions, after approximately 5 cycles, we always get a mixture (enhanced by the presence of walls), and disks can not be separated anymore.

Journal of Physics: Conference Series **166** (2009) 012003 doi:10.1088/1742-6596/166/1/012003

3.3. Design duo

Despite the inability to separate particles for the design without funnel, we note that, for the first cycles, this design separates a greater amount of disks compared with the funnel design. This is clearly indicated by the knee in figure 4 corresponding to the design without funnel. For this reason, we will try to avoid the above mentioned border problem, designing a new device that will segregate particles comming from a combination of the two designs analyzed previously.

Basically, we want to maintain the higher segregation capacity of the design without funnel, but also preventing an excessive diffusion of the disks when they move down the board. This double effect is obtained by building the following "design duo". First, it consists of a series of four BGB without funnel. Then we insert a BGB with funnel, which is followed by a new series of four BGB without funnel. After this, another BGB with funnel is inserted, and so on.

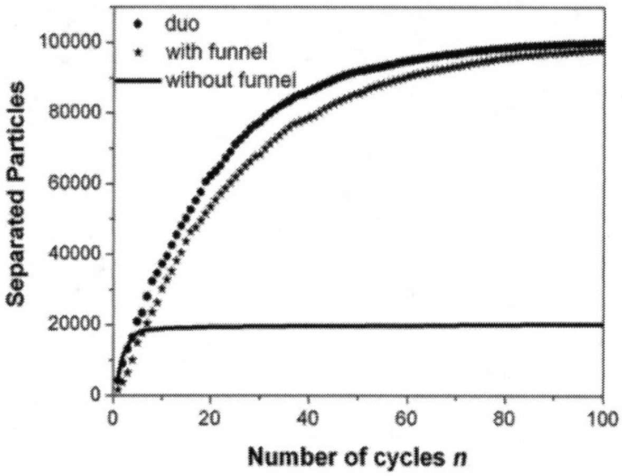

Figure 5. Cumulative number of disks separated vs. number of cycles for designs duo, with funnel and without funnel for mixtures with Δ=0.6.

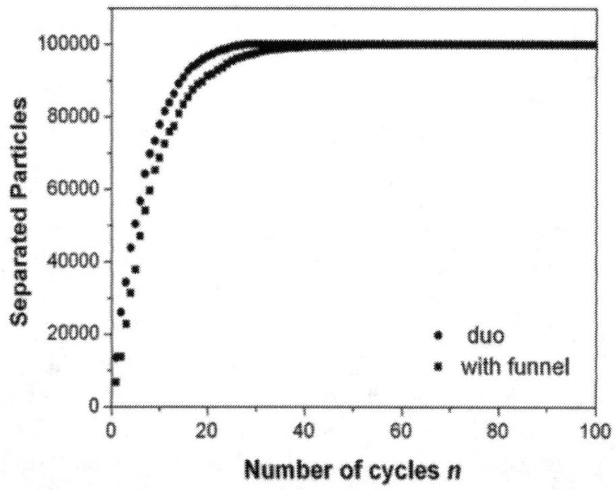

Figure 6. Cumulative number of separated disks vs. number of cycles for designs duo and with funnel and Δ=0.55.

Figure 5 shows the results for the three designs: duo, with funnel and without funnel, for a mixture with Δ=0.6. The amount of disks separated with the duo design is always higher than those corresponding to the previous designs.

A similar behavior is observed in figure 6, for a mixture with Δ=0.55. Again, we observe that the duo design has the greatest ability to keep apart disks. In this case, we only show duo and funnel designs because the poor performance obtained for the design without funnel.

In this way, the new design leaves aside the effects of lateral walls, thus obtaining a better mechanism for particles segregation.

4. Effect of the obstacle density

In order to check whether a different obstacle density in BGB can even improve the segregation obtained by the duo design, we have performed a set of runs changing the density of pins in the central part of the board. The aim of this change is to provoke more diffusion and eventually more segregation among particles of different species.

In figure 7, we present the results obtained for a the density of obstacles in the central part of the table which is twice the density of the rest. The fraction of the total length of the table occupied by higher density of obstacles is 0.11, i.e., 11% of the total area of pins. We found that there is no improvement due to this change of density, as shown in the figure.

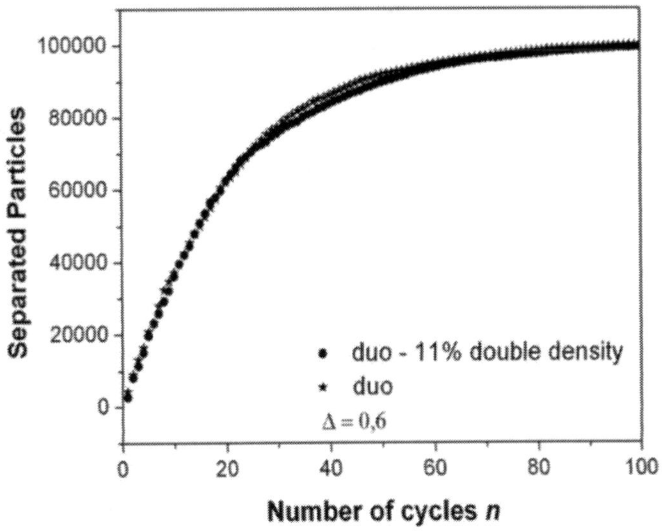

Figure 7. Accumulated number of separated disks vs. number of cycles for designs duo and duo with 11% double density of pins in central part of the table and Δ= 0.6.

We have further increased the fraction area of higher density of obstacles to 14.5% and still obtained the same results. Figure 8 shows the comparison to the original duo design and we see no improvements.

As clearly shown in the above figures, the pin density change does not make better segregation of particles. The increment of obstacles affects identically the behavior of both species and, thus, the probability of finding mixed particles in the collectors is the same as for the case of uniform density tables.

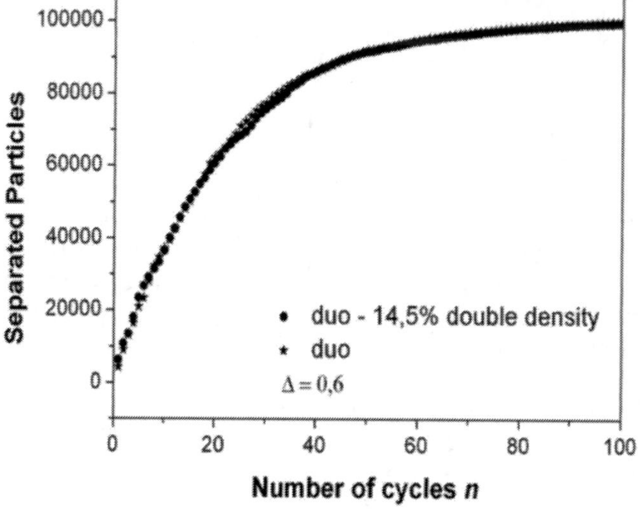

Figure 8. Accumulated number of separated disks vs. number of cycles for designs duo and duo with 14.5% double density of pins in central part of the table and Δ= 0.6.

5. Conclusions

In this paper, we have shown that a BGB can be used to separate particles with different bouncing properties. This result is totally novel taking into account that these kind of apparatus is commonly used to mix particles and has never been tried in the opposite way.

In particular, different designs were tried to search for the best configuration of BGBs to achieve the desired segregation effect.

We have found that a design with funnel notably improves the segregation capability of a series of BGBs. Nevertheless, the "without" funnel design proved to be better at the first cycles of injection, but its performance was too poor for further cycling due to border effects.

The combination of those effects, gave rise to the duo design. This last design was found to be the best one to separate particles under the conditions studied in the present work.

We attempt to improve even more the results obtained with the duo design by increasing the density of pins in the central part of the BGB. It was proven that no effect is produced by this new configuration.

It still remains to be established the relationship between Δ and N through a systematic study as the one performed here for different bouncing relationships.

In the same way, it would be useful to determine the number of lateral collectors that can be drained in each cycle of a series of BGBs to assure a given degree of purity in the separated particles.

References

[1] Fan L T 2001 Bulk-solids Mixing Overview *Handbook of Conveying and Handling of Particulate Solids* 647–658.

[2] Harnby N, Edwards M F and Nienow A W 1985 Mixing in the Process Industries *Butterworths Series in Chemical Engineering*.

[3] Kaye B H 1987 Powder mixing *Powder Technology Series*.

[4] Moran B, Hoover W G and Bestiale S 1987 Diffusion in a periodic Lorentz gas *J. Stat. Phys.*

(Historical Archive) **48** 709.

[5] Hoover W G, Moran B, Hoover C G and Evans W J 1988 Irreversibility in the Galton board via conservative classical and quantum hamiltonian and gaussian dynamics *Phys. Lett. A* **133**(3) 114–120.

[6] Lue A and Brenner H 1993 Phase flow and statistical structure of Galton-board systems *Phys. Rev. E* **47**, 3128.

[7] Rosato A D, Blackmore D, Buckley L, Oshman C and Johnson M 2004 Experimental, simulation and nonlinear dynamics analysis of Galton's board *Int. J. Nonlinear Sci. Num. Simul.* **5** (4) 289.

[8] Bruno L, Ippolito I and Calvo A 2001Granular mixing in a Galton board *Granular Matter* **3** 83–86.

[9] Bruno L 2002 Difusión y Mezcla en Medios Granulares, Ph.D. Thesis *Universidad de Buenos Aires* (Argentina) *and Université de Rennes* 1 (Francia).

[10] Bruno L, Calvo A and Ippolito I 2003 Dispersive flow of disks through a two dimensional Galton board *Eur. Phys. J.* E **11** 131–140.

[11] Bruno L, Calvo A and Ippolito I 2003 Granular mixing and diffusion: a 3D Study *Int. J. Heat Technol* **21** 1.

[12] Benito J G 2007 Estudio numérico y experimental de mezcladores bidimensionales para materiales granulares secos - Final work for bachelor grade *Universidad Nacional de San Luis* (Argentina).

[13] Benito J G, Meglio G, Ippolito I, Re M and Vidales A M 2007 Exit Distribution function crossover in a Galton Board *Granular Matter* **9** 159 -168.

[14] Benito J G, Ippolito I and Vidales A M 2008 Improving mixture of grains by using bi-dimensional Galton boards *Physica A* **387** 5371–5380.

Journal of Physics: Conference Series **166** (2009) 012004 doi:10.1088/1742-6596/166/1/012004

PIV Technique applied to granular flows in hoppers

F Ulissi [1], **I Ippolito**[1,2] **and A Calvo**[1]
[1] Grupo de Medios Porosos, Facultad de Ingeniería, Univ.de Buenos Aires, Paseo Colón 850, 1063, Buenos Aires, Argentina.
[2] CONICET, Av. Rivadavia 1917 – 1033, Buenos Aires, Argentina.

Email: fulissi@gmail.com; iippoli@fi.uba.ar

Abstract. In this work we report a technical method based on PIV, usually used in fluid dynamics, to measure velocity profiles of granular materials in silos and hoppers. The particles inside the hopper were filmed by a digital video camera. Finally, we measure the velocity profile of granular materials inside a quasi-two dimensional hopper and compare the data with some existing models.

1. Introduction

Granular flow in hoppers, silos and similar containers is important in many technologies such as the processing of minerals, grains, chemicals and pharmaceuticals, feeding of furnaces fuelled by solids and mining. In industry, technical solutions to achieve the desired type of flow are empirical, because these materials are difficult to visualize in experimental devices.

The knowledge of the velocity field inside the hopper or silo is important in order to predict the type and flow rate, and for testing the existing models for velocity profiles. Two of the most common flow problems experienced in an improperly designed silo are no-flow and erratic flow. No-flow from a silo can be due to either arching (bridging) or ratholing. Arching occurs when an obstruction in the shape of a bridge forms above the outlet of a hopper and prevents any further material discharge. It can be an interlocking arch, where the particles mechanically lock to form the obstruction, or a cohesive arch. An interlocking arch occurs when the particles are large compared to the outlet size of the hopper. A cohesive arch occurs when particles pack together to form an obstruction [1][2].

In this paper, we will focus on the velocity field measurements in quasi-two dimensional silos applying the Particle Image Velocimetry technique in continuous dry granular flow. As the name suggests, PIV records the position over time of small tracer particles introduced into the flow to extract a discrete velocity field. PIV method knows its origin in the measurement of velocity fields in different fluids. The basic requirements for a PIV system are an optically transparent test-section, an illuminating light source (laser in fluids), a recording medium (CCD camera), and a computer for image processing [3].

[1] Eng. Facundo Ulissi

2. Method of measurement

2.1. PIV Technique: basics

This method allows to obtain the displacement of a group of particles called "markers" that are inserted in the flow. In fluids, the markers are very small non miscible particles (Kerosene or PVC's filings) that join the fluid so that they should not affect the flow.

Figure 1 shows a scheme of the basic general arrangement of devices for applying PIV technique on fluids. The CCD camera only sees the particles in a plane of the fluid that are illuminated by means of a laser sheet that crosses cylindrical lens. In granular materials the images only shows the motion of material against the hopper walls.

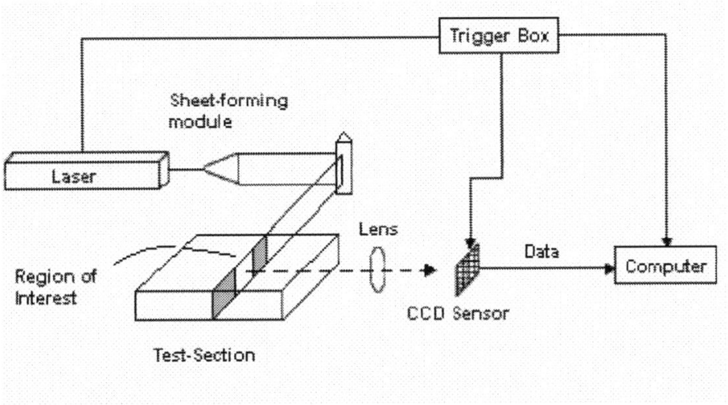

Figure 1. Basic requirements for a PIV system [3].

The basic procedure to apply the PIV method is: a) dividing the obtained images of a movie in basic units called "interrogation area", typically of 32x32 pixels (figure 2). b) Obtaining the medium displacement for all the markers into this area by identifying the maximum of a function called correlation function, applied between the correlation areas of two successive images of a video.

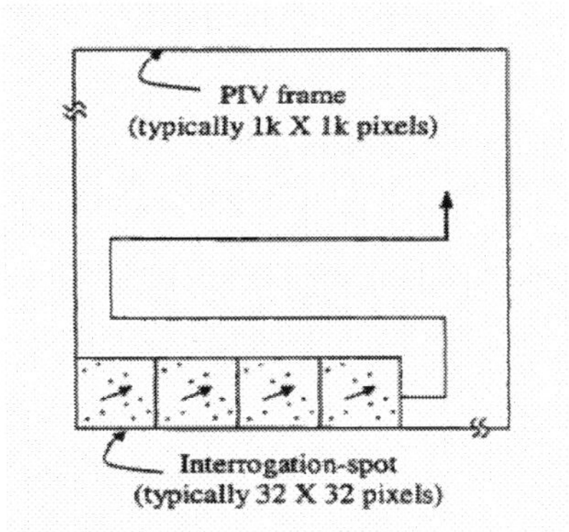

Figure 2. Generic image divided into interrogation areas [3].

Basically the correlation function ϕ measures the "coincidence" between two images. In this paper we use the cross-correlation algorithm as shown in equation (1), to obtain the displacement field by convolution between portions of two successive images of the experiment.

$$\phi_{fg}(m,n) = \sum_{k=-\infty}^{k=\infty} \sum_{l=-\infty}^{l=\infty} f(k,l)\, g(k+m, l+n) \qquad (1)$$

In equation (1) "f" and "g" functions indicate the grey level of each interrogation area that corresponds to analogue centered positions of "t" and "t+Δt" images. In our experiments, sizes of 32x32 pixels and 8x8 pixels were used for "g" and "f" respectively. The displacement of a group of tracers that are confined into an interrogation area is the displacement of "f" against "g" (starting at centered position) in which the cross correlation function reaches a maximum value. This maximum is commonly called "true correlation". Physically, the "true correlation" represents the number of coincidences among tracers between "f" and "g". Then, the discrete velocity field can be found by dividing the displacement obtained by the time between the two images of a pair (Δt).
We call also "mean peaks" to those peak values that are smaller than the "true correlation peak" and the ratio between true correlation and random correlation peaks is the "Signal to Noise Ratio", or S2NR.

2.2. Control parameters and data validation
We have observed that calibration tips for PIV applied in fluid experiments are also valid for granular materials. In order to calibrate this method, we list the main tips.
The cross correlation algorithm cannot work if the minimum quantity of particle tracers into the interrogation area at image 1 are not into the analogue interrogation area at image 2. To avoid this phenomenon called "loss of pairs" it is necessary to control the interrogation size area (d_I), the time between images (Δt), and the tracer´s density.
It is necessary to assure that the real displacement will not exceed the interrogation area dimensions.
In this case, the measurement is not valid because one of the two positions of a particle is missing. The maximum particle displacement $\overline{\Delta X}$ recommended [3] is 0.25 times the size of the interrogation area in pixels (d_I), as shown in equation (2).

$$\Delta\overline{X} \leq \frac{1}{4} d_I \qquad (2)$$

In our cases, we have choosen $\overline{\Delta X} \approx 8$ pixels in average. The time between images was selected taking into account the following tips:

- To avoid the phenomenon of "loss of pairs" due to excessive time between images.
- To avoid velocity errors due to excessive time between images that causes that the detected velocity is not quasi-instantaneous.
- To avoid reaching the low limit of time interval detectable by the CCD camera.
- To avoid measuring low displacements that can be confused with other correlation values (low Signal to Noise Ratio).

Equation (3) suggests an expression to estimate the time interval between images that considers those tips [3].

$$\Delta t \leq \frac{\Delta\overline{X}}{M\,U} \qquad (3)$$

In equation (3) M is a spatial scale factor in pixels/mm and U is the average expected flow speed. In our experiments: U ≈ 20 mm/sec and M = 5 pixels/mm; then, Δt ≈ 0.08 sec. Based on equations (2) and (3), by experimental calibration, we finally use Δt = 0.05 sec and Δt = 0.07 sec (each one computed from different estimated velocities U) depending on the velocity pattern across the flow.
The tracer particle density requires to be also controlled. If the number of tracer particles is increased, the number of real correlations also will be increased and therefore, the relation S2NR and then the reliability of the measurement. The minimum recommended tracer density in an interrogation area is at least 5 particles [4]. Figure 3 shows the effect of seeding density in PIV measurements on fluids.

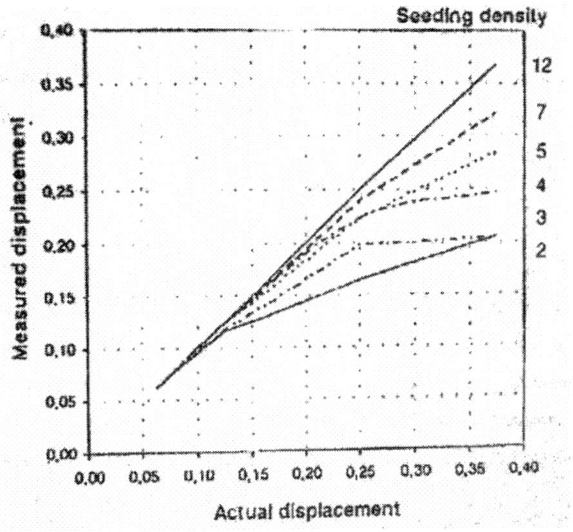

Figure 3. Effect of tracer particle density on the measurements [4].

In our case (figure 4), tracer density is bigger than 5, because vacancies can be also taken into account as tracers. Typically, there are 10 or more tracer particles inside each interrogation area.

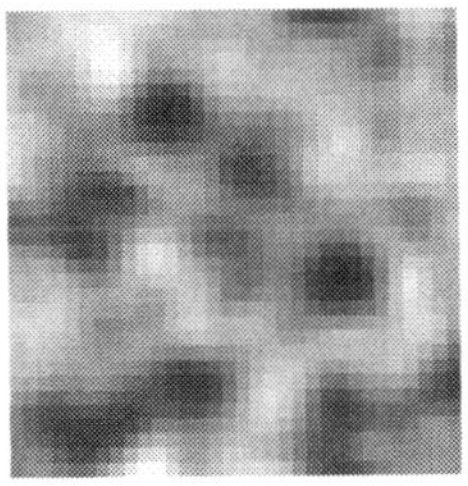

Figure 4. Tracer density on granular measurements in a 32x32 pixels interrogation area (4X magnification).

As it has been described, the PIV method provides typically a grid of independent vectors with a density of 32x32 pixels. In our application, the interrogation is performed with an "overlap" between

adjacent interrogation areas of 50%. In this sense, the center of each interrogation area is displaced $d_I/2$ respect to its neighbour, in both X and Y directions. As a consequence, the total number of vectors increases by a factor of 4 and the detections are not independent [3].
In order to validate the measurement, it is essential to know all the values of the equation (1) in the correlation plane and compare this information to the maximum correlation (position of the measured displacement), according to equation (4).

$$\frac{h_1}{h_2} \geq k \tag{4}$$

The relation expressed in equation (4) is the mentioned "Signal to Noise Ratio" where h_1 is the maximum of the correlation values and h_2 is the comparison parameter, which can be an average value in the correlation plane ("peak to mean" validation) or a second maximum one ("peak to peak" validation). In this work we will use the first one, and the recommended value for k is 3 [5]. In our experiments these values are greater, close to 19 in average which satisfies equation (4).

3. Velocity field measurement

3.1. Experimental set up

The experimental quasi-two dimensional hopper consists of two acrylic plates of 400mm side and 5mm of thickness, separated by thickness rods. The width between plates can be adjusted by varying the rods width. The lateral walls are made of two acrylic tilt rods and the angle between the rods and the horizontal can be varied (figure 5).

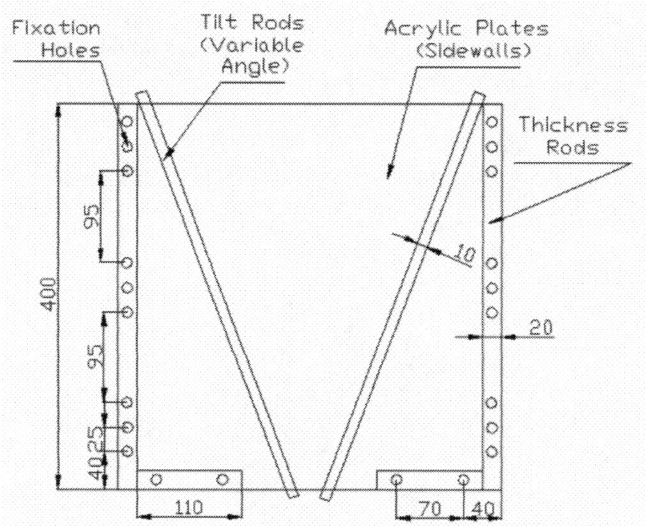

Figure 5. Basic scheme of experimental hopper.

The thickness of the hopper was fixed in 10 mm and the discharge hole in 5 mm in order to optimize the calibration of PIV method. The lighting system consists of four 18W fluorescent tubes placed in front of vertical planes. A scheme of the set-up is shown in figure 6. Special care was taken to horizontal and vertical align the different parts of the general arrangement.

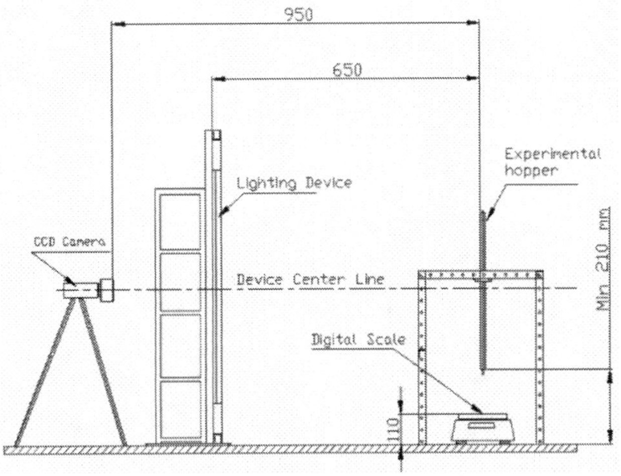

Figure 6. Sketch of the experimental set-up.

3.2. Experimental procedure

The characteristics of grains used in the experiments are listed in Table 1, where Φ_m is the mean size of the grains, s is the standard deviation of sizes and C_v is the relative dispersion of sizes to the mean size.

Table 1. General characteristics of granular materials.

Grains	Mean Size Φ_m [mm]	Standard Deviation s [mm]	Relative Dispersion C_v [%]
0,5 mm Glass Beads	0,5	0,0	8,60%
1 mm Glass Beads	1,1	0,1	7,80%
Type S Quartz	1,0	0,2	23%
Type B Quartz	1,4	0,3	23,90%
Type S Sand	0,6	0,2	27,20%

Depending on the illumination, the grains can be seen as translucent or opaque to prevent from carrying out the PIV method as usually described for fluids. So, we replace the laser plane by conventional light illuminating the frontal plane of the hopper. The markers are some dyed grains, but the interstitial vacancies between grains make up markers too. Typically, we achieve good results by dying 15% of the total mass of the grains in the hopper. The lighting system serves to enhance the contrast between the markers (vacancies and grains) and the granular matrix.

A filling procedure was developed and controlled to achieve a regular charge that is kept constant in all the experiments. The exit orifice is suddenly opened and steady state flow is established before acquiring the images for PIV technique.

We focus the attention on a rectangular region of the hopper of 200mm x 100mm above the orifice with a resolution of 1000 x 500 pixels approximately. The CCD camera acquires 20 frames per second using an external trigger and the exposure time is 5 msec which corresponds to a 10% of the time between images. This exposure time provides a displacement error within 0.1 mm.

Journal of Physics: Conference Series **166** (2009) 012004 doi:10.1088/1742-6596/166/1/012004

The experimental recorded videos are composed of 1200 images approximately and therefore, the maximum discharge time is about 60 seconds. To enhance the contrast and to distinguish boundaries of particle tracers, we apply two convolution filters. For smoothing the image noise, we first apply a Gaussian filter with $\sigma = 0.5$ and a kernel size equal to a half grain size for each image (typically 5x5 pixels). After that, we apply an average filter with the same kernel size. The resulting image is less noisy and the boundary of grains more clear to apply then the velocimetry technique (Figure 7).

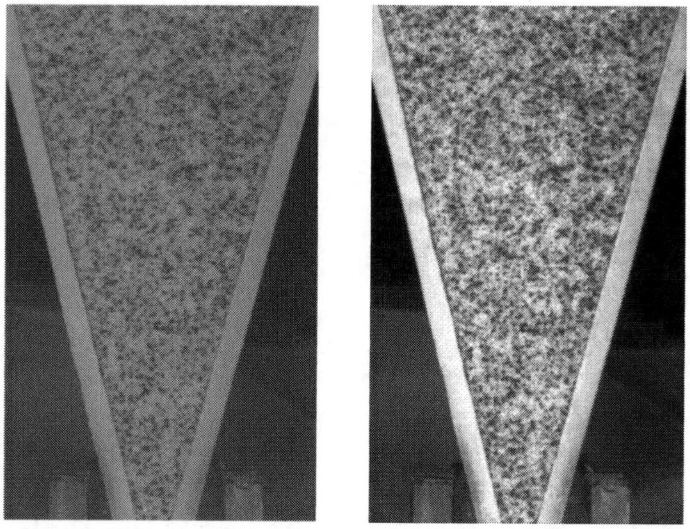

Figure 7. Images after applying Gaussian and Average filters.

3.3. Preliminary results

By applying the PIV technique controlling the mentioned parameters, we achieve satisfactory results. As said before, we achieve a 32x32 pixels resolution in the velocity field. Figure 8 shows the vertical velocity field on a hopper image.

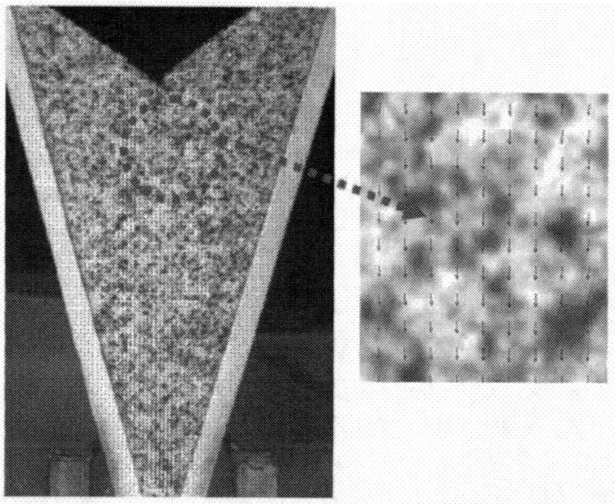

Figure 8. Experimental vertical velocity field.

Figure 9 shows the same results in chromatic scale. The graphic was obtained plotting the vertical velocity values for each spatial position in the hopper. Black regions represent stagnant zones and white regions represent high velocity zones in the hopper. Note that the maximum velocity value is located in the middle of hopper and increases near to discharge.

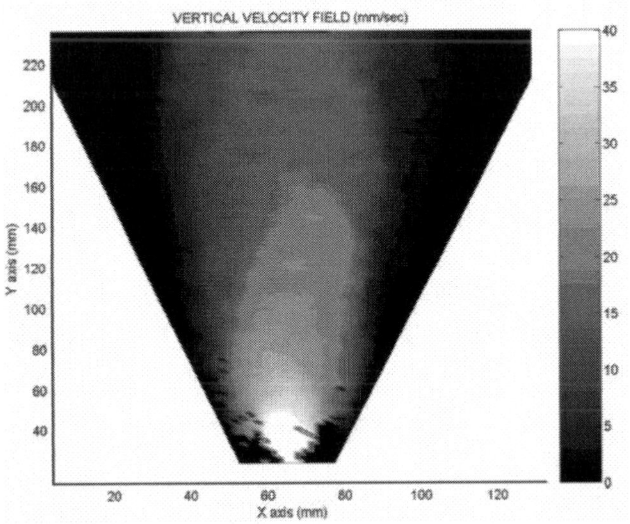

Figure 9. Vertical velocity field, represented by a chromatic scale.

4. Conclusions

In summary, we have used a PIV technique to achieve good resolution in measuring a velocity field on granular materials confined in silos or hoppers. We show that PIV method is reliable and allows us to measure in a good precision velocity profiles. In that sense, we describe a simple, cheap and versatile experimental device that allows to obtain several results in the granular material studies.

Furthermore, we describe the principal characteristics of the method and its control parameters, and we obtained its regulation values to achieve reliable results with granular materials.

5. References

[1] Duran J 1997 *Sables poudres et grains,* (Eyrolles) ISBN 2-212-05831-4.
[2] Nedderman R 1992 *Statics and kinematics of granular materials* ed. (Cambridge University Press) ISBN 0-521-40435-5.
[3] Prasad A 2000 Particle image velocimetry *Current Science* **79** 51–60.
[4] Keane R and Adrian R 1992 Theory of cross-correlation analysis of PIV images *Applied Scientific Research* **49** 191–215.
[5] Westerweel J 1993 Digital particle image velocimetry *Ph.D. Thesis Delft University Press.*

Influence of rheology on buoyancy driven instabilities of miscible displacements in 2D micromodels

M V D'Angelo[1], H Auradou[2], M Rosen[1] and J P Hulin[2]

[1] Univ. Buenos Aires, CONICET. Grupo de Medios Porosos, Facultad de Ingeniería, Paseo Colon 850, 1063 Buenos-Aires, Argentina.

[2] Univ. Pierre et Marie Curie-Paris 6, Univ. Paris-Sud, CNRS. Lab. FAST, Bat 502, Campus Universitaire, F-91405 Orsay, France.

E-mail: vdangelo@fi.uba.ar

E-mail: auradou@fast.u-psud.fr

Abstract. The stability of miscible displacements of Newtonian and shear-thinning fluids of slightly different densities ($\Delta\rho/\rho \approx 3 \times 10^{-4}$) with a mean flow velocity U is investigated in a $2D$ transparent network of channels (average width = 0.33 mm). Concentration maps providing information at both the global and local scale are obtained through optical absorption measurements and compared in gravitationally stable and unstable vertical flow configurations; the influence of buoyant flows of typical velocity U_g is characterized by the gravity number $N_g = U_g/|U|$. For $N_g < 0.2$, the spreading of the mean relative concentration profile is diffusive for both types of rheologies and characterized by a single dispersivity value $l_d = D/U$. For the Newtonian water-glycerol solution, l_d is only the same in the stable and unstable configurations for $|N_g| < 0.01$. For $0.01 < N_g < 0.2$, l_d is increased by buoyancy in the unstable configuration and increasingly large front structures are observed on the concentration maps; for $N_g > 0.2$, front spreading is not diffusive any more. In the stable configuration, in contrast, the front is flattened by buoyancy for $N_g < -0.01$ and l_d reaches values of the order of the length of individual channels. For the shear thinning water-polymer solution, both the concentration maps and the value of l_d are the same in the stable and unstable configurations over the full range of U values investigated: this stabilization is explained by their high effective viscosity at low shear rates keeping N_g below the instability threshold even at the lowest velocities.

1. Introduction

Miscible fluid displacements in porous media are found in many environmental, water supply and industrial processes [1, 2, 3]. The characteristics of these processes, such as the width and the geometry of the mixing front, are often influenced by contrasts between properties of the displacing and displaced fluids such as their density. For unstable density contrast configurations, gravity driven instabilities may appear and broaden the displacement front leading to an early breakthrough of the displacing fluid: this is the case, for instance, when a volume of dense pollutant percolates into a saturated porous medium from its upper surface [4, 5, 6, 7, 8].

The appearance of these instabilities may have harmful consequences, both by enhancing the spreading of pollutants and by reducing the efficiency of industrial processes. The objective of the present work is to demonstrate how such instabilities may be blocked by using shear thinning fluids while, for a same mean given flow velocity, they appear (and may even control the displacement process) for Newtonian fluids.

A key parameter is the characteristic velocity U_g of the buoyant flow components induced by the difference between the hydrostatic pressure gradients in the two fluids which can be estimated from Darcy's law:

$$U_g = -k\frac{\Delta\rho g}{\mu}, \tag{1}$$

where $\Delta\rho$ is the difference between the densities of the upper and lower fluid, k is the permeability of the medium and μ the viscosity: with this definition, U_g is positive if oriented upward. When a natural or externally induced mean flow of velocity U is present, the relative magnitude of U_g and U is a key characteristic of the tracer transport process. In the present problem, it is characterized by the gravity number:

$$N_g = \frac{U_g}{U} = -\frac{\Delta\rho g k}{\mu U} \tag{2}$$

In the high velocity limit ($|N_g| << 1$), tracer transport is controlled by hydrodynamical dispersion [1]. If $N_g > 1$ (low velocity and unstable large density contrast configuration), large fingers corresponding to Rayleigh-Darcy instabilities appear in the displacement front. Finally, for $N_g < -1$, the stabilizing density contrast reduces the dispersion.

The present paper is focused on the domain $0 < |N_g| < 0.2$ including the intermediate case of moderate buoyancy effects. Practically, this corresponds to the release of a small amount of pollutant in a porous medium initially saturated by a clean fluid and inducing only small density contrasts (here $\Delta\rho/\rho \simeq 3 \times 10^{-4}$): in this case U_g is always small compared to U (so that $|N_g| < 1$). We have studied particularly the influence of buoyancy driven flows on the structure and development of the mixing zone: an important issue is whether the variation of the front width Δx remains diffusive ($\Delta x = 2\sqrt{Dt}$ in which the dispersion coefficient D characterizes the mixing process) or whether it is convective ($\Delta x \propto t$).

High resolution optical measurements have been performed in a vertical transparent $2D$ network of channels of random widths [9] with dye added to one of the fluids. The experiments combine visualizations at the pore scale providing a local description of the mixing process and measurements of the average of the relative concentration of the fluids across the width of the model.

Experiments are performed at different mean displacement velocities U in both stable and unstable density contrast configurations. Comparing displacement processes in the two configurations at a same velocity allows one to detect and characterize the development and influence of buoyancy driven flows. The present experiments will demonstrate that an "ideal" tracer situation (*i.e.* negligible buoyancy effects with $U_g << U$ [10]) may be reached by adding a polymer to the solutions: this increases indeed very much their effective viscosity, particularly at low velocities U.

2. Experimental set-up and procedure

The experimental system and the technique for analyzing the data have already been described in reference [10] and the setup is shown in Fig. 1. The model is a network of 140×140 channels with a mesh size $d = 0.67$ mm and a depth equal to 0.5 mm. The width a of the channels takes 7 values between 0.1 and 0.6 mm with a log-normal distribution and a mean $\overline{a} = 0.33$ mm. The permeability of the network is $k = 3 \times 10^{-9}$ m^2. The model is vertical with its open sides horizontal; its lower end is initially slightly dipped into a bath of the initial fluid of density ρ_1 which is sucked uniformly at the top of the model by a syringe pump to saturate the medium. After the saturation, the bath surface is lowered and the initial fluid is replaced by the second one of density ρ_2 without perturbing the fluid inside the model. The bath surface is then raised and fluid is pumped again at the top of the model to induce the miscible displacement. This allows one to obtain a front of the second fluid which is initially perfectly flat.

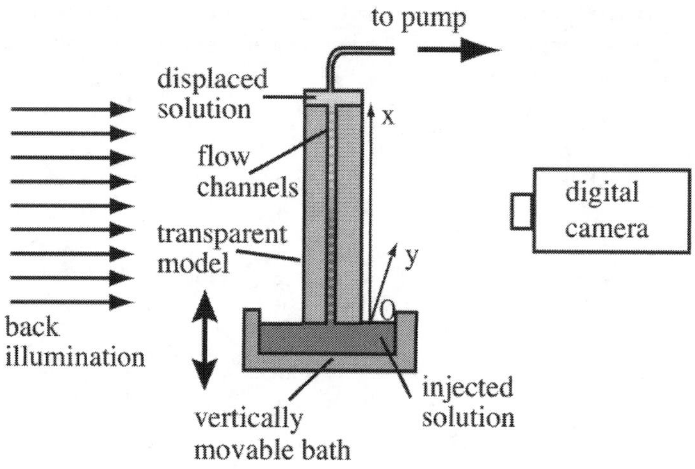

Figure 1. Experimental setup for miscible displacement measurements in 2D micromodels. Flow is vertical and parallel to the x axis. The width of the channel network is horizontal and parallel to the y axis.

Here, the absolute value $|\Delta\rho|$ of the density difference between the two fluids is kept constant with $|\Delta\rho| = |\rho_1 - \rho_2| = 0.3$ kg.m^{-3} and is due to the small amount of Water Blue dye added to one of the fluid. Since the density varies linearly with the concentration of dye, measuring this concentration optically also provides the value of the local density. For each flow rate and pair of fluids used, both stable ($\Delta\rho > 0$ and $N_g < 0$) and unstable ($\Delta\rho < 0$ and $N_g > 0$) density contrast configurations are studied by swapping fluids 1 and 2. The mean flow velocity ranges from 5×10^{-3} to 2.5 mm.s^{-1}.

The fluids used in the experiments are either Newtonian water-glycerol mixtures or shear thinning water-polymer solutions. The Newtonian solution is obtained by mixing 60% in weight of glycerol in pure water. Its viscosity is equal to $\mu = 10^{-2}$ Pa.s at 20 ^{o}C and the buoyancy induced velocity $|U_g|$ is 10^{-3} mm.s^{-1}. At the lowest imposed flow velocity ($U = 5\times10^{-3}$ mm.s^{-1} so that $N_g = U_g/U = 0.2$), the time necessary to saturate completely the network is $t \simeq 3\times10^4$ s. During this time lapse, the characteristic length $U_g t$ characterizing the size of the gravitational instabilities is 28 mm for the water glycerol mixture: this is much larger than the length d of the individual channels of the network and a noticeable influence of gravity on the structure of the displacement fronts is thus expected.

In an effort to reduce the gravitational instabilities, glycerol was replaced in the solutions by 1000 ppm of scleroglucan (high molecular weight polymer from Sanofi Bioindustries). At low shear rates, these solutions reach a constant high viscosity ($\mu \approx 4.5$ Pa.s) and the corresponding velocity of buoyancy driven flows (see Eq.2) is $|U_g| = 2 \times 10^{-6}$ mm.s^{-1}. At the lowest imposed flow velocity used here ($U = 5\times10^{-3}$ mm.s^{-1}), one has then $N_g = 4\times10^{-4}$ and the characteristic size of the gravitational instabilities is less than 0.6 mm, *i.e.* smaller than the pore size d; one expects therefore that, for this solution, gravitational instabilities will be damped and dye will behave like an "ideal" tracer.

The model is illuminated from the back by means of a fluorescent light panel and images are acquired by a 12 bits high stability digital camera with a 1030 × 1300 pixels resolution (pixel size = 0.16 mm). Typically, 100 images are recorded for each experimental run at time intervals between 2.5 and 700 s. The images are then translated into maps of the relative concentration distribution of the two fluids in the model using a calibration procedure already described in refs. [10, 13] and widely used [6].

3. Experimental results

3.1. Qualitative observations.

Figure 2 displays relative concentration maps obtained using these different fluid couples in both stable and unstable density contrast configurations. The water-glycerol experiments displayed in

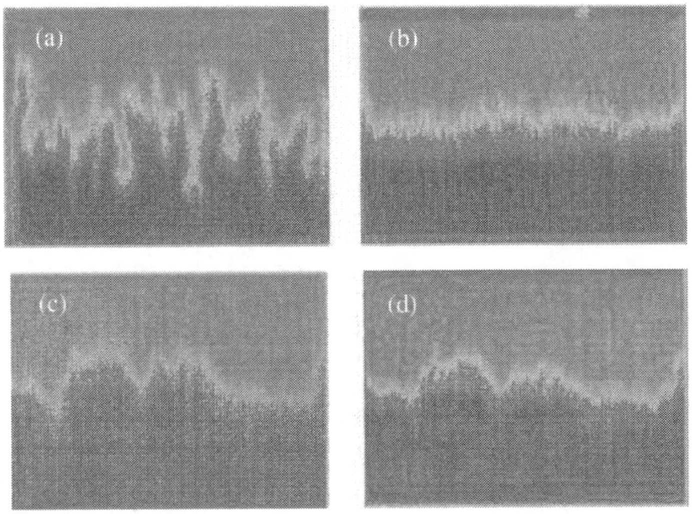

Figure 2. Relative concentration maps for experiments using the water-glycerol (a,b) and polymer solution (c,d) at respective Péclet numbers $Pe = 120$ and $Pe = 25$ (see Fig. 5). Figures (a,c) and (b,d) correspond respectively to unstable ($N_g > 0$) and stable ($N_g < 0$) density contrast configurations.

Fig. 2a and b correspond to a gravity number $|N_g| = 0.04$. Even though this value is significantly lower than 1 and buoyancy might be expected to play a minor role, its effect on the mixing fronts (yellow contours in the figures) is clearly visible. In the unstable configuration (case a), these contours display large fingers increasing the width of the front in the direction parallel to the mean flow. As will be discussed in the next section, the dispersion of the dye tracer is enhanced although it can still be described by the classical convection-dispersion equation. In section 4, the structure of the mixing zone will be analyzed from the point of view of the geometry of the iso-concentration fronts ($C = 0.5$). In the gravitationally stable case (Fig. 2b), gravity flattens the mixing front resulting in a reduced dispersion. Similar pictures obtained during experiments using water-scleroglucan solution are also displayed in Figure 2. Although the velocity U is similar to that corresponding to Figs. 2 a and b, no instability or flattening of the front due to buoyancy appears in either the stable and unstable configuration (cases (c) and (d)).

Globally, these results agree qualitatively with macroscopic dispersion measurements performed on $3D$ packings of glass beads [14] using conductivity tracers detected at the outlet of the samples. In these experiments, buoyancy driven instabilities are also observed at low velocities for Newtonian water-glycerol solutions in a gravitationally unstable configuration but do not appear for water-scleroglucan solutions. With respect to this latter work, the present optical measurements provide additional information on the variation with the control parameters and with time of the width of the mixing zone and of the different geometrical features of the front. We examine first now the variation of the global dispersion characteristics as a function of the experimental parameters and of the configuration of the fluids.

3.2. Quantitative global dispersion measurement procedure

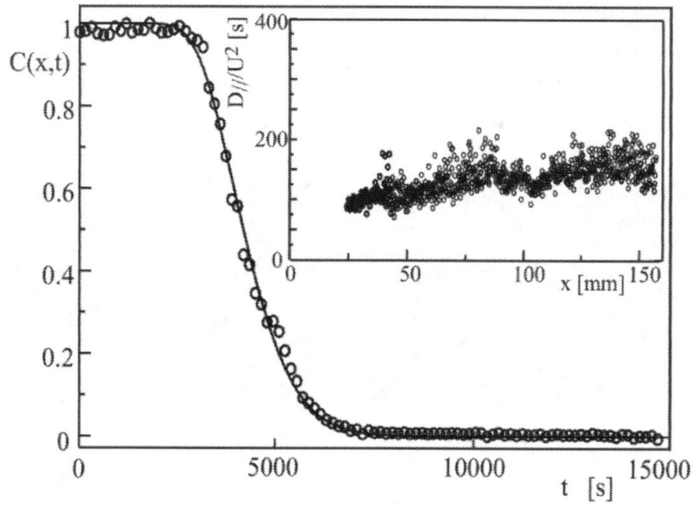

Figure 3. Variation of the normalized mean concentration $C(x,t)$ as a function of the normalized time $(t/\bar{t})$ for $x = 104$ mm and for a displacement experiment using the water-glycerol mixture in a gravitationally unstable configuration ($U = 0.025$ mm.s^{-1} – $N_g = 0.04$). Continuous line: fit by Equation (3) with $\bar{t} = 4350$ s and $D_{\parallel}/U^2 = 110$ s. Inset: variation of $D_{\parallel}/U^2$ obtained by fitting the mean concentration variation $C(t)$ by Eq. (3) as a function of the distance x from the inlet.

The procedure for determining a global dispersion coefficient from the concentration maps is described in detail in reference [10]. A mean value $C(x,t)$ of the relative local concentration $c(x,y,t)$ of the heavy (dyed) fluid at a distance x from the inlet is first determined by averaging the value of $c(x,y,t)$ for individual pixels over a window of width $\Delta y = 140$ mm perpendicular to the flow and located in the central part of the model (only pixels located inside the pore volume are included in the average).

Figure 3 displays the variations with time of $C(x,t)$ for $x = 104$ mm from the inlet. This variation is well fitted (continuous line) by the following solution of the convection diffusion equation when an initial step-like concentration variation and a constant concentration in the direction perpendicular to the mean flow are assumed:

$$C(x,t) = \frac{1}{2}\left[1 - \frac{N_g}{|N_g|}\, erf\left(\frac{t-\bar{t}}{\sqrt{4\frac{D_{\parallel}}{U^2}t}}\right)\right] \tag{3}$$

Adding the factor $N_g/|N_g| = \pm 1$ makes the equation usable in both the gravitationally stable and unstable density contrast configurations (respectively $N_g < 0$ and $N_g > 0$). This fit provides the values of the mean transit time $\bar{t}$ and of the ratio $D_{\parallel}/U^2 = \bar{\Delta}t^2/(2\bar{t})$ in which $\bar{\Delta}t^2$ is the centered second moment.

The parameter $\bar{t}$ is always observed to vary linearly with time, implying that the mixing front moves at a constant velocity. Moreover, but for $N_g < 0.2$, linear regressions on the data indicate that the front velocity is (within $\pm 3\%$) equal to the value estimated from the flow rate. The inset of Figure 3 displays the variation of the second parameter $D_{\parallel}/U^2$ as function of the distance x. $D_{\parallel}/U^2$ increases slightly at first with x and levels off for $x \sim 50$ mm; for $x > 50$ mm, it fluctuates around a constant value which is referred to in the following as D/U^2. In this "plateau" regime and at all velocities, the standard deviation of the second fitting parameter $D_{\|}/U^2$ is always less than 2%. These results prove that, except for $N_g = 0.2$, front spreading is diffusive with an effective width increasing as the square root of time: this is in agreement with previous studies [17, 6] in which a transition between dispersive and convective spreading was reported for $N_g \simeq 0.30$. One may therefore assume that D/U^2 characterizes globally the spreading of the front parallel to the mean flow.

3.3. Dispersion measurement results

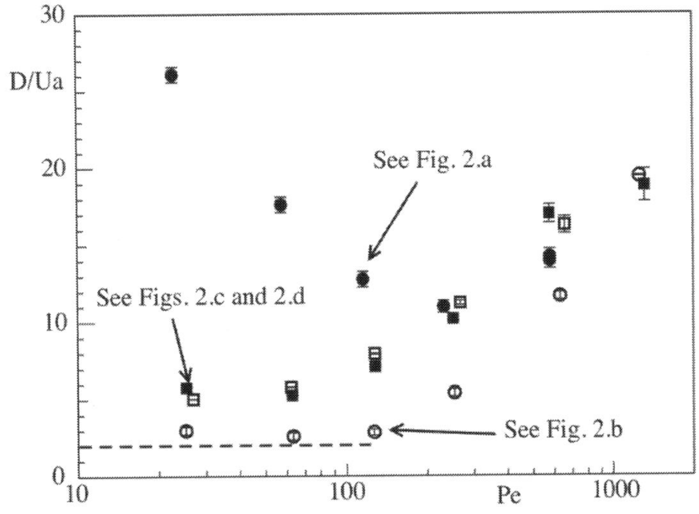

Figure 4. Variation of the dispersivity D/U, as a function of the Péclet number $Pe = U\bar{a}/D_m$ for two fluids of different rheologies: ($\circ, \bullet$) water-glycerol mixture; ($\square, \blacksquare$) 1000 ppm polymer solution in gravitationally stable ($\circ, \square, N_g < 0$) and unstable ($\bullet, \blacksquare, N_g > 0$) configurations

The present section is devoted to the variation of D (or rather of the normalized dispersivity $l_d/\bar{a} = D/(U\bar{a})$) as a function of the Péclet number $Pe = U\bar{a}/D_m$ in the stable and unstable configurations: these variations are displayed in Figure 4 for the different solutions investigated.

For water-glycerol solutions and at low Pe's, the values of $l_d/\bar{a}$ in the gravitationally unstable configuration ($\bullet$) are significantly larger than in the stable one ($\circ$). This difference reflects the increase of the global front thickness parallel to the flow due to the development of finger-like structures at low velocities in the first case (Fig. 2a) while the front is flatter in the stable case (Fig. 2b). At the lowest Pe value ($Pe \simeq 25$) for which the dispersivity can be measured in both configurations, there is nearly a factor of 10 between the values of $l_d/\bar{a}$ for $N_g > 0$ and $N_g < 0$. For $N_g < 0$ and at low Pe's, the normalized dispersivity $l_d/\bar{a}$ reaches a limiting value (1 mm) of about twice the mesh size of the lattice (horizontal dashed line in Fig. 4).

For displacements in the same range of Pe values but using a 1000 ppm water-polymer solution, the variations of $l_d/\bar{a}$ with Pe are instead the same for both $N_g > 0$ and $N_g < 0$: this could be expected in view of the very similar maps displayed in Figs. 2c-d for the two configurations. This coincidence confirms that the high viscosity of these shear thinning solutions at low shear rates prevents the development of fingering instabilities in the gravitationally unstable configuration. Quantitatively, this is consistent with the low value $N_g \simeq 5\ 10^{-4}$ estimated for the lowest experimental velocities by taking the viscosity of the solutions equal to the low shear rate limit $\mu \sim 4.5$ Pa.s. Similarly, in the stable configuration, $l_d/\bar{a}$ is not reduced and the front is not flattened by buoyancy, unlike for the water-glycerol solutions.

At higher velocities ($U > 0.1$ mm.s^{-1} or $Pe > 630$), l_d takes similar values for both $N_g > 0$ and < 0 for all fluid pairs and increases significantly with Pe. For the polymer solution, $l_d/\bar{a}$ varies as $Pe^{0.35}$ (like at lower Péclet numbers) and, for the water-glycerol solution, it increases as $Pe^{0.5}$. Both behaviors differ from the slow increase observed in such $3D$ media as homogeneous monodisperse grain packings. Such variations have been studied experimentally in ref. [10] and numerically in ref. [11]: they have been shown to reflect the combined effects of geometrical dispersion due to the disorder of the velocity field and of Taylor dispersion due to the velocity profiles in individual channels. The latter becomes important at high Péclet numbers due to reduced mixing by transverse molecular diffusion at the junctions [12]: as a result, the correlation lengths of the velocity of the tracer particles along their trajectories (and, therefore, the dispersivity l_d) become larger.

In all cases, the dispersivity variations reported above combine several types of effects. A first one is the local spreading of the concentration variation front which may be deduced from concentration profiles along a single line of pixels parallel to x. A second one is the variations of the geometry of the front resulting from variations of the mean transit time between neighbouring flow paths. These latter variations are now investigated in the present system down to small lengths scales by taking advantage of the $2D$ geometry of the model and of the high resolution of the imaging system.

4. Spatial structure of the displacement fronts for unstable flows

In porous media, the front structures associated to heterogeneities have been shown experimentally to be strongly amplified in unstable density contrast configurations [16]. This latter effect is studied here by comparing front geometries observed for different values of N_g. At a time t, the front is assumed to correspond in the relative concentration map $c(x, y, t)$ to the set of points (or iso-concentration contour) for which $c(x, y, t) = 0.5$; as mentioned above, c is the local relative concentration on the pixel of coordinates (x, y). Examples of such contours determined at four gravity numbers $0.02 \leq N_g \leq 0.2$ are displayed in Fig. 5 for both solutions used in the present work at low and high velocities. For the water-glycerol solution and for $N_g = 0.2$ which corresponds to the lowest flow velocity investigated, front spreading, as mentioned above, is not dispersive. In this latter case, $U_g \sim U$ so that the displacement is dominantly controlled by buoyancy driven flows: they are localized in a few fingers soaring up in the model while the front displacement is almost zero in other regions (Fig. 5a). At higher mean flow velocities ($N_g < 0.2$), the global spreading of the concentration profile is again dispersive (see Sec. 3.3): the large rugosities of the front disappear as the velocity increases as can be seen Fig. 5b. For the shear-thining water-scleroglucan solution, no instability is visible, even at the lowest U value which, in Fig. 5c is of the same order of magnitude as in Fig. 5a: as mentioned above, this results from the high effective viscosity at these velocities which leads to a much lower value of $N_g \simeq 0.00045$. For this solution, increasing the velocity leads to similar large scale structures of the front as shown in Fig. 5d; its local roughness is however of larger amplitude which reflects an enhancement of local spatial velocity contrasts [10].

5. Discussion and conclusion

The present experiments on have demonstrated that buoyancy driven instabilities of the vertical miscible displacement of two fluids of slightly different densities could be prevented by using shear thinning fluids: the stabilization results from the high viscosity of these latter solutions at low shear rates. These results agree qualitatively with those of previous investigations on a $3D$ bead packing which used only tracer concentration measurements at the outlet of the sample [14].

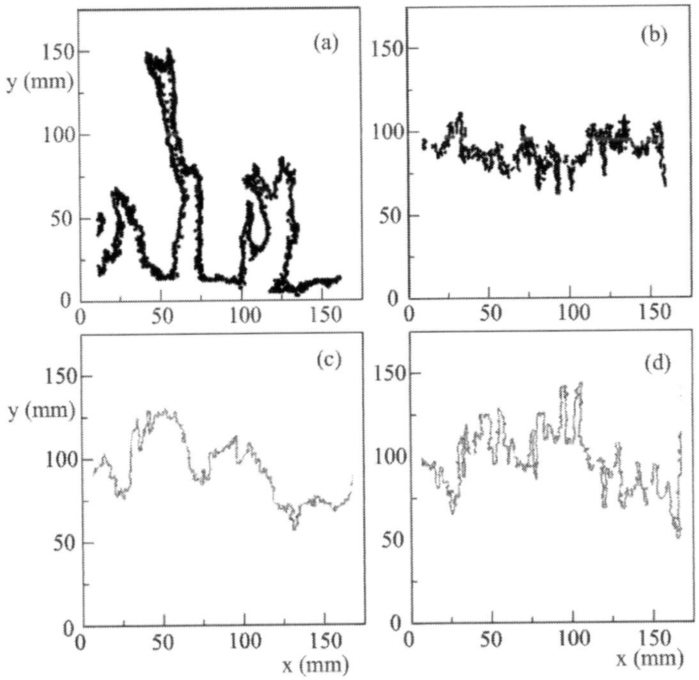

Figure 5. Isoconcentration contours $c(x, y, t) = 0.5$ for mixing fronts of Newtonian and shear thinning fluids in a gravitationally unstable flow configuration (injected volume of fluid = half the pore volume of the model). a), b): water-glycerol solution. c), d): water-scleroglucan solution. (a) $U = 0.005$ mm.s^{-1} ($N_g = 0.2$), (b) $U = 0.05$ mm.s^{-1} ($N_g = 0.02$), (c) $U = 0.005$ mm.s^{-1} ($N_g = 0.00045$)(d) $U = 0.125$ mm.s^{-1}. Flow is upward on the figure and $\vec{g}$ is oriented downward.

In contrast, for Newtonian water-glycerol solutions in a gravitationally unstable density configuration, fingering instabilities are observed even for low velocities and small $\Delta\rho$'s. In a stable situation, the effect of buoyancy is, instead, a clear flattening of the front at low flow velocities.

Quantitatively, the variation with time and with distance of the mean concentration is, almost always, well fitted by solutions of a classical convection-dispersion equation. For the water-glycerol solutions, the corresponding dispersivity $l_d = D/U$ in the unstable configuration becomes larger than in the stable one below a threshold velocity; the difference increases at still lower velocities for which the amplitude of the instabilities is observed to be very large on the concentration maps (at the lowest velocity investigated, front spreading is not even dispersive). The values of the threshold velocities at which the dispersivities in the stable and unstable cases become different suggest that the ratio $N_g = U_g/U = -\Delta\rho g k/\mu U$ of the characteristic velocity U_g of buoyancy driven motions and of the mean flow velocity U is the suitable parameter for predicting the onset of the instabilities.

For $N_g < 0.2$, the present measurements demonstrate that, even when instabilities are present, front spreading is dispersive all along the length of the model parallel to the flow with a front width increasing as the square root of distance. Front spreading results here from the combined effects of the structure of the velocity field (sole mechanism in the passive tracer case) that controls the mixing of the fluids and of buoyancy driven flows; the latter either reduce or enhance dispersion depending on the stable or unstable configuration of the fluids.

At low Péclet numbers or equivalently for $N_g > 0.2$, the front displays large spikes and troughs and their dynamics cannot be analyzed from linear stability theories. An approach taking into account the combined influence of longitudinal buoyancy forces parallel to the mean flow and of the exchange of solute between the instability fingers and the surrounding fluid is needed to account for the dependence of the dispersivity on the Péclet number ([15]). For the shear-thinning water-polymer solutions, $|N_g|$ is always less than 10^{-4}: N_g is then always lower than the threshold value and the dispersivities measured in the stable and unstable configurations are indeed the same.

Preventing such instabilities may be relevant both to the control of the propagation of pollutants and to the improvement of the efficiency of displacement processes in petroleum or chemical engineering. Compared to achieving the same result by means of a Newtonian high viscosity fluids, shear thinning fluids allow one to avoid both an excessive energy dissipation at high shear rates (their effective viscosity is reduced) and to leave the molecular diffusion coefficient practically unchanged.

Acknowledgments
We thank C. Zarcone and the "Institut de Mécanique des Fluides de Toulouse" for providing us with the micromodel used in these experiments and R. Pidoux for manufacturing the experimental set-up. This work is part of the ECOS Sud program A03-E02 and of the CNRS-CONICET Franco-Argentinian "Programme International de Cooperation Scientifique" (PICS n^o2178).

References

[1] J. Bear, Dynamics of Fluids in Porous Media, Elsevier Publishing Co., New York (1972).
[2] F.A.L. Dullien, Porous Media, Fluid Transport and Pore Structure, 2nd edition, Academic Press, New York (1991).
[3] M. Sahimi, Flow and Transport in Porous Media and Fractured Rock, VCH, Weinheim, Chap. 12 (1995).
[4] R.A. Schincariol, F.W. Schwartz and C.A. Mendoza, Water Resour. Res. **33** (1), 31–41 (1997).
[5] M. Beinhorn, P. Dietrich and O. Kolditz, J. Contam. Hydrol. **81**, 89–105 (2005).
[6] T. Menand and A.W. Woods, Water Resour. Res. **41**, W05014 (2005).
[7] K. Johannsen, S. Oswald, R. Held and W. Kinzelbach, Adv. Water Resour. **29**, 1690–1704 (2006).
[8] T. C. Flowers and J. R. Hunt, Water Resour. Res. **43**, W01407 (2007).
[9] R. Lenormand, C. Zarcone and A. Sarr, J. Fluid Mech. **135**, 337–353 (1983).
[10] V. D'Angelo, H. Auradou, C. Allain, J-P Hulin, Phys. Fluids **19**, 033103 (2007).
[11] C. Bruderer and Y. Bernabé, Water Resour. Res. **37**, 897–908 (2001).
[12] Y. Park, J.R. de Dreuzy, K. Lee and B. Berkowitz, Water Resour. Res. **37**, 2493–2501 (2001).
[13] A. Boschan, H. Auradou, I. Ippolito, R. Chertcoff and J.P. Hulin, Water Resour. Res. **43**, W03438 (2007).
[14] M.A. Freytes, A. d'Onofrio, M. Rosen, C. Allain, J.P. Hulin, Physica A **290**, 286-304 (2001).
[15] V. D'Angelo, H. Auradou, C. Allain, M. Rosen, J-P Hulin, Phys. Fluids **20** 034107 (2008).
[16] V. Kretz, P. Berest, J.P. Hulin, D. Salin, Water Resour. Res. **39**, 1032-1041 (2003).
[17] H.H. Liu and J.H. Dane, J. Contam. Hydrol. **23**, 233–243 (1996).

X Meeting on Recent Advances in the Physics of Fluids and their Applications IOP Publishing
Journal of Physics: Conference Series **166** (2009) 012006 doi:10.1088/1742-6596/166/1/012006

Numerical study of unstable reaction-diffusion fronts in porous media

D Müller[1], A D'Onofrio[1,2] and M Cachile[1,2,3]

(1) Grupo de Medios Porosos. Facultad de Ingeniería. Universidad de Buenos Aires. Paseo Colón 850 (1063). Buenos Aires. Argentina, (2) CONICET

E-mail: (3) `mcachil@fi.uba.ar`

Abstract. Numerical simulations using a lattice-gas method (FHP-III) were performed. The method was modified for simulating two miscible fluids into a rectangular cell with obstacles (porous medium) and without obstacles (Hele-Shaw cell). An external force can be applied to each fluid. A suitable combination of this force and initial conditions can develop an unstable interface front (Rayleigh-Taylor instability). A chemical reaction between the fluids was also introduced, with a parameter that controls the reaction rate. The influence of the parameters that control the force and the chemical reaction on the development of the instability was studied.

1. Introduction

Usually the numerical resolution of a fluid flow is based in discretizing the relevant equations in several ways (finite elements, spectral techniques, etc.). The lattice methods, such as lattice-gas, use a different approach: the key idea is to use pseudo particles moving in a discrete lattice with a small number of possible velocities, constrained by the geometry of the lattice itself. In many real systems, the behaviour at macroscopic scales does not depend on the details of the dynamics at the microscopic scales. In these cases, it appears as a feasible option to approximate the macroscopic phenomena by using simplified rules applied at a much smaller scales.

Although at microscopic (lattice) scales the interparticle dynamics is extremely simplified, the method reproduces the correct behavior of a fluid at larger scales. Under some hypothesis, this method satisfies the incompressible Navier-Stokes equation[1, 2].

The main advantage in using this method is that it is not required the use of the macroscopic equations of the problem, since only elementary microscopic interactions between particles are involved.

At the same time, it is relatively easy to define rather complex boundaries and contours, which makes this method suitable to study flows in porous media [2, 3].

The lattice-gas method can be applied in problems with two fluids, with different mass densities. If the density gradient through the interface is opposite to the gravity, there will appear density-driven instabilities [4, 5]. It has been shown that this situation can be simulated by the lattice-gas method by applying a different force (*weight*) for each species [6]. This scheme can be applied not only to a difference in density, but to any other problem in which each species is under the influence of a different force.

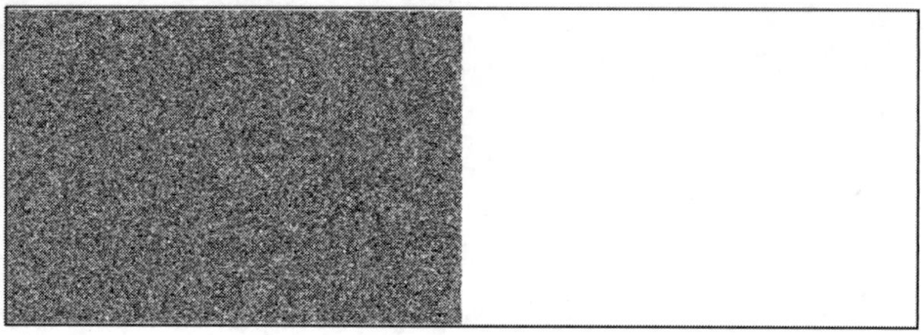

Figure 1. Typical initial condition. For clarity, only green particles are shown.

In literature, the porous media used in this kind of simulations are usually composed of one-node obstacles, known as *scattering nodes* [7]. These obstacles reproduce quite well the effect of the porous media on the flow, but with a different permeability - porosity relationship of real porous media.

In addition, the model was modified to simulate a chemical reaction, where one species is transformed into the other. It was done in a way that reproduces an autocatalytic reaction, previously studied numerically [8] as well as by experimental methods [9, 10].

The main goal of this work is to analyze the influence of the external force and chemical reaction on the development of the unstable front between the two fluids, inside Hele-Shaw cells and porous media.

2. Model Set Up

A variation of the basic FHP-III model of seven velocities [2] has been set up, modified in order to simulate two fluids.

Particles of both species are identical, except in a scalar magnitude we called *colour* used to distinguish each species. The colours were chosen to be red and green, from RGB colour model of images.

2.1. Species assignment

During the collisions, the particles are treated in the same way regardless their species, which implies the need of a redistribution after each collision. A random redistribution, uncorrelated with the initial condition was used, which ensures that both species behave as miscible.

2.2. Initial and boundary conditions

On the solid boundaries, including the outer walls as well as the obstacles which form the porous media, we used viscous collisions, *i. e.* when a particle collide with an obstacle, its velocity is reversed. In this way, in the neighbourhood of solid boundaries, as in a real viscous fluids with no-slip condition, the fluid has essentially null velocity.

Two kind of boundaries were used: closed cells, used only in diffusion simulations, where the four walls are solid and periodic cells, where left and right walls (perpendicular to the force) are solid, while top and bottom walls are open.

The simulations were performed with an initial condition that consists in a uniform distribution of particles in the cell. The green particles are placed at the left half of the cell, while the red ones are in the other half of the cell. The concentration of both species is identical. In figure 1 is shown this initial condition.

The particle concentration was controlled by the average filling fraction per node, which is kept constant in each simulation. Concentration $c = 1$ is obtained with 7 particles per node, values of c ranging from 0.15 up to 0.99 were investigated.

2.3. External Force
The method used by Kadanoff [6] has been implemented. The particle velocities after a collision are reoriented towards the direction of the force. The only possible velocity changes are those parallel to the force ($+x$ in this case) occuring with a probability u_f. This value controls the force strength. Values of u_f ranging from 10^{-4} to 10^{-2} were studied. This force is applied only to the green particles (initially placed at the left) and is pointing to the right, favoring the appearence of density-driven instabilities.

2.4. Obstacles
Two kind of configurations were investigated: Hele-Shaw cells (without obstacles) and porous media (with obstacles). The obstacles which form the porous media are obtained by placing, at random positions circles of constant diameter d_c, with the condition that the minimum distance between any pair of centers is not smaller than $1.5d_c$. We investigated values of d_c between 1 and 10 lattice units. The number of circles has also been varied, ranging from 700 to 3000 circles per 10^6 nodes.

2.5. Chemical Reaction
With the aim of simulating the behaviour of an autocatalytic reaction between the species, a tranformation of one species into the other was implemented through a random change of the color (red to green) of one particle at each node. This is done with a probability of the net reaction rate p_q determined by the product $p_q = Q\ c_r\ c_v^2$, where Q is a chemical constant, and c_r and c_v are the number of red and green particles at the node. The values of Q used were between 10^{-5} and 10^{-2}.

3. Results
3.1. Diffusion
The governing equation, with no external forces is [8]:

$$\frac{\partial c_i(x,t)}{\partial t} + \overline{u}_i \cdot \overline{\nabla} c_i(x,t) = D\nabla^2 c_i(x,t) + q_i c_i(x,t) \tag{1}$$

Where x y t are the position and time, D is the diffusion coefficient, q_i is the reaction rate and c_i is the concentration of the species i. The solution of equation (1), without chemical reaction, and with a concentration step of c_0 placed at x_0 as initial condition, is given, for all x by the error function, erf [11]:

$$c(x,t) = \frac{c_0}{2}\left(1 - erf\left(\frac{x - x_0}{2\sqrt{Dt}}\right)\right) \tag{2}$$

This study of the diffusion has been done in matrices of 1000 x 2000 nodes, the average particle concentration, kept constant in each simulation, ranged from 0.15 to 0.99.

The concentration profiles for each species, averaged by column were fitted using equation (2) and then D was found for each concentration.

The same study was done using porous media, in order to analyze its influence on the diffusion coefficient. Porous media formed by circles of constant radius (between 1 and 10 nodes), different obstacle concentration (between 700 and 3000 circles per 10^6 nodes) and particle concentration between 0.15 an 0.99 were investigated. These results are shown in figure 2.

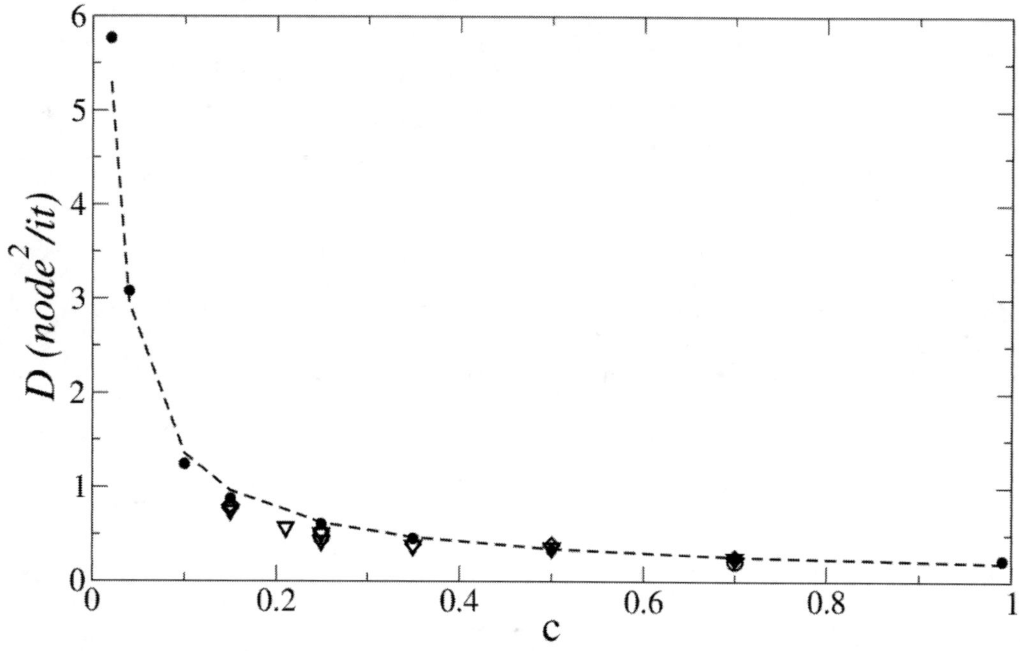

Figure 2. Diffusion coefficient as a funtion of particle concentration. Markers represent different porous media.

3.2. Chemical Reaction

The chemical reaction used in the code transforms red particles into green ones. Since in the in the initial condition green particles are at the left, the front advances to the right. Without forces, this movement is expected to be stable, *i.e*, the front should remain straight This behaviour has been observed in our simulations.

The position of the front can be determined by plotting the profile along the horizontal axis and then finding the mean-concentration point x_f where $c(x_f, t) = c_0/2$. Plotting x_f as a function of time, the front velocity is found. After a short transient, the velocity attains a constant value, V_f, shown in figure 3.

4. Instabilities

Using the same initial conditions of the previous study, simulations with the addition of an uniform force, applied to the particles at the left (green) were run.

A short time Fourier analysis of the interface between the two species was developed to study the growth of the instabilities as a function of the wavelength. This study has shown that there exist a range of wavelengths where instabilities can grow, as shown in figure 7.

Porous media were also added, and the radius as well as the concentration of the obstacles were varied.

In figure 8 is shown the behaviour of the growth rate (in iterations^{-1}) when the diameter of the obstacles is changed, while its number is kept constant in 1500 circles $/10^6$ nodes.

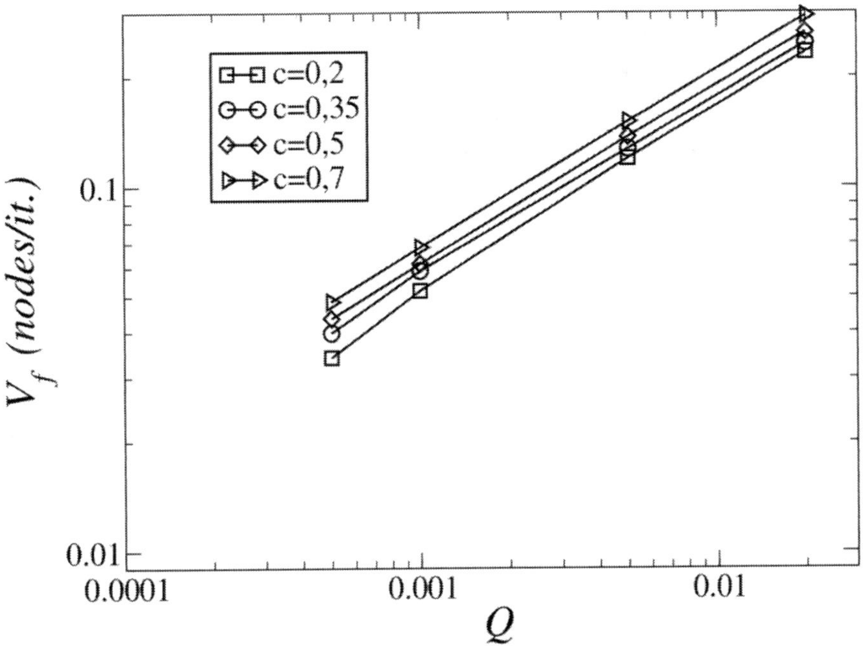

Figure 3. Reaction front velocity V_f as a function of the chemical parameter Q and different particle concentrations.

In figures 4 and 5 examples of this instability are shown for different configurations with and without porous media.

When the instability occurs under a chemical reaction, a change in the shape of the interface is observed, coherent with experimental works [10] and other numerical simulations [8] as shown in figure 6. An influence of the chemical parameter Q on the development of the instability has been observed [8, 13]. As Q increases, the instability becomes smaller, as shown in figure 9.

5. Analysis and discussion

Without forces, nor chemical reaction, the system evolves diffusively, *i.e.* when the initial condition is a concentration step, each species has a concentration profile that follows equation 2 not only in space, but also has the time variation given by the same equation.

The study on the diffusion has shown that the diffusion coefficient varies as $1/c$, coherently with results found in gaseous systems, in an important range of concentrations as shown in figure 2. This result is not contradictory with the previous one, since local concentration does not appart significantly from the average value.

It has been found that the diffusion coefficient is not modified by the presence of porous media, as shown in figure 2. This fact makes it independent on the presence of obstacles in the integration domain.

The dispersion relation of the instability at short times shows an interval of wavelengths in which instabilities can grow. It can be seen that the wavelengths in the center of that interval

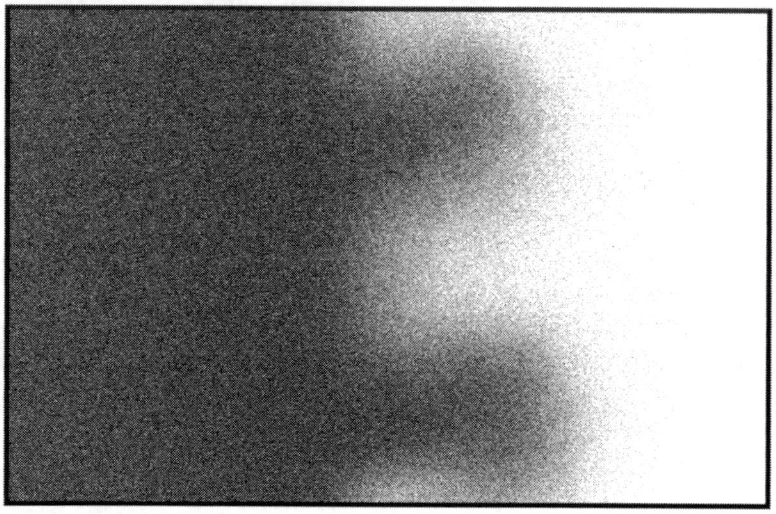

Figure 4. Unstable evolution. 10.000 iterations, no porous medium, concentration 0.7, force parameter: 8 x10^{-4}.

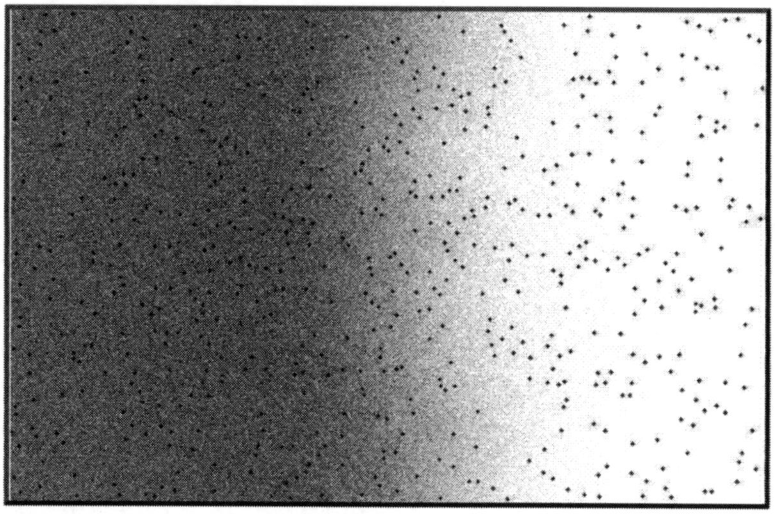

Figure 5. Identical conditions as in figure 4 , but with a porous medium of 1500 obstacles /10^6 nodes.

grow faster, as in the case of Rayleigh-Taylor instability [4, 5]. This fact shows that the model reproduces the essential characteristics of that instability in Hele-Shaw cells. Similar behaviour has been obtained when a porous medium is included in the cell.

With the addition of porous media, instabilities are dimmed, and also some shorter wavelengths become stable. Although this study is not completed, the tendency observed on the influence that the porous media have on the instabilities is very close to the observed in experiments in Hele-Shaw cells [13].

In absence of external forces and with the addition of a chemical reaction, a stable front which moves at constant velocity has been obtained. This velocity depends on the chemical parameter

Figure 6. Identical conditions as in figure 4, but with the inclusion of a chemical parameter $Q = 3\text{x}10^{-4}$.

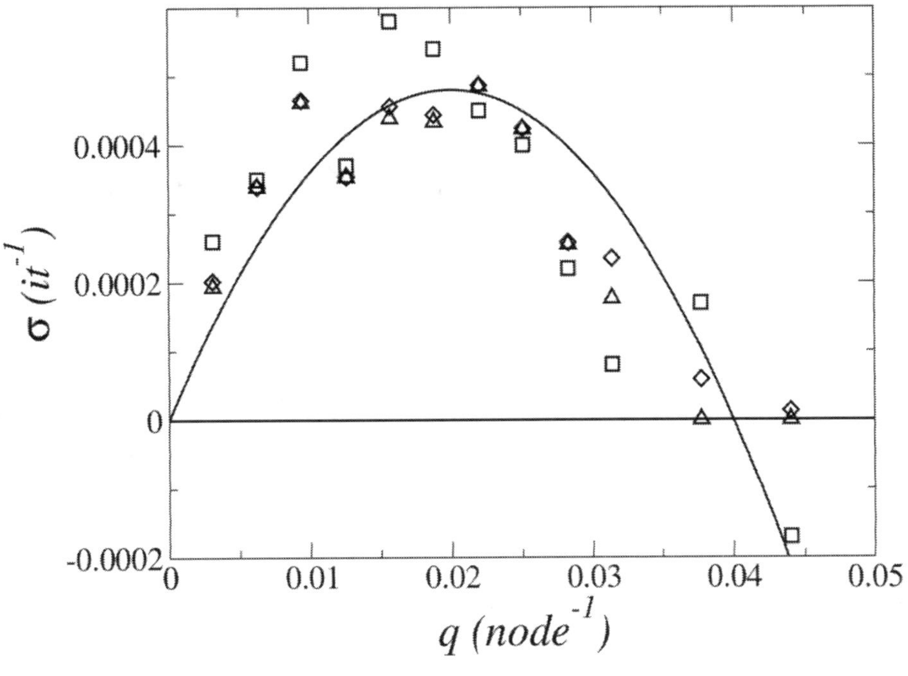

Figure 7. Dispersion relation for the Hele-Shaw cell. The full line is to guide the eye. Markers represent different runs.

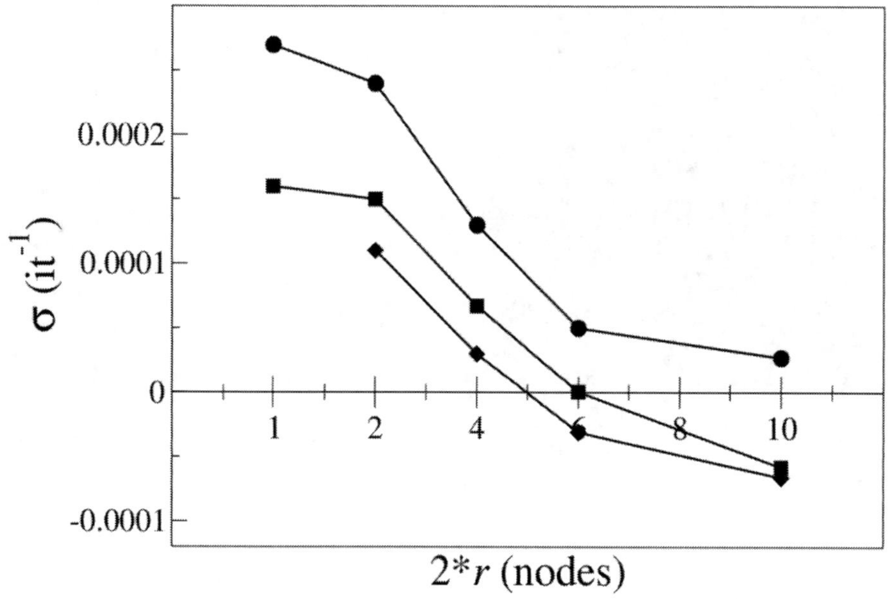

Figure 8. Growth rate dependence as a function of the diameter of the obstacles. for porous media with 1500 obstacles /10^6 nodes. The force parameter is, from up to down: 8, 4 *and* 2 x10^{-4}, wavenumber $q = 0.03\ nodes^{-1}$.

as $V_f \propto \sqrt{Q}$, coherently with experimental results from Horvath [9].

By adding a chemical reaction, the morphology of the instability is clearly modified, as shown in figure 9. This results together with the reduction in the amplitude of the instability because of the presence of a chemical reaction, are coherent with previous experimental works [10, 13].

The results of these simulations show that instabilities grow slower when: a) the diameter of the obstacles (of the porous medium) is kept constant and more obstacles are included, b) the number of obstacles is kept constant and the diameter is increased and c) when chemical parameter Q is increased.

This preliminary study shows that the Lattice gas method with two species is a suitable computational method for the study of the flow in porous media when studying two miscible fluids and under a chemical reaction.

Acknowledgments

This work has been financed by I062 project of the University of Buenos Aires.

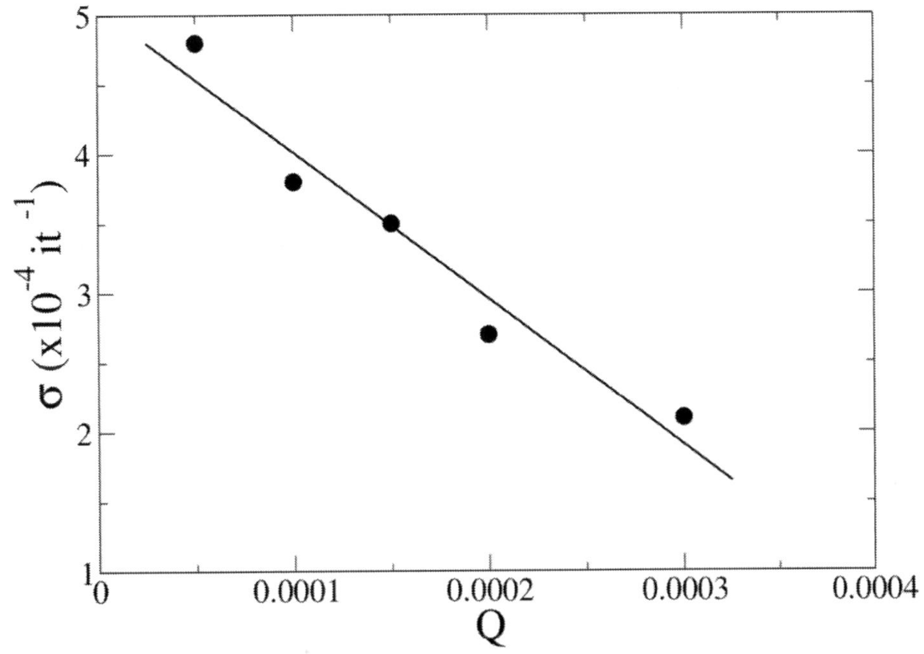

Figure 9. Growth rate σ, corresponding to $\lambda = 800$ nodes, as a function of the chemical parameter Q. The straight line is the linear regression.

Journal of Physics: Conference Series **166** (2009) 012006 doi:10.1088/1742-6596/166/1/012006

References

[1] Frisch U, Hasslacher B and Pomeau Y 1986 *Phys. Rev. Lett.* **56** 1505
[2] Rothman D and Zaleski S 1997 *Lattice Gas automata: simple models of complex hydrodynamics* (Cambridge: Cambridge University Press)
[3] Rivet J P and Boon J P 2001 *Lattice gas hydrodynamics* (Cambridge: Cambridge University Press)
[4] Chandrasekhar S 1961 *Hydrodynamic and Hydromagnetic Stability* (Oxford: Oxford)
[5] Manickam O and Homsy G M 1995 *J. Fluid Mech.* **288** 75
[6] Kadanoff L McNamara G and Zanetti G 1987 *Complex Systems I* 791
[7] Kohring G A 1991 *J. Stat Phys.* **63** 411
[8] Yang J D'Onofrio A Kalliadasis S and De Wit A 2002 *J. Chem. Phys.* **117** 20
[9] Toth A Horvath D and Siska A 1997 *J. Chem. Soc., Faraday Trans.* **93** 73
[10] Horvath D Toth A 1998 *J. Chem. Phys.* **108** 1447
[11] Jost W 1960 *Diffusion in Solids, Liquids, Gases* (New York: Academic Press)
[12] Manickam O HomsyG M 1995 *J. Fluid Mech.* **288** 75
[13] Garcia Casado G Tofaletti L Müller D and D'Onofrio A 2007 *J. Chem. Phys* **126** 114502

Journal of Physics: Conference Series **166** (2009) 012007 doi:10.1088/1742-6596/166/1/012007

Synthetic Schlieren: Determination of the density gradient generated by internal waves propagating in a stratified fluid

L P Thomas[†1], **B M Marino**[†] **and S B Dalziel**[‡]
† Instituto de Física *Arroyo Seco*, Universidad Nacional del Centro de la Provincia de Buenos Aires, Pinto 399, 7000 Tandil, Argentina
‡ Department of Applied Mathematics and Theoretical Physics, University of Cambridge, Wilberforce Road, Cambridge CB3 0WA, United Kingdom

E-mail: *lthomas@exa.unicen.edu.ar, bmarino@exa.unicen.edu.ar, s.dalziel@damtp.cam.ac.uk*

Abstract. The synthetic schlieren technique and a convenient image processing are employed to attain a new insight into the visualization of internal waves and to make quantitative precise measurements of the bidimensional distributions of very small density gradient variations, harmonics distributions patterns, etc., in a stratified fluid. The laboratory implementation of this technique is much simpler than that of the optical methods as classical schlieren and interferometry, and makes possible to obtain useful information in situations in which shadowgraph gives only qualitative results. In addition, much greater domains than those with optical methods may be analyzed, and high optical quality windows are not required. Its application is illustrated by means of the visualization of the internal gravity waves generated by an oscillating body and thermals.

1. Introduction

The information related to density fluctuations in a stratified fluid obtained in many situations has contributed to improve the understanding of the flows produced there. The available techniques range from the introduction of passive tracers, such as dying or neutrally buoyant, to shadowgraph, schlieren and interferometric techniques which depend on the refractive index variation. Examples of how to perform these visualizations [1-3] and how to extract the quantitative information of interest from them [4-5] may be found within the literature.

In particular Dalziel *et al.* [6] reported a number of techniques for obtaining both high quality visualizations and accurate measurements of the density field, based on the same optical principles as the classical schlieren method but with the advantage of being easier to setup and adaptable to much larger domains. They outlined four related techniques: qualitative synthetic schlieren, line refractometry, dot tracking refractometry and pattern matching refractometry. The first one has two operational modes: "line mode" and "dot mode", and allow the basic qualitative visualization of the density variations with respect to an unperturbed initial state of the fluid. Together with the other three quantitative techniques, it conforms the "synthetic schlieren" diagnostic, that is relatively cheap to setup and provides a powerful method to visualize, for example, bidimensional internal waves of small magnitude and measure the associated density fields.

[1] To whom any correspondence should be addressed.

Journal of Physics: Conference Series **166** (2009) 012007 doi:10.1088/1742-6596/166/1/012007

Synthetic schlieren allows analyzing the density fluctuations that are too small to be visualized by applying the techniques related to the concentration of a passive dying (e.g. light induced by fluorescein or light absorption by dying), and those in which the curvature of the beams due to density variations is out of the range of application of the common interferometry. The name reflects the relationship with classical schlieren, although it must be noted that the digital processing is used to form the "synthetized" schlieren image, instead of the knife edge in the focus of a lens or a convergent mirror.

A particular experimental challenge is the visualization of the internal waves generated in a stratified medium. Basically, if a fluid element is perturbed vertically in a stratified fluid, it experiences a buoyancy force directed to restore it to its original position. This buoyancy force, in combination with the inertia of the fluid element, acts as a simple harmonic oscillator with a frequency (the buoyancy or Brunt-Väisälä frequency)

$$N^2 = -\frac{g}{\rho_0}\frac{d\rho}{dz},$$

where ρ is the fluid density, ρ_0 is a reference density, g is the gravity and z is the coordinate oriented vertically upwards. If the element oscillation occurs in a direction at an angle ϕ to the vertical, the restoring force is reduced, allowing the formation of waves of angular frequency

$$\omega = \pm N\cos\phi .$$

This is the dispersion relationship for small amplitude waves in the Boussinesq approximation. The fluid motion and group velocity occurs at an angle ϕ to the vertical, while the phase velocity is perpendicular to them (the sign of the vertical component of the phase velocity is opposite to that of the group velocity).

Therefore, a linearly stratified fluid (N: constant) is able to transmit waves of frequencies $0 \leq \omega \leq N$ corresponding to $0 < \phi < \pi/2$. Mowbray & Rarity [7] demonstrated theoretically and experimentally (using classical schlieren) that for the waves forced by an oscillating cylinder, the fluid motion was confined within a region of fluid resembling a Saint Andrew's Cross. Other authors have subsequently used different experimental techniques to analyze these waves (for example Peters [8] and Merzkirch and Peters [9] used interferometry). A recent work by Hurley & Keady [10] has provided additional theoretical details of the structure of this wave field for oscillating cylinders of a finite size in a viscous fluid. The first experimental results confirming their prediction were provided by Sutherland *et al.* [11] by means of the vertical oscillation of a circular cylinder in a linear stratification.

In this work we present the results obtained with an optimized application of the synthetic schlieren technique and the subsequent image processing to determine the density gradient field in two different situations. In the following we review briefly others schlieren techniques in order to distinguish the comparative advantages of the synthetic schlieren, and describe the experimental setup and the methodology that introduces a new procedure to process the images using the specialized software *Digi*Flow. As an example, the results corresponding to the bidimensional internal waves and their harmonics generated by the periodical oscillation of a cylindrical body in a stratified medium are presented in Section 3. Up to our knowledge, the intensities and phases of the internal waves corresponding to the higher harmonics have not been yet quantified. In Section 4 another interesting application of the synthetic schlieren diagnostic is presented, which arises when the fluctuations field of the density gradient produced by the displacement of a thermal in a stratified fluid is looked for.

2. Experimental description

2.1. Classic schlieren techniques

The schlieren techniques ([7], [12]) employ the same optical principle, namely that a beam of light is bent in regions with refractive index changes. As the refractive index of a stratified fluid varies with density, the beams of light passing through it to angles close to the horizontal are bent depending on the density gradient. When internal gravity waves propagate in a stratified fluid, isopycnal (constant

Journal of Physics: Conference Series **166** (2009) 012007 doi:10.1088/1742-6596/166/1/012007

density) surfaces are displaced resulting in regions where the local density gradient changes with respect to the background gradient and the path of a nearly horizontal light beam passing through the medium is deflected [11].

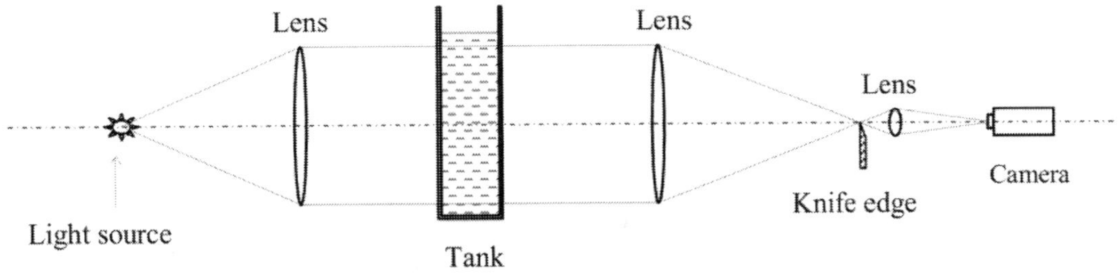

Figure 1. Optical setup for applying classical schlieren method.

In the classical schlieren technique, whose typical setup is illustrated in figure 1, a convergent lens or a parabolic mirror is employed to create beams parallel to those emitted from a light source. After passing through the test section of the tank containing the stratified fluid the rays are focused by a second lens or a parabolic mirror, while the edge of a "knife" in the focus stops the rays deviated from the parallelism in the test section. This technique allows determining the first derivative of the density variations with respect to (for example) a state with constant initial gradient [12]. Practical difficulties with this technique are associated with the careful setting up and high cost of the lens and parabolic mirrors. In addition, high quality images can only formed when the apparatus itself is optically good, and the need for a point source of illumination limits the intensity of illumination available.

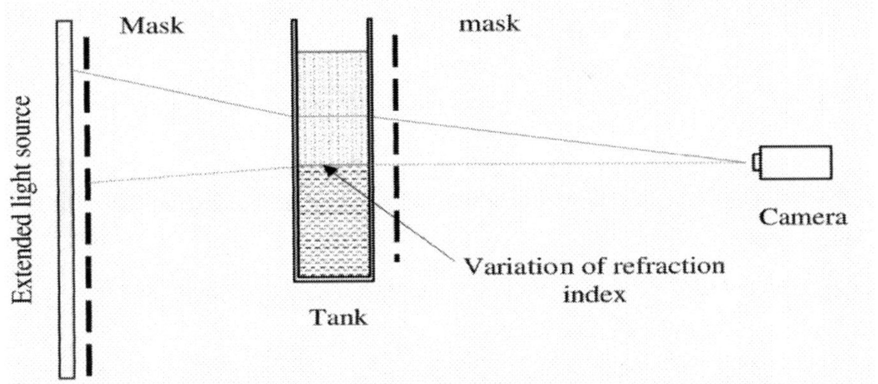

Figure 2: Moiré fringe method optical setup.

The Moiré fringe method [13], illustrated in figure 2, operates in a similar manner to classical schlieren but the pair of parabolic mirrors and the knife edge are replaced for a pair of accurately aligned masks that consist of a set of parallel lines. They are normally aligned so that 50% of the light passing through the first mask located at a side of the test section is stopped by the second "analyzer" mask located on the opposite side of the test section. Due to camera parallax, the lines on the mask in front of the tank are more closely spaced than those at the rear of the tank. The Moiré fringe method is cheaper to implement and may be scaled up to cover larger domains more easily than classical schlieren because expensive parabolic mirrors are not required. The main difficulty of this method is that the alignment between the masks is critical and non-trivial, especially if light entering the camera is not parallel or if the stratification is non-linear so that the lines spacing of the mask are not related by a constant scale factor.

2.2. Synthetic schlieren

The synthetic schlieren method overcomes the above mentioned difficulties. It needs neither lens nor good quality mirrors, the positioning and alignment of the elements are no critical, and it may be scaled to any dimension without raising the cost or making difficult the laboratory implementation. An analyzer digital "virtual" mask is used, which allows measuring quantitatively the density variations with respect to an arbitrary equilibrium state. Measurements are made in two directions simultaneously, so that the knife edge (classical schlieren) or the masks (Moiré fringe) must not be rotated to obtain the gradients in the perpendicular direction.

The figure 3 illustrates the typical arrangement of the experimental setup for applying the synthetic schlieren technique. The beams of light generated in a diffuse light source pass through a mask and the tank containing the stratified fluid forming small angles to the horizontal, and are captured by a video camera located at a distance that is longer than the size of the test section. The solid line represents the path of the beam deviated when the density gradient is greater than the equilibrium gradient.

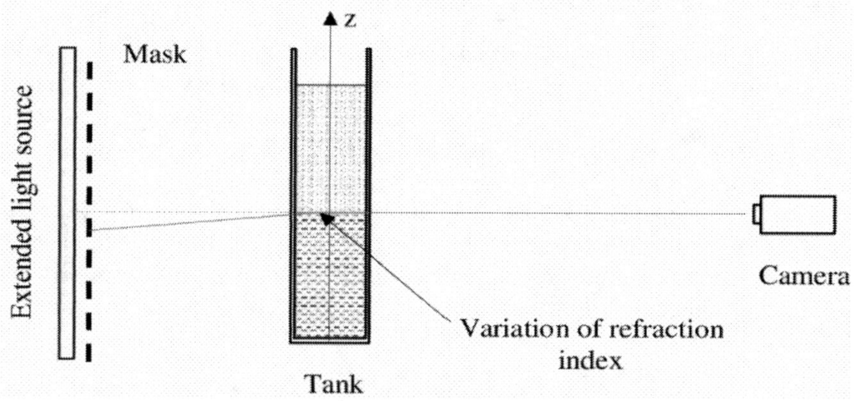

Figure 3: Synthetic schlieren experimental setup.

A white and black camera JAI CVM4+MCL, with resolution 1372×1024 pixels of 8 bits operating up to 24 images/s, is employed to acquire the images that are stored directly in a PC for processing digitally later. The camera is placed at 3.00 m far away from the tank to minimize the angle (< 3°) at which the rays are captured by the camera. A fabric with small circular holes of 1 mm diameter is used as a mask. A reference images sequence of the initial situation, that replaces the mask of the Moiré fringe method, is saved. Then any small change in the refraction index gradient inside the environment produces the deflection of the light beams. During each experiment, the deviations of the intensities with respect to the reference image corresponding to each mask hole are detected and quantified. *DigiFlow* software [14] allows the images to be processed easily obtaining qualitative information on-line. Quantitative results are obtained with an additional processing in which the light intensity deviations are converted to changes of the vertical and horizontal density gradients, provided the density and the refraction index of the stratified fluid follow an almost linear relationship [15].

The particular features related to the implementation of this diagnostic are introduced en the following two sections showing concrete applications.

3. Internal waves generated by an oscillating object

By using the double-bucket method, a 0.30 m deep, 0.91 m long and 0.08 m wide tank is filled with 0.275 m deep uniformly stratified salt water. The maximum density of the fluid, measured with a Reichert-TS refractometer and a 5500 Antor-Paar densimeter, varies between 1.0645 and 1.0225. Then the buoyancy frequency is $N^2 = 1.43 \pm 0.07\ s^{-2}$. A cylindrical body of radius 0.02 m with its centre at 0.15 m over the tank bottom is immersed with its axis aligned horizontally and spanning the full width of the tank. It is suspended from a thin metallic supporting arm which is driven by an adjustable speed eccentric cam, forcing the body to undergo sinusoidal vertical oscillations (see figure 4). The amplitude of the body oscillation (that is, the maximum distance between the lower and higher positions of the body axis) is 0.0081±0.0001 m. The flow around the body is laminar for the oscillation frequencies and amplitudes generated.

Figure 4: Experimental setup to generate internal waves by means of an oscillating body.

After the completion of an experiment, a segment of a vertical column of pixels passing through the lowest point of the oscillating cylinder was extracted from the images sequence to form a time series image as that shown in figure 5. The upper and lower zones clearly distinguished in the figure correspond to the body and the fluid, respectively. The vertical position of the lowest part of the body, $h = h(t)$, is obtained from the contour between these zones. The fundamental frequency ω_0 is determined by means of: (a) the best fit sinusoidal curve in the table (t,h), and (b) the spectral analysis of frequencies of the time series. The uncertainty in each case and the difference between both values are smaller than 1%.

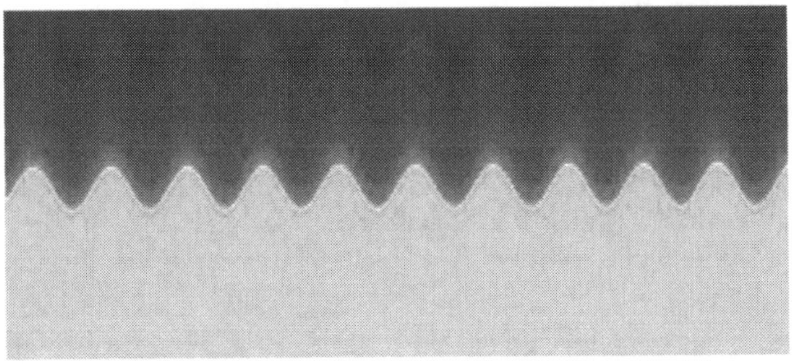

Figure 5: Time series of a fraction of a column of pixels of the images. The interface between the two colours indicates the height of a point belonging to the lower contour of the oscillating body as a function of time

3.1. Internal waves visualization

Figure 6 shows the perturbations at a given time generated by a cylinder oscillating with a frequency $\omega_0 \cong 0.298\pm0.003$ s^{-1}. Applying the synthetic schlieren technique, quantitative information of the variation of the instantaneous density gradient field $\rho^{-1}\partial\rho / \partial y$ is obtained starting from the intensity distribution of the images.

As mentioned in Section 1, internal waves are evidenced by means of fluid elements oscillations

Figure 6: Intensity of the vertical component of the density gradient in arbitrary units for a cylinder oscillating in a stratified fluid. The colour scale (below) ranges between black (negative values) and white (positive values).

whose direction forms an angle $\phi = \cos^{-1} \omega / N$ to the vertical. In the case reported here, four beams at angle ϕ are seen far from the body. Each of these beams is composed by two thinner beams of opposite phases forming the known St. Andrew Cross. Due to the symmetry of the body cross-section and since the motion axis coincides with the direction of stratification, the intensity distribution looks symmetrical with respect to vertical and horizontal axis passing through the centre of the oscillation. This suggests that the used experimental setup avoids spurious horizontal motions that might give place to flow asymmetries.

The beams evolve as expected [7], that is, they appear above (below) the cylinder with increasing intensities, then move down (up) following a direction that is perpendicular to the respective beam while their intensities decrease, and finally vanish. The angle ϕ and the direction of the motion of the beams are related to the oscillation frequency as anticipated by theory.

3.2 Higher harmonics visualization

The angular fundamental frequency ω_0 of the oscillations generates distributions varying in time as those showed in figure 6. An important experimental difficulty arises when the deviations from these distributions are wanted to visualize. The internal waves corresponding to the harmonics are also difficult to visualize because of their very small intensities (typically of one order of magnitude smaller than those of the amplitude of waves with main frequency ω_0) and their superposition with the waves of frequency ω_0. We use the fact that the density gradient in each point of the region of interest varies in time with a given frequency. This information is found in the time variations of the intensity distributions that appear in the sequences of images registered during n periods of oscillation.

Therefore, for each pixel located at (x,y) in the image, we calculate the complex parameter

$$\Re(x,y) = 2\int_0^{nT} \left.\frac{1}{\rho}\frac{\partial \rho}{\partial y}\right|_{(x,y)} e^{i\omega t}\, dt ,$$

where $\omega = \omega_0, 2\omega_0, 3\omega_0, \ldots$ are the fundamental frequency and its harmonics. Thus the value of $|\Re|$ is maximum if the frequency of the variations of the pixel intensity coincides with ω, and is minimum if it is different from ω.

Figures 7 and 8 show the images corresponding to the modulus and phase of $\Re(x,y)$, respectively, calculated with the corresponding sequence of images of the instantaneous density gradient as those shown in figure 6. The white regions in figure 7 are associated with the maximum amplitude of the time variations of the density gradient, while the light blue ones, for which $|\Re| \approx 0$, do not vary significantly with time at the fundamental frequency.

Figure 7. Modulus of $\Re$ at frequency ω_0 for the same case of figure 6. The false colour scale is also the same.

Figure 8. Phase of $\Re$ at frequency ω_0 for the same case of figure 6. The false colour scale is also the same.

Figure 9. Modulus of $\Re$ at frequency $2\omega_0$ for the same case of figure 6. The intensity scale is amplified by a factor 5 with respect to the values of figure 7.

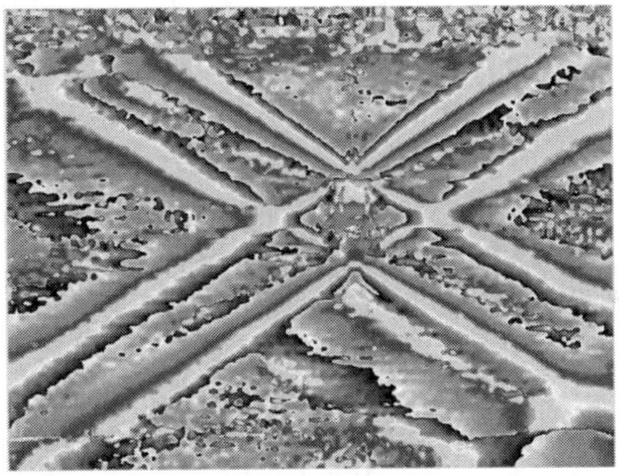

Figure 10. Phase of $\Re$ at frequency $2\omega_0$ for the same case of figure 6.

Figure 8 shows a 180° change of phase at the horizontal axis of symmetry resulting that the inferior half-plane is nearly a mirror reflection of the superior one but with an opposite phase. Then the density gradients at points located below the oscillating objects are in the opposite direction with respect to the mirror points located above. This may also be observed in figure 6.

Thus, from figures 7 and 8 it is possible to determine the intensities and phases corresponding to oscillations with frequency ω_0 of each point (x,y) located in the plane of observation. This information is more useful and contains major precision than that obtained from figure 6, because only the oscillations with the frequency of interest is involved, and a sequence of more than 100 images is used to obtain it.

Figure 9 show the modulus of $\Re$ for the first harmonics ($\omega = 2\omega_0$) obtained using the same images sequence that in figure 6. The beams now form a greater angle $\phi_1 = arccos(2\omega_0 / N)$ to the vertical when compared with those of figure 6. It is found that the harmonic waves are generated in the same regions around the body where the waves of frequency ω_0 are produced. The amplitude of the beams is about one fifth of that corresponding to the beams formed at the main frequency. The distribution of the phase of the function $\Re$ for the first harmonics is presented in figure 10 and shows the same main features of figure 8.

Figures 11 and 12 show the modulus and phase of $\Re$, respectively, for the second harmonic ($\omega = 3\omega_0$) obtained using the same images sequence than that employed to obtain the distribution in figure 6. The intensity scale is amplified 5 times with respect to that of figure 9. Although tenuous, the beams corresponding to this harmonics may be distinguished forming an angle $\phi_2 = arccos(3\omega_0 / N)$ to the vertical, superposed to the beams belonging to the main and first harmonic frequencies. This superposition comes from an incomplete filtering applied to the calculus of $\Re$ in the digital processing. Then it may be inferred the very small intensity of the oscillations associated with the second harmonic and the practical limit of the method for visualizing them. Better results would require images with a better space and time resolutions, and a sequence of a greater number of images.

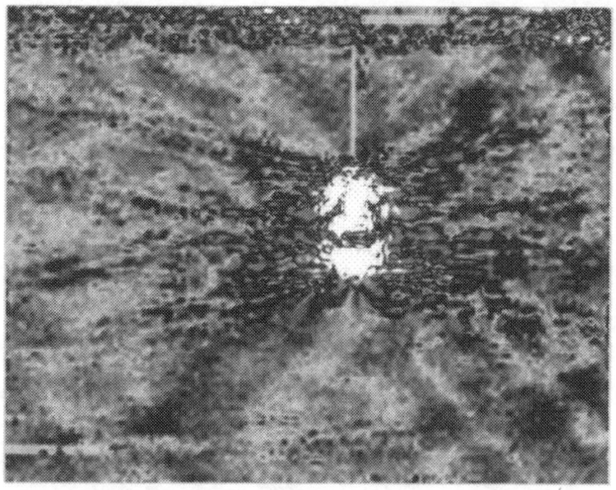

Figure 11. Modulus of $\Re$ for the frequency $3\omega_0$ (or second harmonics) for the same case of figure 6.

Figure 12. Phase of $\Re$ for the frequency $3\omega_0$ for the same case of figure 6

4. Internal waves generated by thermals

The experiments were performed in a 75 cm long, 29.5 cm wide and 38 cm high tank of transparent walls. It is filled up to a deep of 34 cm with linearly stratified salt water prepared by means of the double-bucket method, with a density ρ decreasing from the bottom to the fluid free surface where $\rho_0 \approx 0.998$ g/cm^{-3}. The value of $(1/\rho_0)\Delta\rho/\Delta z$, with z as the vertical coordinate, ranged in the experiments from 0 (no stratified fluid) up to 0.002 cm^{-1}.

The thermals were released from a Perspex cylinder with an interior diameter of 3.7 cm as shown in figure 13. The rounded lower end of the tube was covered with a stretched latex membrane. The tube was then immersed completely in the stratified medium slowly with its upper end at about 3 cm below the water free surface. A saline solution of known volume and density is added inside the tube using a pipette, minimizing the mixing during this process. The experiment starts when the membrane is broken using a needle, releasing the salt water in a negligible time. A series of experiments was carried out combining different initial volumes $3 < V_i < 10$ ml and densities $1.09 < \rho_i < 1.2$ g/cm^{-3} (saturated salt water) of the dense fluid released, and the stratification of the environment.

Figure 14 shows a series of images corresponding to a run. A diagnostic based on the attenuation of the light [16] by the presence of a calibrated quantity of dye in the salt water provides visualization, and synthetic schlieren technique allows observing the perturbations generated by the descending motion of the thermal in the stratified surrounding. The tank was illuminated from behind using a fluorescent strip lights and a diffusing sheet to provide nearly uniform back-lighting. The light attenuation due to the dyed water allows the fluid density variations to be observed. The thermal descends from rest under the unique influence of density difference between the fluids while its volume increases continuously due to mixing between the fluids. The total vertical momentum increases progressively under gravity action but density and velocity decrease due to the entrainment of external fluid. Since the total excess of buoyancy is reduced as the thermal descends, the dense fluid density may balance that of the surrounding fluid at an intermediate height level of the tank, stopping its vertical motion and then extending axially at that height.

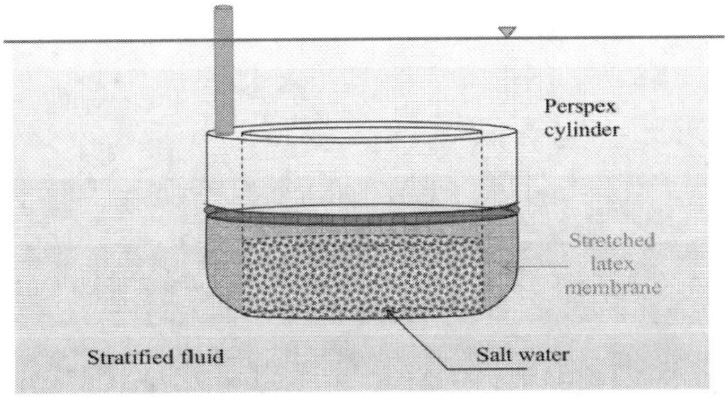

Figure 13. Experimental device for releasing thermals.

Figure 15 shows the distribution of the variations of the density gradient in the surrounding fluid obtained by synthetic schlieren. It is observed that, from the release of the thermal, internal waves forming beams with different angles to vertical are produced. The horizontal beams correspond to internal waves of very small frequency. As the thermal evolves, the angle of the others beams increases indicating the decreasing of the frequency of the waves excited in the medium. This effect is associated with the decreasing of the thermal velocity. Eventually new vertical beams are generated implying the formation of new internal waves. The symmetry of the dense blob, and also that of the internal waves, is lost gradually giving place to the formation of a great number of beams at different angles at the end of the run.

As appreciated, the synthetic schlieren technique allows measuring without inconvenient the small density fluctuations produced when a dense transparent fluid moves and mixes in another transparent fluid. The difference between figures 14 and 15 makes evident the extreme sensitivity of this diagnostic.

5. Conclusions

A non-intrusive optical technique suitable for determining the density gradient fluctuations in two-dimensional unsteady flows is employed to visualize and measure quantitatively the amplitude of the bidimensional internal waves generated in a stratified medium by an oscillating body and the displacement of a thermal. Combining the application of this diagnostic with the image processing employing the specialized software *Digi*Flow, the intensity changes of small lit circles are related to the deviation of light beams, and density gradient $\rho^{-1}d\rho/dx$ and $\rho^{-1}d\rho/dy$ is determined with respect to an initial reference image.

The methodology is more powerful, useful and versatile than other available techniques of visualization due to its low cost of implementation, and because it allows working in a larger spatial domain and provides non-intrusive measurements of the amplitude of the quasi-bidimensional wave field. The resolution of the experimental setup depends on the distances between the mask and the tank and between the camera and the tank (see figure 3), the characteristics of the camera employed and the density of points in the mask. Thus, in the arrangement described in Section 2.2 it is possible to measure changes of the density gradient so small as 1% of the ambient density gradient.

A key point of the images processing, integrated to an adequate software, is the on-line visualization of the results in a qualitative way, the later estimation of the density gradients in two directions and the possibility of performing more sophisticated processing as that described in Section 3.2. Therefore, the methodology introduced allows obtaining results that are hard to acquire using the traditional techniques (as optical schlieren, interferometry, etc.). The results may be transformed directly in a matrix of data to be compared with results of theoretical models and to detect experimentally new aspects of the physical process involved.

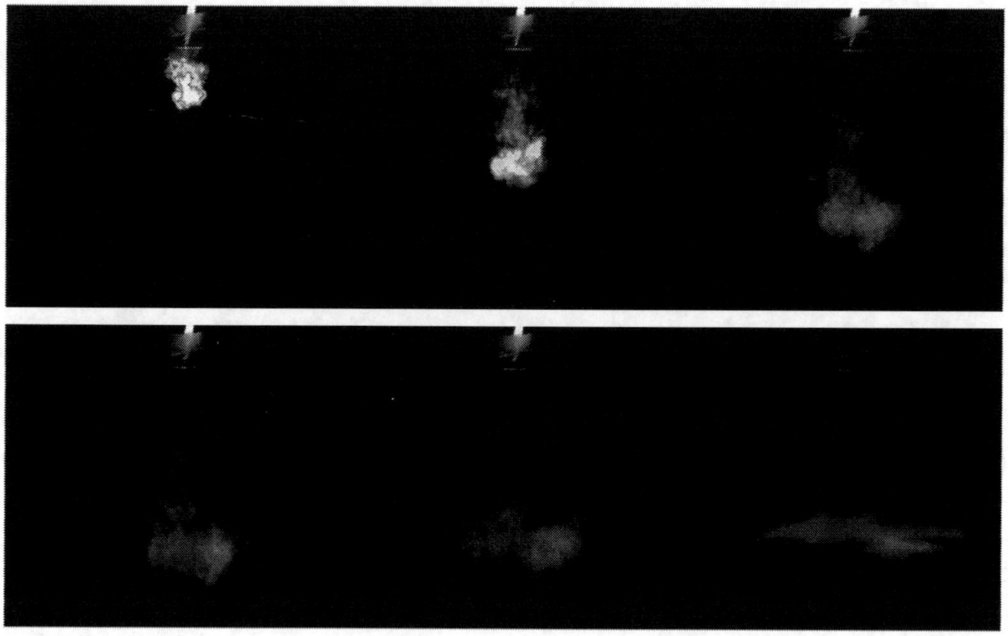

Figure 14. Sequence of images in false color obtained by using the diagnostics based on the light attenuation for a thermal with V_i = 5 ml and ρ_i = 1.2 g/cm^{-3} in a stratified medium ($\Delta\rho/\Delta z(1/\rho_0)$ = 0.0005).

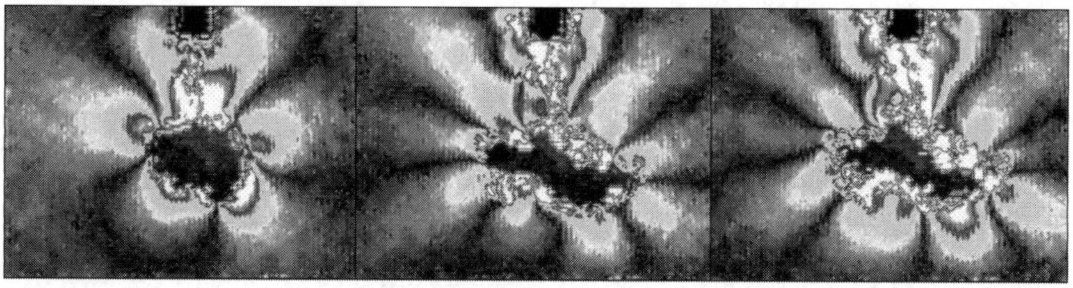

Figure 15. Variations of $(1/\rho_0)\Delta\rho/\Delta z$ in the ambient fluid generated by the thermal of figure 14 and visualized using synthetic schlieren.

Acknowledgments
This work was supported by CONICET and ANPCyT (Argentina) under Grants PIP 5893, and 34088/05 and 1185/06, respectively.

References

[1] van Dyke M. 1982 *An album of fluid motion* (Stanford: Parabolic Press).
[2] Thomas L P, Dalziel S B and Marino B M 2003 The structure of the head of an inertial gravity current determined by Particle tracking Velocimetry. *Expts. Fluids* **34** 708-16.

[3] Thomas L P, Marino B M and Dalziel S B 2003 Measurement of density distribution in a fluid layer by light-induced fluorescence in non-rectangular cross section channels. *Int. J. Heat & Technology* **21** 143-7.

[4] Irvin B R and Ross J 1991 Observation of pattern evolution in thermal convection with high-resolution quantitative schlieren imaging. *Phys. Fluids* A **3** 1699-710.

[5] Greenberg P S, Klimek R B and Buchele D R 1995 Quantitative rainbow schlieren deflectometry. *Appl. Optics* **34** 3810-22.

[6] Dalziel S B, Hughes G O and Sutherland B R 2000 Whole-field density measurements by "synthetic schlieren". *Expts. Fluids* **28** 322-35.

[7] Mowbray D E and Rarity S H 1967 A theoretical and experimental investigation of the phase configuration of internal waves of small amplitude in a density stratified fluid. *J. Fluid Mech.* **28** 1-16.

[8] Peters F 1985 Schlieren interferometry applied to a gravity wave in a density-stratified fluid. *Expts. Fluids* **3** 261-9.

[9] Merzkirch W and Peters F 1992 Optical visualization of internal gravity waves in stratified fluid. *Optics and Lasers in Engng.* **16** 411-25.

[10] Hurley D G and Keady G 1997 The generation of internal waves by vibrating elliptic cylinders. Part 1: Approximate viscous solution. *J. Fluid Mech.* **351** 119-38.

[11] Sutherland B R, Dalziel S B, Hughes G O and Linden P F 1999 Visualization and measurement of internal waves by synthetic schlieren.Part 1: Vertically oscillating cylinder. *J. Fluid Mech.* **390** 93-126.

[12] Hecht E and Zajac A 1974 *Optics* (Reading, MA: Addison-Wesley Publ. Co.).

[13] Sakai S 1990 Visualization of internal waves by Moiré Method. *Kashika-Joho* **10** 65-8.

[14] Dalziel S B 2006 *Digi*Flow *User Guide*. http://www.damtp.cam.ac.uk/lab/digiflow/.

[15] Weast R C 1981 *Handbook of chemistry and physics* (Cleveland, OH: C.R.C. Press).

[16] Cenedese C and Dalziel S B 1998 *Concentration and depth fields determined by the light transmitted through a dyed solution*; Proc. 8th International Symposium on Flow Visualization, 1-4 September, Sorrento, Italy. (ed. G M Carlomagno & I Grant) Paper 061, 1.

Local flows in the Quequén Grande River Estuary, Argentina

M G Pereyra, L P Thomas[1] and B M Marino
Instituto de Física *Arroyo Seco*, Universidad Nacional del Centro de la Provincia de Buenos Aires, Pinto 399, 7000 Tandil, Argentina

E-mail: mpereyra@exa.unicen.edu.ar , lthomas@exa.unicen.edu.ar , bmarino@exa.unicen.edu.ar

Abstract. Quequén Grande River is one of the most important water courses of Buenos Aires Province due to the economic and strategic significance of its estuary, where the populous Quequén-Necochea area and Quequén Port are located. The minor Las Cascadas falls, at 15 km from the place where the river meets the sea is the point where the maximum tidal propagation is detected marking the head of the estuary. Artificial dredging is needed to insure the adequate navigability conditions in the Quequén harbour, which has induced a highly stratified water column in the last 2 km of the estuary. Thus, an abrupt step is established at the head of the harbour, implying a much reduced water circulation and in some cases nonexistent, producing strong reductive and even anoxic conditions. The foot of the step is a sediment and organic matter trap and becomes an interesting place of study. The goal of this article is to present the information obtained with Doppler sonar at the neighborhood of the step, which allows distinguishing local turbidity currents that may influence the deposition patterns of the sediments.

1. Introduction

Estuaries are dynamic systems because of the noticeable interaction between the discharge of fresh water and the saline intrusion, which generates the variations of the salinity distribution. Sometimes the mix of the waters produces homogeneous environments, while others estuaries are characterized by the presence of well differenced layers.

The estuaries constitute ideal places to settle harbours, which gives place to two remarkable effects for the particulate material flows: 1) frequent dredging is usually needed to remove the accumulated sediments and provide adequate navigability conditions, and 2) the circulation of ships generates local currents that are important enough to change the patterns of sedimentation.

The Quequén Grande River, one of the most significant water courses in the south-west of Buenos Province, is 173 km long and born in the Tandilia Sierras range. Its drainage basin is located in a highly developed farming zone covering a total area of 9370 km^2 (figure 1) in which the conditions to form small falls and rapids along the river are provided. The last minor falls, named Las Cascadas, located at 15 km from the mouth are the point where the maximum tidal propagation is detected, and mark the head of the estuary following the estuarine definition criteria proposed by Perillo [1]. The monthly mean river runoff varied throughout the year since a minimum of 5.3 m^3/s in February to a maximum of 11.4 m^3/s in November according the files obtained at the Las Cascadas Gauging station. Nevertheless, large flash floods have been registered occasionally; among them the most important

[1] To whom any correspondence should be addressed.

occurred in 1905, 1913, 1915, 1998, 2002 and the largest one ever in 1980, which reached over 200 m^3/s and destroyed three major bridges in the estuarine zone [2].

The Quequén harbour is located in the last 2 km of the estuary and covers both margins modifying the original geomorphology of the river mouth. Because of its activity, mainly related to the exportation of grains coming from an extensive hinterland, the harbour is the second largest deep water system in Argentina. To provide deep water conditions for the harbour activities, this part of the estuary is kept at 12 m depth by continuous dredging. However, further upstream the thalweg has a depth of 2–4 m. In this way, an artificial step is formed and separates the estuary in two parts, being one of them notoriously affected by the tide [2].

The formation of this abrupt depth jump in the harbour head induces flows with different behaviour and a low or even null circulation in the deep lower layer where velocities are extremely small, restricting the estuarine circulation only to the depth of the crest of the step. Thus any sediment particle or contaminant that drops below the interface becomes trapped and settle at the foot of the step. Based on Perillo *et al.* [2] results, this sector of the harbour is reductive and even anoxic at times, with both reduced circulation and, obviously, oxygenation of the deeper water. Thus, this particular zone of accumulation of particulate material becomes an interesting place of study.

The river itself, and mainly the estuary, has a meandering pattern which originally ended in a coastal plain dominated by a coastal dune system on the Necochea side and a cliffy coast with extended beaches on the Quequén side. However, these features have been lost with the development of the harbour in 1908 and the construction of the jetties in 1915. Due to the prevailing SW–NE littoral drift along the coast, the effect of the jetties was to accumulate sand on the Necochea side and to erode the Quequén beaches. Although the construction of the jetties was prompted by the embankment of the harbour mouth by littoral drift and designed to reduce wave activity within, the results along the years have been very poor [2].

Quequén harbour presents singular man-made structures that have produced major consequences altering the geomorphology and circulation in the estuary. First of all, both jetties, especially the South one, have modified the littoral circulation and sediment transport pattern. Originally, the mouth of the estuary was deflected to the north by a spit that was cut periodically by seasonal peak runoff. The presence of the North jetty reduces the width of the mouth to 165 m developing a basin with low circulation.

Perillo *et al.* [2] consider the estuary as a microtidal and partly mixed system from the step to the head, while in the last 2 km the water column is highly stratified meaning that the upper 1-3 m of the water column layer is fresh water and a halocline (or interface of density) develops reaching salinities of over 30 salinity practical units being homogeneous down to the bottom [3]. According to the morphogenetic classification [1], it is considered as a coastal-plain primary estuary. The bottom sediments along the thalweg are constituted by sands, silty sands and silty clays with mean grain size decreasing inland while the bottom sediments within the estuary reflect the influence of the Pleistocene loess sediments of the adjacent area as most of the materials found have a related mineralogical composition [4]. However, at the head of the harbour, and due to its particular dynamics, sediments are very fine and the conditions are highly reductive [5].

There are very few hydrographic studies on the Quequén Grande River estuary and consequently the basic properties of the geomorphology, its physical oceanography and the relation with the mixing processes are poorly understood. Perillo *et al.* [2] reported a number of salient features of the estuarine dynamics, especially those relative to salinity, temperature and current distributions and tidal behaviour. Within the period April 1994–August 1998, six surveys covering all seasons and spring–neap conditions were carried out in the estuary to provide basic information to characterize the system. Sediment samples were taken at a series of cross-sections along the estuary to describe the general sediment distribution. All measurements were performed before the modification and extension of the jetties started at 2003. Developing gravity currents or arrested wedges have not been detected at all. A better understanding of both, geomorphology and hydrography of the estuary will certainly be useful in solving its pollution and dredging problems.

The preliminary results of a two day-campaign are introduced here, whose purpose was to study the local flows around the step neighbourhood and their influence on the deposition of the suspended material. In the next Section, the methodology is described and the relevant information obtained is introduced. Then some of the results are shown and briefly discussed; finally the conclusions are presented.

Figure 1: Geographic location of the Quequén Grande River estuary.

2. Methology

The field work was performed between the midday on 27 and the dawn on 28 December (2007) during a tide cycle. The figure 2 shows the area where 45 runs were made and that includes the harbour zone located between the step and the jetties. As an example, one of these trajectories is marked with a red line, together a reference frame that indicates the distances to the North and East from the starting point.

An *Workhorse* Acoustic Doppler Current Profiler (or ADCP), provided by *RDInstruments*, and a Global Positioning System (GPS) were employed. The ADCP, attached to a side of the *IADO 4* boat, measures absolute velocities in three dimensions and maps the bed giving a bathymetry. In addition, the intensity of the reflected acoustic signal (or backscatter) provides information about the suspended material, being possible to determine the evolution of the sediments transport by a given current.

During the time of the campaign, the suction dredger *Beachway* was operating in the harbour zone. This vessel is 95m long and 16 m wide, has two powerful engines of 810 kW which suck through a pipe of 0.80 m diameter and a storing volume of 3580 m^3. Measurements were made during the time in which the dredger went offshore to unload the sediments, which allowed visualizing flows generated under particular conditions as described in the next Section.

The values of the air temperature, humidity and atmospheric pressure, and wind intensity and direction every 10 min were provided by the meteorological station located in Quequén Port. According to the data reported in Table I, the weather conditions prevailing during the measurements days were stable and the sky was clean. The maximum temperature was reached at 14:00 and its minimum at 00:00. The maximum wind intensity was 32.2 km/h registered between 15:00 and 15:30 on 27 December. According to the wind rose shown in figure 3, it is may be inferred that the East direction is the most frequent and in this direction the most intense intensity was registered.

Figure 2: Run N° 44 wholly performed in the 2 km long harbour zone.

Table 1. Data registered at the meteorological station of the *Consorcio de Gestión de Puerto Quequén* on 27-28 December

Mean temperature	19.9 °C
Maximum temperature	26.9 °C
Minimum temperature	15.5
Mean wind velocity	21.06 km/h
Maximum mean wind	29.2 km/h
Main wind direction	E
Rain	0 mm
Mean humidity	47%
Atmospheric pressure	1012.23 bar

The figure 4 shows the variation of the tide level during the days of the survey. The height difference reached 2 m and the main period was of about 12 hours (according to a microtidal estuary and a semidiurnal tide). The interference of the different components of the tide causes a significant variation of the maximum and minimum values that coincide with those of the astronomical tide tables provided by the Servicio Hidrológico Nacional, suggesting that the wind did not affect the measurements. The maximum height was 1.75 m at 9:00 on the day 28. Since the dawn of day 28 the wind was more variable, tending later to blow from the south-east, that is perpendicularly to the coast; consistently, the tide height increased about 0.30 m with respect to the astronomic tide. The mean height of the waves was approximately constant, which is consistent with the wind intensity registered.

Journal of Physics: Conference Series **166** (2009) 012008 doi:10.1088/1742-6596/166/1/012008

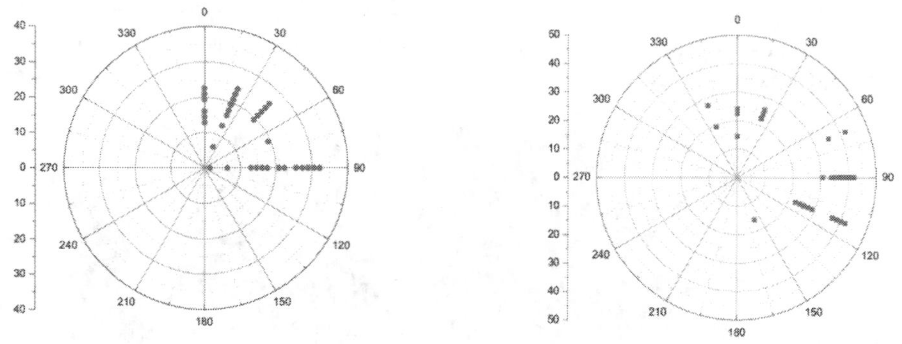

Figure 3: Wind rose for the days December 27 (left) and 28 (right). The wind direction 0° corresponds to the north, 90° to the east, etc. The wind intensity is given in km/h.

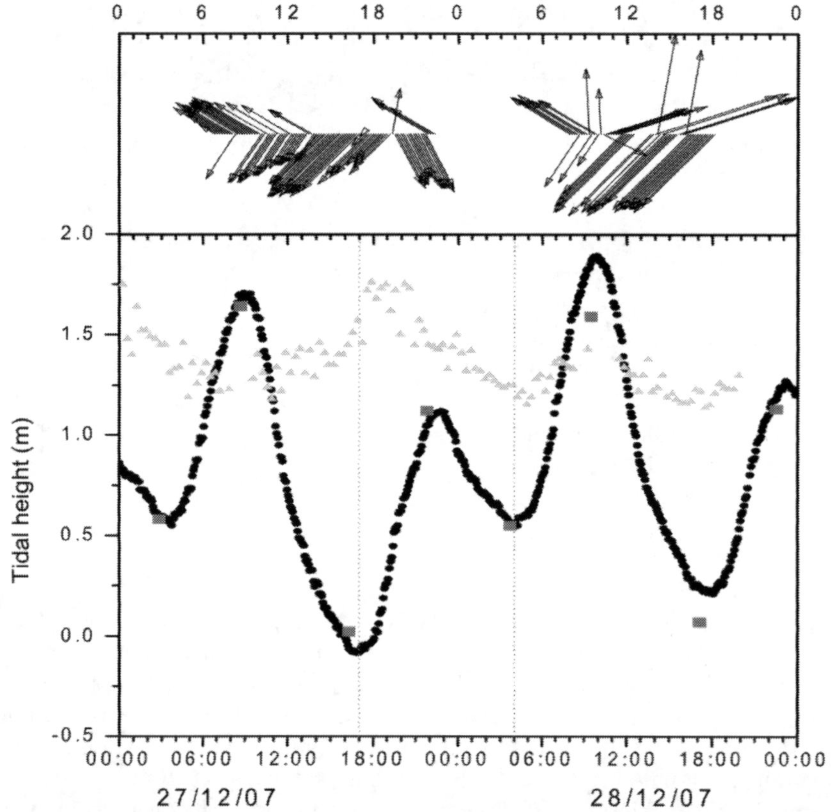

Figure 4. Tide height registered by the gauge of the *Consorcio de Gestión de Puerto Quequén* (circles). The squares represent the height forecasted by the Servicio de Hidrografía Naval and the triangles represent the main wave height. The wind intensity and direction (the north is upwards) is showed above.

3. Results and discussion

The most interesting findings come from the analysis of the intensity distribution of the ADCP backscatter signal. Particularly, a remarkable sedimentation pattern is observed, which may be understood taking into account the dredging operation and the movement of others ships during the runs. The dredger sucks sediments from the estuary bed and simultaneously generates turbulent flows with suspended sediments, thus increasing the water density in the dredging zone. The density difference between the fluids due to the suspended particles generates horizontal flows on the bed known as turbidity currents and whose dynamics is well described by the gravity currents theory [6].

Briefly, a gravity (or density) current on a rigid bottom is characterized by a frontal zone (or head) of height h that is greater than that of the following flow, and moves forwards with a velocity V that depends on the relative density difference $\Delta\rho/\rho$ and the thickness h:

$$V = F\sqrt{g'h} \tag{1}$$

where $F \approx 1$ is the Froude number, $g' = g\ \Delta\rho/\rho$ is the reduced gravity with g: gravity acceleration, and $h = \varepsilon h_0$ with $h_0 \approx 12$ m: total depth of the estuary water column and $0 < \varepsilon < 1$. The density difference between the salt and fresh waters is $\Delta\rho/\rho = 0.025$ for 3.5% of dissolved salts (sea water). If it is possible to measure V and to estimate the values of g' and h independently, the matching between both sides of Eq. (1) suggests the presence of gravity currents.

In the following the backscattering results obtained in some of the 45 trajectories performed during a tidal cycle are presented. The time in which each run occurred is indicated in a tide evolution graph, allowing determining the stage of the flood or ebb tide at that moment. The figure 5 shows the backscatter-intensity distributions near the step obtained on a part of the trajectories initiated at 16:39 and 17:06. The deep blue zone indicates a denser fluid moving to the right. From the leading mark and the time difference it is possible to estimate $V = \Delta x/\Delta t = 1.65$ m/s, while $F\sqrt{g'h} \approx 1.41$ m/s. The figure 6 shows the distributions corresponding to the trajectories initiated at 19:38 and 20:24, giving $V = 1.97$ m/s and $F\sqrt{g'h} \approx 1.55$ m/s. The acoustic signal dispersion corresponding to the trajectories initiated at 20:32 and 20:35 (figure 7) shows a gravity current head moving forwards the step with velocity $V = 2.18$ m/s that is close to the value of $F\sqrt{g'h} \approx 1.73$ m/s.

The trajectories started at 23:20 and 23:30 in figure 8 show a practically arrested gravity current. Its front velocity results $V = -0.15$ m/s despite $F\sqrt{g'h} \approx 1.55$ m/s. In this case the tidal flow towards the sea causes the retreat of the turbidity current when it was advancing to the step.

During the trajectories corresponding to initial times 01:21 and 01:30 a gravity current is observed (figure 9) again traveling forward the step with velocity $V = 2.3$ m/s and $F\sqrt{g'h} \approx 1.73$ m/s.

Other kind of flows also was detected. The backscattering registered during the runs initiated at 0:23 and 0:31 (figure 10) suggest the presence of an homogeneous fluid without sediments on the right (red colour), separated from another fluid containing a significant quantity of suspended material through the whole water column. Here, the sediments are more concentrated near the bottom but no gravity current seems to be produced. The velocity of the separation plane between the two fluids is insignificant.

The figure 11 shows a bore [6] reaching the step. In the present case, the typical undulations of this kind of flow in a stratified medium are observed behind the frontal zone of the current suggesting a turbulent bore. When an intense enough turbidity current meets a wall, a hydraulic jump as that observed in the figure 12 propagates away from the wall after the interaction.

The velocity values obtained from figures 5-9 allow estimating the time $T = L/V$ that the turbidity current needs to travel along the harbor zone of the estuary. Table 2 presents the values of T for a length $L \approx 2250$ m obtaining times of the order of 20 min.

Journal of Physics: Conference Series **166** (2009) 012008 doi:10.1088/1742-6596/166/1/012008

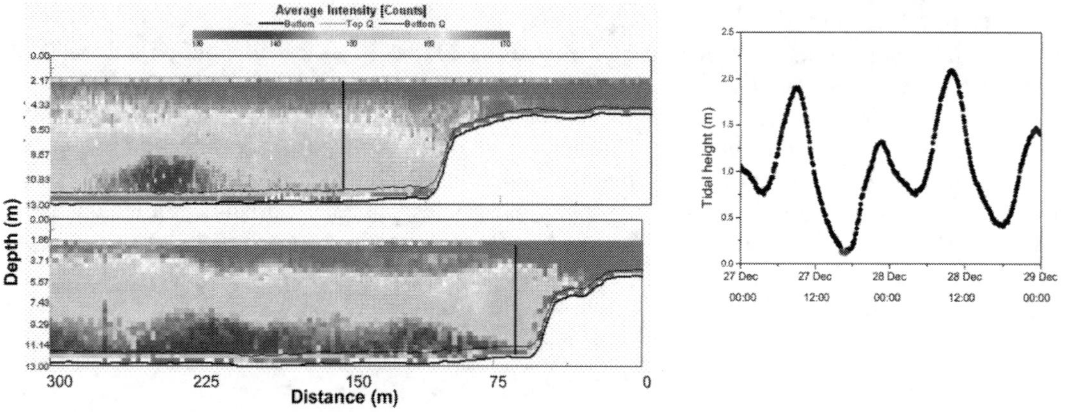

Figure 5: On the left, intensity registered by the ADCP in a part of the trajectories N° 2 and 5. The vertical lines indicate the gravity current front positions. On the right, the initial time of the trajectories (red point) are marked on the tide variation graph.

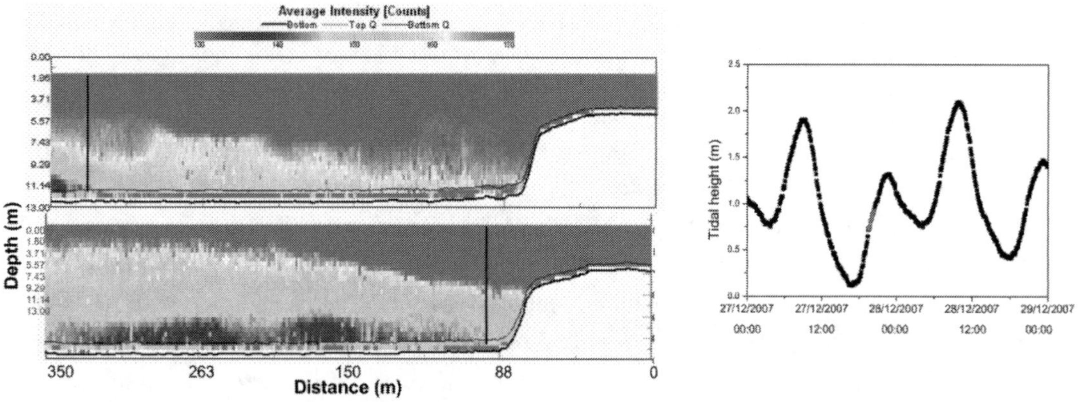

Figure 6: Idem figure 5 for trajectories N° 9 and 13.

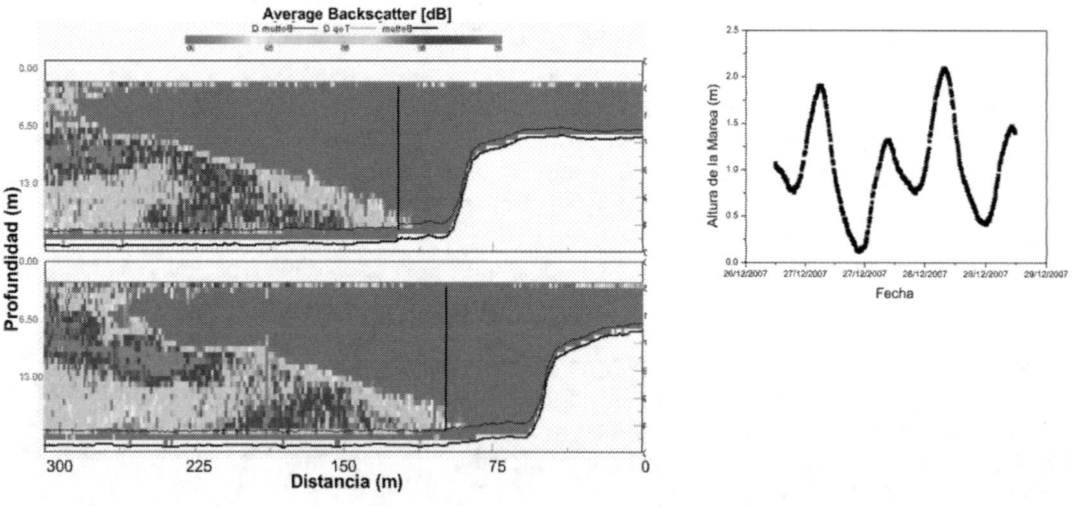

Figure 7: Idem figure 5 for trajectories N° 15 and 16.

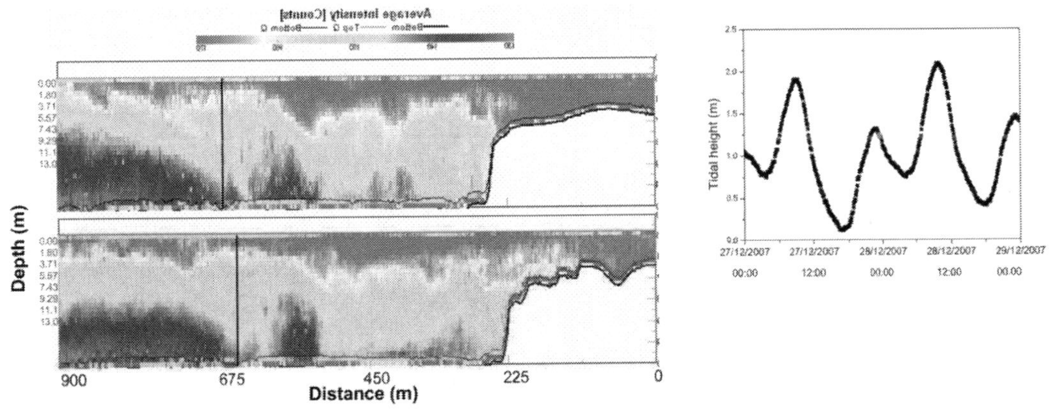

Figure 8: Idem figure 5 for trajectories N° 25 and 26. An arrested turbidity current is evident.

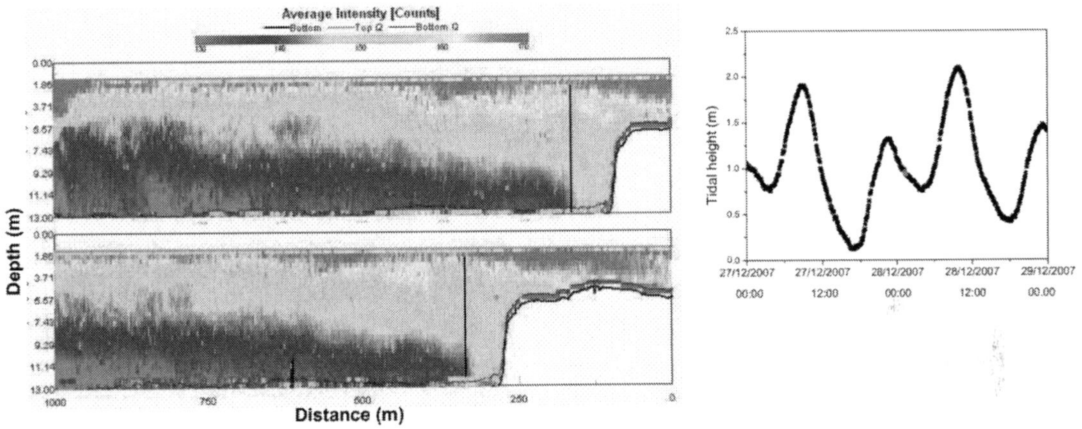

Figure 9: Idem figure 5 for trajectories N° 37 and 38

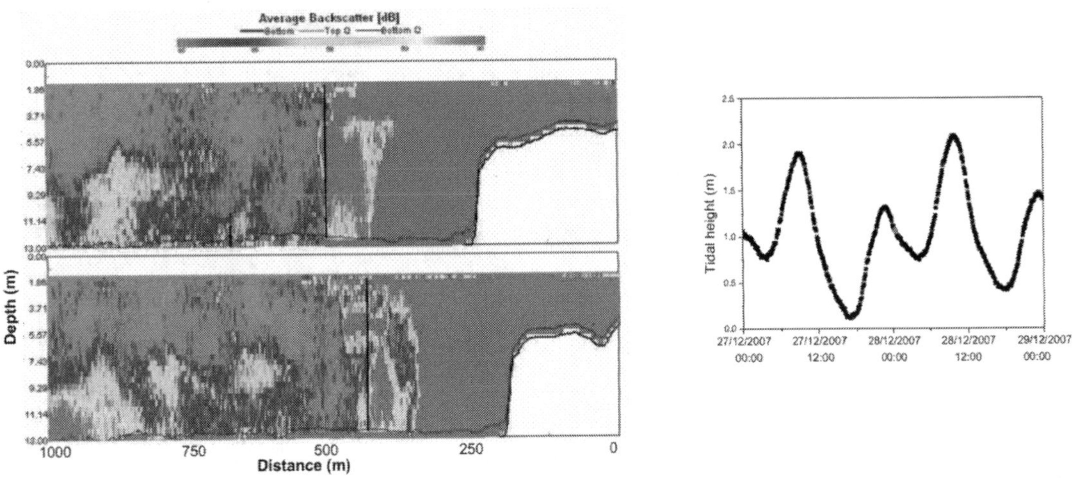

Figure 10: Backscattering registered by the ADCP in a part of the trajectories N° 28 and 29. It is clear the separation between a zone with a high concentration of suspended materials and another of an homogeneous fluid without sediments

motion of other ships were clearly identified and their velocities of displacement suggest circulation times that coincide with those of the oscillations reported by Perillo *et al.* (2005).

Perillo *et al.* [2] also suggested the occurrence of at least two macro vortexes during floods at the harbour, with opposite components of velocity in depth and surface, and such vortexes may create special conditions for the sedimentation process in that place. However, the turbidity currents observed this time might be responsible for such a velocity change.

Thus, the measurements made and the information acquired allow visualizing many interesting aspects of the local flows and the movement of the suspended material in the harbour sector of the estuary. There is a complex interaction between the displacing of vessels, the river discharge, the tide and the sediments transport. Then, conditions for creating an intense sedimentation area at the foot of the step exist and more hydrodynamic information is needed to understand the varied processes that occur.

Acknowledgments

Authors thank to the *Consorcio de Gestión del Puerto Quequén* for supplying the data from their monitoring stations, and to Lic. Ernesto Alberdi (IADO) for his valuable collaboration in the ADCP operation and information processing. This work was financial supported by the *Programa Institucional de Ambiente*, UNCPBA, and performed in the frame of PICT ANPCyT 1185/06.

References

[1] Perillo G M E 1995 *Definition and geomorphologic classifications of estuaries.* In Perillo G M E (ed.) *Geomorphology and sedimentology of estuaries, Development in Sedimentology* (Amsterdam: Elsevier Science BV), **53** 17-47.

[2] Perillo G M E, Pérez D E, Piccolo M C, Palma E D and Cuadrado D G 2005 Geomorphology and physical characteristics of a human impacted estuary: Quequen Grande River Estuary, Argentina. *Estuarine, Coastal and Shelf Science* **62** 301-12.

[3] Piccolo M C 1998 *Oceanography of the Western South Atlantic continental shelf from 33S to 55S.* In: Robinson, A.R., Brink, K.H. (Eds.), *Coastal Oceanography: The Sea Series.* (John Wiley & Sons), pp. 253-271.

[4] Wright R 1968 Miliolidae (Foraminiferos) recientes del estuario del río Quequén Grande (provincia de Buenos Aires). *Hidrobiología* **2** 225-56.

[5] Piccolo M C and Perillo G M E 1999 *Estuaries of Argentina: a review.* In: Perillo G M E, Piccolo M C and Pino Quivira M (Eds.), *Estuaries of South America: Their Geomorphology and Dynamics*, Environmental Science Series (Berlin: Springer-Verlag), pp. 101–132.

[6] Marino B M and Thomas L P 2007 *Corrientes de gravedad* (Tandil: Independencia Gráfica y Editora).

Journal of Physics: Conference Series **166** (2009) 012009 doi:10.1088/1742-6596/166/1/012009

Stability of a finite-length rivulet under partial wetting conditions

Javier A. Diez, Alejandro G. González
Instituto de Física Arroyo Seco,
Universidad Nacional del Centro de la Provincia de Buenos Aires - CONICET, Pinto 399, 7000, Tandil, Argentina

Lou Kondic
Department of Mathematical Sciences
Center for Applied Mathematics and Statistics
New Jersey Institute of Technology, Newark, NJ 07102

Abstract. We study the stability of a finite-length fluid rivulet at rest on a partially wetting surface. We consider the problem by including the intermolecular force (van der Waals interaction) within the framework of the lubrication approximation. The results are validated by comparison with numerical simulations of the full nonlinear equation. For finite length rivulets, we show that the distance between drops after breakup is very close to the wavelength of maximum growth rate predicted by the linear theory for infinite rivulets. Finally, we compare theoretical and numerical results with reported experimental data.

1. Introduction

The evolution of partially wetting thin liquid films on solid substrates and the subsequent pattern formation is a subject of growing importance. The interest on the subject is raised by a variety of important applications such as thin films in micro- and nano-fluidics [1, 2], as well as templating or lithographic patterning and various types of coatings [3, 4, 5]. Many of the fundamental dewetting studies have focused on continuous liquid or polymeric [6, 7] thin films. More recently, the dewetting characteristics of continuous thin metal films were studied [8]. In all of these examples, the films are often unstable and break up, leading to the formation of dry spots. One important question is what are the mechanisms involved in the dewetting and breakup of unstable thin films. The instability has been found to be either driven by nucleation at defect sites which are not spatially correlated, or via spinodal-type of instabilities resulting in a spatial correlation [9, 8, 10].

In this work we consider the instability of finite-length fluid rivulets placed on a horizontal substrate under partial wetting conditions (see Fig. 1). The case of an infinite rivulet has previously been considered by other authors assuming sharp contact lines and circular cross section shapes. Among the most relevant papers on the subject we mention the works by Davis [11], Langbein [12], Roy & Schwartz [13], Sekimoto et al. [14] and Brochard-Wyart et al. [7]. Here, we perform a stability analysis by considering van der Waals forces which diffuse the contact line into a thin precursor film and also consider the effects of gravity. Thus, the cross section of the rivulet cannot be described as a circle either because of the non sharp contact line or because of the influence of gravity.

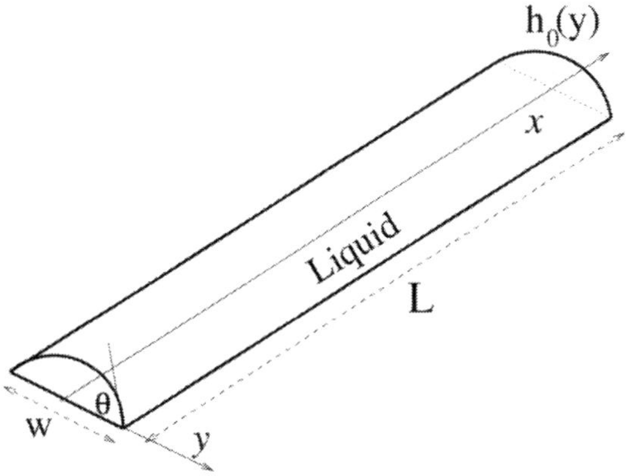

Figure 1. Schematic of the liquid rivulet placed on a horizontal substrate.

One significant difference to previous approaches is that intermolecular forces are taken into account. Here, we choose a conjoining/disjoining pressure model. Instead of imposing a fixed contact angle at the contact line where $h \to 0$, we introduce the wettability properties directly in the model so to reach a given free surface slope (apparent contact angle, θ) far from the contact line. To do so, an equilibrium film is required ahead of the macroscopic contact line, whose thickness is very close to that of equilibrium for an infinite flat film.

The paper is organized as follows. The linear stability analysis (LSA) within the framework of lubrication approximation is presented in Section 2. The comparison between the linear theory and the numerical evolution of the full nonlinear equation is given in Section 3, where the case of finite rivulet is also considered. The comparison between experimental data reported elsewhere and the present theoretical and numerical results is discussed in Section 4. Finally, Section 5 summarizes main results and conclusions.

2. Long-wave approximation and linear stability analysis

In the study of thin film flows on solid substrates, the evolution of the fluid thickness, h, is typically described under the framework of lubrication theory. This approach allows to reduce Navier-Stokes equations to a single nonlinear partial differential equation for h. In addition, finite contact angles can be included in the model by accounting for van der Waals forces, as described below.

We note that although lubrication theory is strictly valid only in the problems characterized by vanishing free surface slopes, it has been used commonly in partial wetting conditions, therefore in situations where the contact angle is not necessarily small [15, 16, 17]; see also [18, 19, 20, 21] for further discussion regarding involved issues. This approach has been justified in part by the works which show that even in the case of large contact angles, only relatively small deviations from more complete models result. For example, [22] compares the solutions for the cross section of a rivulet flowing down a plane obtained by solving the complete Navier-Stokes equation with the predictions of the lubrication approximation (see their Table I). For a contact angle of 30°, they find that the differences between the two approaches related with the shape of the free surface are of the order of few percent. Although the accuracy of lubrication approximation is not so good regarding the details of the velocity field, they find that these velocity differences cancel out when the total flux along the rivulet is computed. The issue of appropriateness of the use of lubrication approximation was also discussed earlier [23]. In that work it was shown that there are some differences in the free surface slope between lubrication theory and Stokes formulation, but only very close to the contact line.

Another concern regarding the use of lubrication approximation is that one typically (as we do here) approximates curvature of the free surface by h_{xx}, where x is in-plane coordinate. This issue was considered in [24], where it is shown that use of the complete nonlinear curvature yields only a few percent difference (see Fig.2b in [24]). There have been also some attempts [25, 26] to improve the typical lubrication approximation approach. These works show that including a correction factor to the flux term in the continuity equation extends the limits of its validity. Therefore, it may be appropriate to implement these improvements when precise quantitative results are desired. In the present problem, where we are mainly concerned with the basic mechanisms involved in the dewetting and breakup processes, we expect that standard lubrication approach is sufficient.

The van der Waals forces are included in the formulation of a lubrication model via disjoining pressure $\Pi(h)$ (see for instance [9]). The resulting equation for the fluid thickness, $h = h(x, y, t)$, is (see, e.g., [27, 15, 16])

$$3\mu\frac{\partial h}{\partial t} + \gamma\nabla\cdot\left(h^3\nabla\nabla^2 h\right) + \nabla\cdot\left[h^3\nabla\Pi(h)\right] - \rho g\nabla\cdot\left(h^3\nabla h\right) = 0, \qquad (1)$$

where μ is the viscosity, ρ is the fluid density, γ is the surface tension, and g is the gravity. Here, the first term stands for viscous dissipation and the other three terms account for the driving forces, which are surface tension, van der Waals and gravity force, respectively. The disjoining pressure model that we use

$$\Pi(h) = \kappa f(h) = \kappa\left[\left(\frac{h_*}{h}\right)^n - \left(\frac{h_*}{h}\right)^m\right], \qquad (2)$$

introduces κ (proportional to the Hamaker constant) and the exponents $n > m > 1$ (note that $f(h)$ is a dimensionless function). The first term represents liquid-solid repulsion, while the second term is attractive, leading to a stable film thickness $h = h_*$. Within this model, $\kappa = S/(Mh_*)$, where S is the spreading parameter, and $M = (n-m)/((m-1)(n-1))$ (see [9]). The spreading parameter can be related to the apparent contact angle θ via Laplace-Young condition $S = \gamma(1-\cos\theta)$.

Here, we take x-axis along the rivulet, and y-axis in the transverse direction. By defining the dimensionless variables $\tilde{h} = h/a$, $\tilde{x} = x/a$, $\tilde{y} = y/a$, $\tilde{t} = t/t_c$, with $t_c = 3\mu a/\gamma$, Eq. (1) becomes

$$\frac{\partial h}{\partial t} + \nabla\cdot\left(h^3\nabla\nabla^2 h\right) + K\nabla\cdot\left[h^3 f'\nabla h\right] - \nabla\cdot\left(h^3\nabla h\right) = 0, \qquad (3)$$

where $f' = df/dh$ and we have omitted the 'hat' symbol for simplicity. We define $K = \kappa a/\gamma$, with $a = \sqrt{\gamma/\rho g}$ being the capillary length.

2.1. Steady state (base) solution

We first concentrate on the x-independent solution, $h(y,t)$, so that Eq. (3) reduces to

$$\frac{\partial h}{\partial t} + \frac{\partial}{\partial y}\left[h^3\left(\frac{\partial^3 h}{\partial y^3} + Kf'\frac{\partial h}{\partial y} - \frac{\partial h}{\partial y}\right)\right] = 0. \qquad (4)$$

We consider now $\partial h/\partial t = 0$. After integrating twice the resulting equation and using the boundary conditions $h_0''' = h_0' = 0$ at $y = \pm\infty$, where the primes stand for y-derivatives, we obtain (see e.g. [9]):

$$h_0'' + Kf(h_0) - h_0 + p = 0, \qquad (5)$$

where the constant $p > 0$ is the equilibrium pressure [28, 29] within the fluid. An analysis [28, 30, 9] of the solutions of Eq. (5) shows that there is a range of pressures $p_c < p < p^*$

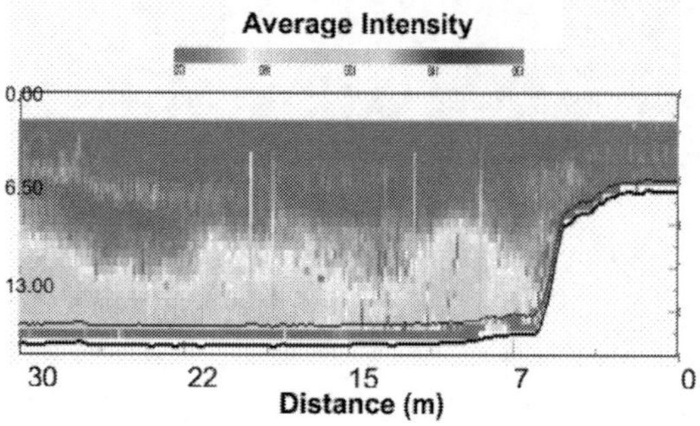

Figure 11: Turbulent bore moving forwards the frontal zone of a turbidity current.

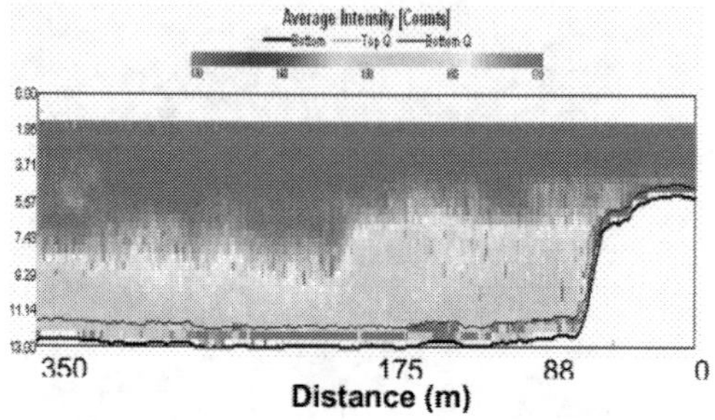

Figure 12: Hydraulic jump produced by the reflexion of the turbidity current at the step wall.

Table 2: Estimated time during which the turbidity currents travel across the analyzed sector of the estuary.

Trajectory N°	Measurement time	*T* (min)
	Intensity	
2 - 5	17	28
9 - 13	20	23
25 - 26	23:30	23
37 - 38	1:30	16
	Backscatter	
15 - 16	20:30	17

4. Conclusions

The measurements performed in this campaign allow detecting for the first time the occurrence of turbidity currents, *bores* and hydraulic jumps in the harbour zone of the Quequén Grande river estuary. It is found that the circulation times of these flows are of the order of 15-30 min. Since the time difference between the detection of two consecutives currents is at least a couple of hours, which is similar to the time employed by the dredger in going to offshore, unloading the sediment and coming back to its work section, and also is similar to the time between the exit of two big ships, the detected turbidity currents seem to be generated each in an individual way by the vessels displacement.

Some interesting relationships arise when the results showed here are compared with those reported by Perillo *et al.* [2] who detected dangerous oscillations over the water surface of about 1 m with respect to that explained by the astronomical tide. As the harbour tidal records used to show the presence of high frequency oscillations, Perillo *et al.* [2] made specific measurements at the two extreme tide gages during August and September 1998 at 1 min interval to evaluate these oscillations. The periods were estimated resulting between 3 and 25 min, although the latter are predominant. Even though normal oscillations have amplitudes of the order of 0.10 m, oscillations of up to 1.5 m often are observed. Normally, the oscillations increase in size suddenly and inexplicably. When they occur, the strong oscillations may induce serious consequences for the navigation of large cargo vessels. The theoretical resonance periods of the surface waves for the harbour vary between 3.3 and 6.4 min, which are well below the typical 25 min found during the field measurements. In the present field work those oscillations did not appear, but turbidity currents generated by the dredger action and the for which there exist bounded-in-height solutions, where $h_{0,min} < h_0 < h_{0,max}$. Within that range, the value of p corresponds to different cross section areas,

$$A = 2\int_0^\infty (h_0(y) - h_{0,min})\, dy, \tag{6}$$

where [28, 9] $h_{0,min} \approx h_* + ph_*/K(n-m)$. Figure 2 shows relationship between p and the corresponding area, A. For the case considered here, with gravity effects included, p cannot be lower than a certain limit, p_c, and, consequently, the maximum drop thickness, $h_{0,max}$, is bounded from above.

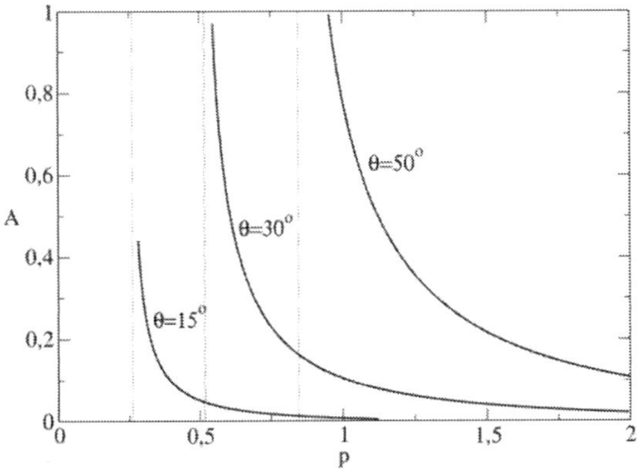

Figure 2. Relationship between pressure, p, in Eq. (5) and the corresponding area, A, of the rivulet cross section for three different contact angles, θ. Here, we use $n = 3$, $m = 2$ and $h_* = 10^{-2}$.

The solution of Eq. (5) is studied in detail in the Appendix B of [9]. The case with no van der Waals forces, i.e. $K = 0$, but supplemented with boundary condition $h = 0$ at the contact lines ($x = \pm w/2$, w being the width of the 1D drop) is also analyzed in that paper. Here, we use both solutions as possibilities to describe the cross section of the rivulets.

2.2. Perturbations and eigenvalue problem

In order to perform the linear stability analysis (LSA) of the transverse thickness profile, $h_0(y)$, with respect to longitudinal perturbations (x-direction), we write:

$$h(x, y, t) = h_0(y) + \epsilon g(y) \exp(\sigma t + ikx) \tag{7}$$

where ϵ is small number, $g(y)$ is the amplitude of the perturbation, $k = 2\pi/\lambda$ is the wave number, and λ is the wavelength.

By replacing Eq. (7) into Eq. (4), we obtain to $O(\epsilon)$ the eigenvalue problem:

$$\mathcal{L}g = -\sigma g$$

where $\mathcal{L}$ is the linear operator defined by:

$$\mathcal{L}g = c_4(y)g_{yyyy} + c_3(y)g_{yyy} + c_2(y)g_{yy} + c_1(y)g_y + c_0(y)g, \tag{8}$$

where the coefficients $c_i(y)$ ($i = 0, ..., 4$) are defined by:

$$\begin{aligned}
c_4(y) &= h_0^3, \\
c_3(y) &= 3h_0^2 h_0', \\
c_2(y) &= -\left(1 + 2k^2 - Kf'\right) h_0^3, \\
c_1(y) &= -h_0^2 \left[3(1 + k^2 - Kf') - 2h_0 K f''\right] h_0' \\
c_0(y) &= c_{0,2}h_0^2 + c_{0,3}h_0^3 + c_{0,4}h_0^4
\end{aligned} \tag{9}$$

and $c_{0,j}$ ($j = 2, 3, 4$) are given by:

$$\begin{aligned} c_{0,2}(y) &= 3h_0'^2 K f'', \\ c_{0,3}(y) &= k^2\left(1 + k^2 - K f'\right) - K f''\left(p + K f\right) + h_0'^2 K f''' \\ c_{0,4}(y) &= K f'' \end{aligned} \tag{10}$$

We use $n = 3$, $m = 2$, and $h_* = 0.01$. The considered contact angles range between $\theta = 30°$ and $\theta = 50°$.

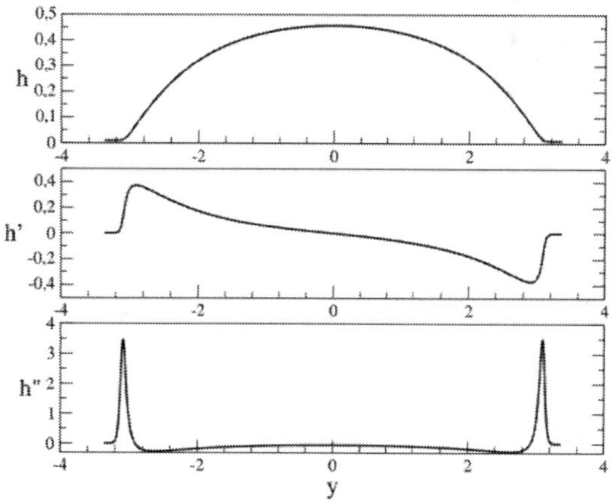

Figure 3. Profiles of (a) thickness, (b) slope and (c) curvature of a rivulet cross section for $A = 2$ and $\theta = 30°$.

Figure 3 shows the thickness, slope and curvature profiles for $A = 2$ and $\theta = 30°$. Figure 4 show the resulting dispersion relation curves. We see that the maximum growth rate as well as its corresponding wavenumber diminish as A is increased.

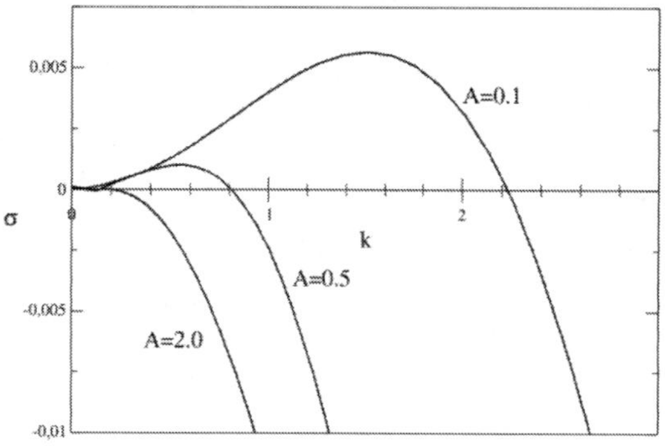

Figure 4. Dispersion relations for the three rivulet cross sections, A, using $\theta = 30°$.

3. 3D numerical simulations

In order to understand the nonlinear effects in the rivulet instability, we perform numerical simulations of the evolution of the rivulet governed by the full nonlinear Eq. (3). The calculations are carried out in a rectangular computational domain defined by $0 \le x \le X$ and $0 \le y \le Y$, which is divided into cells of size $\delta x \times \delta y$ (typically, we use $\delta_x = \delta_y = 0.04$). Equation (3) is discretized in space using a central finite difference scheme. Time discretization is performed using implicit Crank-Nicolson scheme. We note that all the results presented in this paper are fully converged, as verified by grid refinement; more details about numerical issues can be found in [31].

Journal of Physics: Conference Series **166** (2009) 012009 doi:10.1088/1742-6596/166/1/012009

We choose a rivulet with pressure $p = 5$ and, due to the symmetry along the line $x = 0$, we simulate only half of the whole rivulet, i.e. $y > 0$. We take a numerical domain length, L, coincident with the wavelength of maximum growth, $\lambda_m = 1.087$ and impose no-flow (symmetry) conditions at the boundaries, which is equivalent to periodic boundary conditions for this infinite rivulet case. We perturb the free surface of the rivulet at $t = 0$ as prescribed by Eq. (7), and take $\epsilon = 10^{-3}$ and $g(y)$ as given by the corresponding (varicose) eigenfunction. Fig. 5 shows the time evolution of the free surface shape, where the breakup process is clearly observed at the middle of the domain.

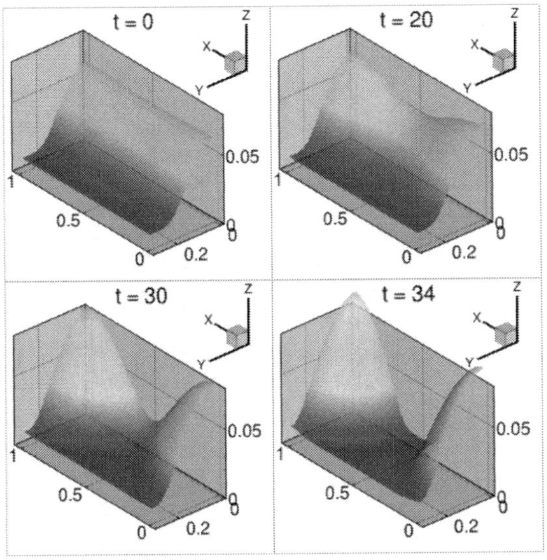

Figure 5. Evolution of a fluid rivulet of length $L = 1.087$ with $p = 5$. We employ a perturbation with wavelength $\lambda_m = 1.087$.

In Fig. 6 we show the amplitude of the perturbation in the middle of the rivulet, i.e., at $x = L/2$ and $y = 0$, measured as the difference between the maximum height of the rivulet and the fluid thickness at that point. For early times, the perturbation follows an exponential law, as expected for a linear regime. The good agreement between the exponent, β, that fits this behavior, and the eigenvalue σ corresponding to $\lambda_m = 1.087$ indicates that the eigenvalue problem describes correctly the early evolution of a given steady initial profile.

An interesting feature of this analysis is that a perturbation different from the corresponding eigenfunction, say by setting $g(y)$ in Eq. (7) equal to unity instead, yields a stable flow, i.e. the amplitude of the perturbation decreases and no rivulet breakup occurs.

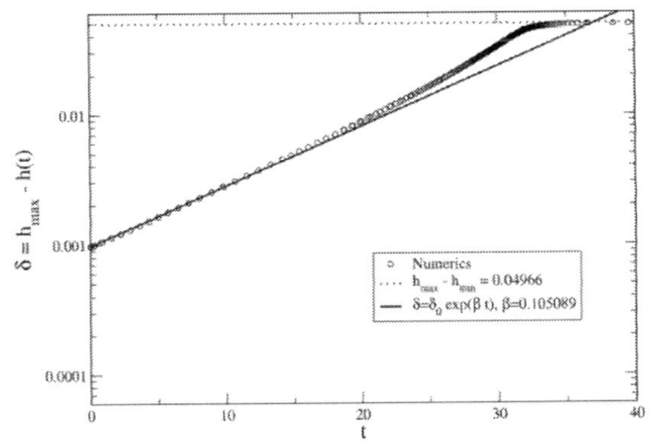

Figure 6. Evolution of the amplitude, δ, of the perturbation at $(x = L/2, y = 0)$ with $L = \lambda_m = 1.087$. Here, $\theta = 50°$ and $p = 5$.

Motivated by the comparison with experimental results in the following section, we here analyze the evolution of finite length rivulets. Thus, we perform the numerical simulation for an initial condition obtained by modulating the steady solution of the rivulet as [9]:

$$h(x,y,0) = h_0(y)\left[\arctan\left(q(x-x_1)\right) - \arctan\left(q(x-x_2)\right)\right]/\pi, \tag{11}$$

where x_1, x_2 are the coordinates of the end points of the rivulet, so that the fluid length is $\ell = x_2 - x_1$. The computational domain has an x-length $L = \ell/2 + 1$, i.e., half length of the rivulet plus a small space for the precursor film ahead of it. The value of q determines the width of the transition region between the unperturbed rivulet of height h_{apex} and the precursor film region. We use a relatively large q (typically, $q = 100$) to obtain an initial condition resembling a step function connecting $h_0(y)$ and h_* at the end points of the bulk.

The evolution of a finite length rivulet for $p = 5$ is shown in Fig. 7. This mechanism of rivulet breakup has previously been studied in connection with the one dimensional (1D) rupture of flat thin films [9, 32]. In those cases, there exist instabilities of two different kinds: spinodal or nucleation instability. These regimes depend on the value of the thickness of the film, with nucleation being relevant for thinner films. The main difference between the final patterns formed by these two mechanisms is the distance, D, between consecutive 1D-drops. In the spinodal regime, D is given by the wavelength, λ_m, of maximum growth rate of the corresponding LSA, while in the nucleation regime D is not correlated with λ_m, but it is determined by the nonlinear evolution of the flow.

In the 3D-rivulet case, its thickness $h_{apex} = h_0(0)$ depends on the value of the pressure, p, inside the rivulet. In particular, for $p = 5$, we find that $D \approx 1.09$, which is very close to λ_m. We find basically the same result for other values of p, so that we conclude:

$$D \approx \lambda_m \tag{12}$$

for different rivulet thicknesses h_{apex}. This indicates that no nucleation regime exists in the rivulet problem and, therefore, the distance between droplets is close to the wavelength of the maximum growth resulting from LSA, λ_m.

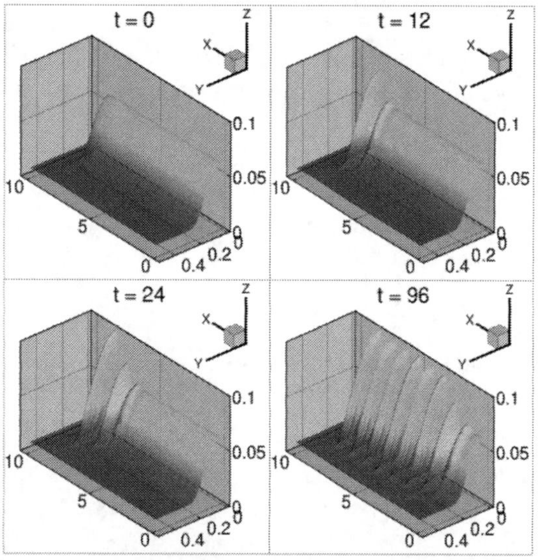

Figure 7. Evolution of a finite rivulet of length $L = 18$ with $p = 5$. The final pattern is formed by droplets separated an average distance $D = 1.03$.

The time needed for surface perturbations, like that used in LSA, Eq. (7), to break up the rivulet is given by $t_s = -\ln\epsilon/\sigma_m$. In this case, we have $\sigma_m = 0.114$, so that surface perturbations of amplitude $\epsilon = 10^{-3}$ lead to breakup only for $t_s = 60$. Thus, for the considered rivulet length, breakup due to dewetting from the rivulet ends proceeds faster than surface perturbations have time to grow.

4. Comparison with experimental results

In this section we aim to compare the LSA results, as well as from 3D-numerical simulations, for the evolution of the rivulet instability with the experiments reported by González et al. [33]. In that work, the rivulet is formed by capturing a thin jet of silicon oil (polydimethysiloxane, PDMS) on a coated glass slide, which is a partially wetting surface for the oil. The evolution of this rivulet shows a repetitive process of length contraction and breakup into droplets (see Fig. 8). Each experiment is characterized by a contact angle, θ, and the cross sectional area of the rivulet, A. When the fluid is placed on the substrate, the initial width, w, of the rivulet is also measured and reported in that work. Thus, the theoretical model used to describe this initial configuration must be able to yield w for given θ and A. Since here we use the capillary length a ($= 0.145$ cm for PDMS), as a characteristic length scale, we write the main experimental results reported in González et al. [33] in units of a (see Table 1). The last column in Table 1 corresponds to the predictions given in [13]. We observe large difference between the experiment and that theoretical result.

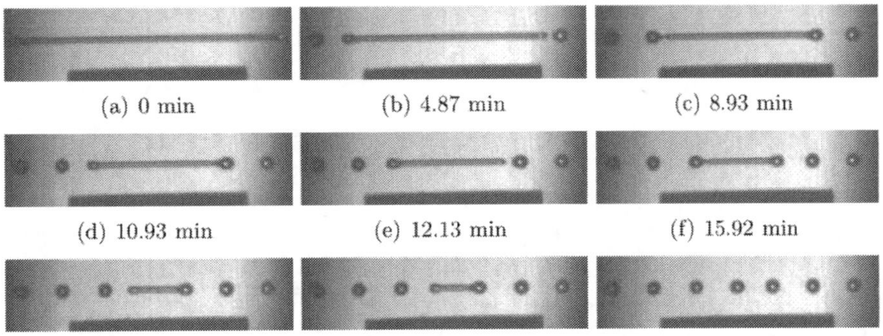

(a) 0 min (b) 4.87 min (c) 8.93 min (d) 10.93 min (e) 12.13 min (f) 15.92 min (g) 17.83 min (h) 19.33 min (i) 22.95 min

Figure 8. Evolution of a silicon oil rivulet of length 7 cm on a substrate covered with EGC–1700 ($\theta_e = 57°$) as reported by González et al. [33] and formation of primary drops. The black segment corresponds to 4.2 cm.

Table 1. The first six columns correspond to the experimental data reported in Table I of González et al. [33], where w is the rivulet width and n is the number of drops in the rivulet of length $L = 50$ (in units of a). The last column shows λ_m predicted by the results in [13].

Exp.	θ	A	w	n	D_{exp}	λ_m [13]
1	57°	0.069	0.517	12	3.501	1.015
2	57°	0.296	1.076	4	12.84	2.113
3	57°	0.316	1.110	3	19.53	2.213
4	63°	0.041	0.352	16	2.802	0.701
5	63°	0.296	0.993	7	6.117	1.980

Next, we compare the values of λ_m given by LSA presented here, with the experimental distance between drops, D_{exp}. To do so, we obtain the base thickness profile by using the solution of Eq. (5) with van der Waals and gravity forces. Since this solution is determined by a fixed value of the pressure, p, inside the drop, we find the corresponding p by relating it with the experimental drop area, A, which is defined as the area for $h > h_*$ in the theoretical profile.

The values of λ_m reported in Table 2 are obtained for these values of p and for θ as given by the experimental data for few different values of the exponents (n, m). These results are closer to the experimental data than the results from [13], but there is still only qualitative agreement. It should be noted that the discrepancy is particularly large in the cases when only few drops are seen, and the statistical fluctuations of the experimental results may be large. We also note that the influence of the exponents (n, m) in van der Waals model is relatively weak.

Table 2. Comparison of D_{exp} with λ_m as obtained from LSA for a cross section with van der Waals and gravity forces by using few different pairs of (n, m).

D_{exp}	$\lambda_m^{(3,2)}$	$\lambda_m^{(4,3)}$
3.501	2.424	2.356
12.84	5.427	4.885
19.53	5.694	5.136
2.802	1.743	1.742
6.117	5.178	4.700

We believe that one reason for the relatively larger discrepancy between the values of D_{exp} and λ_m is the differences between the rivulet widths, w, of the base solution profile used in LSA, and the corresponding measured values reported in Table 1 (see Fig. 9a). These departures come from the inability of this base profile to yield large values of contact angles at the inflection points, as explained in Appendix A. In fact, the experimental values of the contact angle refer to the slope at these points, since they are measured by the maximum deviation of a light beam perpendicular to the substrate. Consequently, when we set θ equal to the values reported in [33], the profiles with van der Waals forces (as those used in LSA for Table 2) exhibit apparent contact angles different to those experimentally measured.

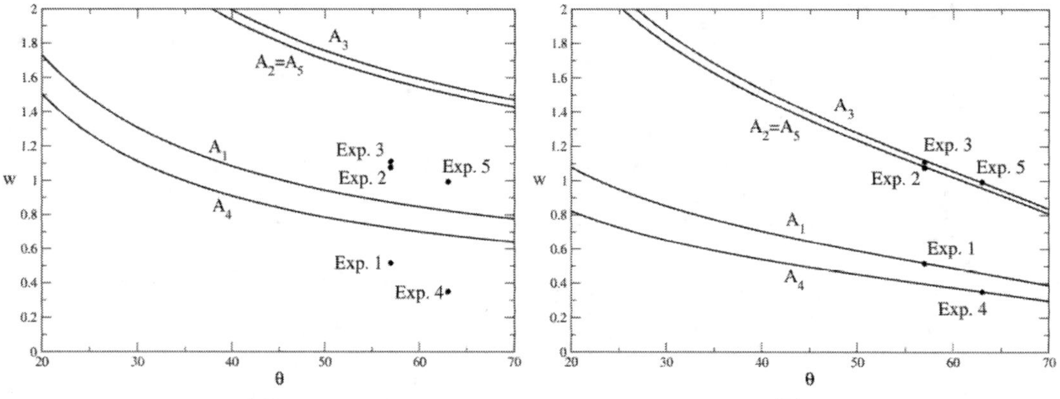

Figure 9. Width of the rivulet, w, as a function of contact angle, θ, as given by the solution: (a) with and (b) without van der Waals forces. The symbols correspond to five experimental cases reported in [33].

On the other hand, we find that the steady solution of a drop *without* considering van der Waals forces (see e.g. Appendix B in [9]) yields very accurate values of w when compared with experiments (Fig. 9b). It is worth noting that the LSA performed by using the base solution without van der Waals forces yields stability for zigzag modes and only marginal instability ($\sigma = 0$) for varicose modes for all $k > 0$.

Thus, we perform three dimensional (3D) numerical simulations with an initial condition as given by the solution *without* van der Waals forces. We start the evolution with the same area, contact angle and width as in the experiments, which assures an initial condition close to the experimental one. The evolution must be performed with the inclusion of van der Waals forces, since they are necessary to account for the partial wetting condition. An example of this evolution is shown in Fig. 10 for experiment 2.

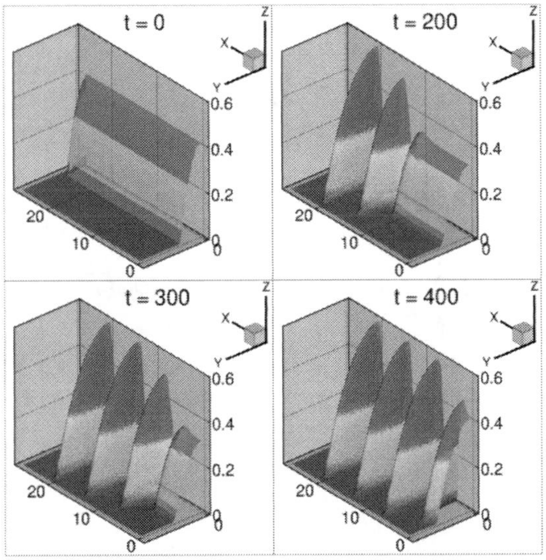

Figure 10. Evolution of experiment 2 of Table 2 with initial condition *without* van der Waals forces, so that A, θ and w are given by the values in second line of Table 2. The final pattern is formed by droplets separated an average distance $D_{num} = 6.6$.

The average distance between drops obtained for all experimental cases reported in [33] are shown as D_{num} in Table 3. Clearly, regardless of the choice of exponents, the large differences with the experimental data where only small number of drops is present, similarly to the results shown in Table 2. It should be also noted that both λ_m and D_{num} are calculated using relatively thick precursor film ($h_* = 0.01$. This large value, required due to significant computational cost involved if thinner precursor were used, may also contribute to inaccuracy of the results.

Table 3. Distance between drops, D_{num}, from numerical simulations for (n, m) equal to $(3, 2)$ and $(4, 3)$ compared with the experimental values from Table 1.

D_{exp}	$D_{num}^{(3,2)}$	$D_{num}^{(4,3)}$
3.501	2.745	3.936
12.84	6.600	9.600
19.53	7.000	7.080
2.802	1.754	2.730
6.117	7.200	7.080

5. Conclusions

The linear stability analysis (LSA) of partially wetting fluid rivulets has proved to be very useful for a deeper understanding of the breakup process. On one hand, it allows to consider simultaneously the effects of wettability and gravity. Therefore, it constitutes an advantage with respect to previous approaches, which are limited to a circular cross section and sharp contact lines with imposed contact angles. A more detailed comparison of our LSA results with those theories is out of the scope of this work and will be presented elsewhere. On the other hand, we have shown that the results of LSA apply also to large perturbations - therefore, the wavelength of maximum growth rate determines the distance between drops after breakup.

The comparison of the predictions obtained from both LSA and 3D numerical simulations has opened some challenging issues. The fact that: (a) for given area and contact angle the width of the rivulet given by the profile with van der Waals forces does not agree with experimental measurements (cf. Fig. 9a), and (b) that relatively large contact angles cannot be reached by the cross section profile with intermolecular forces considered here (see Appendix A), indicate intrinsic limitations of this lubrication approach. Future work is required to fully understand the breakup of a partially wetting rivulet on horizontal substrate.

Appendix A. Rivulet width and apparent contact angle

In this Appendix we show a peculiar feature of the steady state solution for the cross section of the rivulet. The value of the contact angle, θ, that characterizes the wettability of the substrate is introduced in the theory by means of the prefactor K in the van der Waals potential. Let us recall [27, 9] that the dependence of K on θ is obtained by performing a local analysis in the contact line region (which matches onto the precursor film of thickness h_*) and assuming that far away from that location the thickness profile asymptotes to a linear behavior with slope θ. However, the calculation of a profile that includes a maximum thickness at a finite distance from that region does not completely satisfies this assumption. This case unavoidably has an inflection point that connects the region near the precursor with that in the neighborhood of the apex (bulk). Thus, the actual contact angle is basically given by the slope at the inflection point, θ_i.

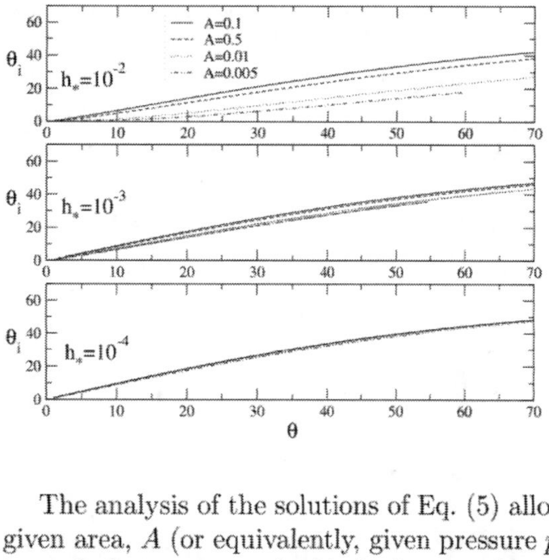

Figure A1. Slope at the inflection point (apparent contact angle) as a function of the contact angle used to determine prefactor K in van der Waals term, for typical areas A's. No gravity effects are considered here.

The analysis of the solutions of Eq. (5) allows to obtain the relationship between θ_i and θ for given area, A (or equivalently, given pressure p) as shown in Fig. A1. We see that, even for very large θ, θ_i cannot reach this high value, but it saturates around 50° (these θ_i's are calculated without including gravitational effects, but these are not of relevance to the present discussion). All the curves collapse into a single one as $h_* \to 0$, thus showing that the difference between θ_i and θ is only due to the fact that the drop width is finite. This result is an important drawback since these large apparent contact angles cannot actually be reached by the theory with the commonly used van der Waals model.

Acknowledgements. J.A.D. and A.G.G. thank support from Consejo Nacional de Investigaciones Científicas y Técnicas de la República Argentina (CONICET) and from Agencia Nacional de Promoción Científica y Tecnológica (ANPCyT) through grant PICTR 0094/02.

References

[1] Stone H, Stroock A and Ajdari A 2004 *Ann. Rev. Fluid Mech.* **36** 381
[2] Stone H A and Kim S 2001 *AIChE J.* **47** 1250
[3] Ruschak K 1999 *Ann. Rev. Fluid Mech.* **17** 65
[4] Kheshgi H S 1997 *Liquid Film Coating* ed Kistler S F and Scheizer P M (London: Chapman and Hall) p 183
[5] Larson R G and Regh T J 1997 *Liquid Film Coating* ed Kistler S F and Scheizer P M (London: Chapman and Hall) p 709
[6] Becker J, Grün G, Seemann R, Mantz H, Jacobs K, Mecke K R and Blossey R 2003 *Nature Mat.* **2** 59
[7] Brochard-Wyart F and Redon C 1992 *Langmuir* **8** 2324
[8] Herminghaus S, Jacobs K, Mecke K, Bischof J, Fery A, Ibn-Elhaj M and Schlagowski S 1998 *Science* **282** 916
[9] Diez J A and Kondic L 2007 *Phys. Fluids* **19** 072107
[10] Favazza C, Kalyanaraman R and Sureshkumar R *Nanotechnology* **17** 4229
[11] Davis S H 1980 *J. Fluid. Mech.* **98** 225
[12] Langbein D 1990 *J. Fluid. Mech.* **213** 251
[13] Roy R V and Schwartz L W 1999 *J. Fluid. Mech.* **391** 293
[14] Sekimoto K, Oguma R and Kawasaki K 1987 *Annals of Physics* **176** 359
[15] Schwartz L W and Eley R R 1998 *J. Colloid Interface Sci.* **202** 173
[16] Schwartz L, Roy R, Eley R and Petrash S 2001 *J. Colloid Interface Sci.* **234** 363
[17] Glasner K B 2003 *Phys. Fluids* **15** 1837
[18] Pismen L M and Rubinstein B Y 2000 *Phys. Fluids* **12** 480
[19] Pismen L M and Pomeau Y 2000 *Phys. Rev. E* **62** 2480
[20] Davis J M and Troian S M 2003 *Phys. Rev. E* **67** 016308
[21] Pismen L M and Thiele U 2006 *Phys. Fluids* **18** 042104
[22] Perazzo C A and Gratton J 2004 *J. Fluid Mech.* **507** 367
[23] Goodwin R and Homsy G M 1991 *Phys. Fluids A.* **3** 515
[24] Münch A and Wagner B 2005 *Physica D* **209** 178
[25] Snoeijer J H 2006 *Physics Fluids* **18** 021701
[26] Hirasaki G J and Yang S Y 2002 *Contact Angle, Wettability and Adhesion* **2** 1–30
[27] Oron A, Davis S H and Bankoff S G 1997 *Rev. Mod. Phys.* **69** 931
[28] Glasner K B and Witelski T P 2003 *Phys. Rev. E* **67** 016302
[29] Bertozzi A L, Grün G and Witelski T P 2001 *Nonlinearity* **14** 1569
[30] Glasner K B and Witelski T P 2005 *Physica D* **209** 80
[31] Diez J and Kondic L 2002 *J. Comp. Phys.* **183** 274
[32] Thiele U, Velarde M G, Neuffer K and Pomeau Y 2001 *Phys. Rev. E* **64** 031602
[33] González A G, Diez J, Gratton R and Gomba J 2007 *Europhysics Letters* **77** 44001

Selective decay in a helicity-injected spheromak

P L García Martínez, R Farengo

Centro Atómico Bariloche and Instituto Balseiro, Av. Bustillo 9500, 8400 Bariloche, Argentina

E-mail: pablogm@cab.cnea.gov.ar

Abstract. The non-linear evolution of several unstable equilibria, representative of helicity-injected spheromak configurations inside a cylindrical flux conserver, is studied by means of three dimensional resistive MHD simulations. These equilibria are force-free ($\nabla \times \mathbf{B} = \lambda(\psi)\mathbf{B}$) but do not correspond to minimum energy states, having linear $\lambda(\psi)$ profiles with negative slope. Several aspects of this process are studied (magnetic energy relaxation, selective helicity decay, relaxed profiles) for different initial λ slopes. The stability threshold predicted by linear theory is recovered. The results show that complete plasma relaxation leading to a uniform λ, is achieved only if the initial profile is hollow enough. The evolution for cases just above the stability threshold is more gentle and does not end in a Taylor state. The final state in these cases has a linear $\lambda(\psi)$ profile, as the initial condition, but with a smaller slope.

1. Introduction

A spheromak is a toroidal confinement configuration having a simply connected bounding surface inside which the magnetic field is produced almost entirely by currents flowing in the plasma. A distinctive feature of spheromaks is that the toroidal magnetic field is nonzero on internal flux surfaces but vanishes at the bounding surface, which implies that no external coils link the plasma [1, 2]. This represents an important advantage from an engineering and economical point of view, but introduces some difficulties related to the sustainment of the configuration against resistive dissipation. Most of the steady-state driven spheromaks have currents flowing through plasma-facing electrodes which are orthogonal to the magnetic axis. In order to drive toroidal current along the magnetic axis, the system relies on magnetic relaxation [3] which necessarily involves non-axisymmetric (three-dimensional) MHD mode activity. According to the theory of magnetic relaxation [3], the configuration should relax to a state of minimum magnetic energy subject to the conservation of magnetic helicity. While this theory qualitatively predicts the observed configurations [4, 5, 6] it does not provide any explanation on how the toroidal current is driven and on the details of the dynamics of the process. Furthermore, in practice the observed configurations are not fully relaxed, but are maintained by the external energy source in a steady state whose energy is larger than the minimum [4].

The physical mechanism underlying Taylor's hypothesis is called "selective decay" [7, 8]. It is an MHD turbulence situation in which, under certain conditions, a system with initial magnetic helicity present evolves with a rapid decay of total energy relative to the magnetic helicity.

In this work, the non-linear evolution of several unstable equilibria, representative of a helicity-injected spheromak inside a cylindrical flux conserver, is studied by means of three dimensional resistive MHD simulations. Through relaxation the system evolves from a state representative of a driven configuration, like the one shown in Fig. 1 (a), to a relaxed (or partially

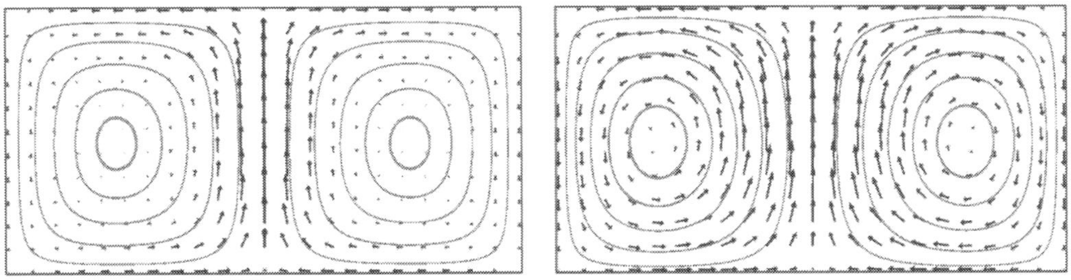

(a) (b)

Figure 1. Magnetic field vectors and poloidal flux contours for a driven configuration with a linear $\lambda(\psi)$ profile (a) and for a relaxed configuration with uniform λ (b).

relaxed) state, more similar to the one shown in Fig. 1 (b), via a selective decay process. It has to be emphasized that no helicity is injected to the system during these simulations. We study the evolution of an isolated configuration from an initial condition that mimics the real spheromak plasma when helicity is being injected to it (sustainment phase).

Spheromak relaxation has been studied using numerical simulations in the past. Katayama and Katsurai [9] simulated a decaying spheromak in a cylindrical flux conserver for several q-profiles and observed the growth of an $n = 1$ mode and poloidal flux generation for the hollow λ profile situation (see section 2). Their study was restricted to unstable modes with relatively high growth rates (larger than $0.01/\tau_A$, where τ_A is the Alfvén time) due to computer limitations, and all the final states obtained in their simulations were quite near the Taylor state. The decay of an unstable spheromak equilibrium in a "bowtie" flux conserver was simulated by Izzo and Jarboe [10]. Using the same initial condition, they varied the Lundquist number over one order of magnitude and identified the minimum value needed for poloidal flux generation. Formation and sustainment in both a cylindrical "flux core" configuration and a coaxial gun were studied by Sovinec *et al* [11].

In this paper, we study the magnetic relaxation event for different values of α, the slope of the linear $\lambda(\psi)$ profile imposed at the beginning of the simulation (see Sec. 2). This slope has a threshold value above which the configuration becomes unstable to kink modes. For unstable cases well above the stability threshold, we observe, as expected, a rapid evolution of the system towards a Taylor state. However, for cases just above the stability threshold the evolution is more gentle and does not end in a uniform λ profile (Taylor state) but in a state having a linear $\lambda(\psi)$ profile with a lower slope, below the stability threshold.

In Section 2, we describe the initial conditions of the simulations. The physical model and the details of the numerical scheme employed are presented in Sec. 3. The simulation results are discussed in Sec. 4 and, finally, the conclusion are summarized in Sec. 5.

2. Initial condition

The initial condition is an axisymmetric force-free ($\mathbf{J} \times \mathbf{B} = 0$), pure magnetic ($\mathbf{u} = 0$) state. When the magnetic field is axisymmetric, the poloidal flux ψ acts as a stream function for the poloidal magnetic field and the force-free condition may be expressed in terms of the Grad-Shafranov equation for ψ,

$$\Delta^*\psi + \lambda(\psi) \int_0^{\psi} \lambda(\hat{\psi})\partial\hat{\psi} = 0, \tag{1}$$

where $\Delta^* = \frac{\partial^2}{\partial r^2} - \frac{1}{r}\frac{\partial}{\partial r} + \frac{\partial^2}{\partial z^2}$, and $\lambda = \mu_0 \mathbf{J}\cdot\mathbf{B}/B^2$. The special case $\lambda(\psi) = \lambda_0$ (constant λ-profile) corresponds to the Taylor state which has the minimum magnetic energy for a given helicity. Experimental observations showed that driven spheromaks are well described by solutions of eq. (1) using the linear relationship,

$$\lambda(\psi) = \bar{\lambda}[1 + \alpha(2\tilde{\psi} - 1)], \tag{2}$$

where $\bar{\lambda}$ is the average of λ, $\tilde{\psi} = \psi/\psi_{max}$ is the normalized poloidal flux, the constant α controls the slope of the λ-profile and its sign determines whether the current profile is hollow (negative α) or peaked (positive α). Hollow current profiles are representative of the sustainment phase of a helicity-injected spheromak having a $\lambda(\psi)$ profile peaked towards the outer flux surfaces. The linear ideal MHD stability analysis shows that this type of configurations becomes unstable to an $n = 1$ mode (kink) when α is below -0.5 [4, 12].

In section 4 we present results from the non-linear evolution of solutions to eq. (1), using the linear (in ψ) λ-profile (2) and the values given in table 1 for α and $\bar{\lambda}$. The domain is a cylindrical flux conserver ($\psi|_{\partial\Omega} = 0$, which implies $(\mathbf{B}\cdot\mathbf{n})|_{\partial\Omega} = 0$) with $h/a = 1$ (where, h is the cylinder height, and a its radius). To this pure magnetic initial condition we add a small perturbation to the axial component of the velocity (u_z), having toroidal wavenumber $n = 1$ and an energy $\sim 5 \times 10^{-4} \mathrm{W}_0$, where W_0 is the initial magnetic energy.

Table 1. α and $\bar{\lambda}$ values for the simulations presented in Sec. 4.

α	$\bar{\lambda}$
0.0	4.955
-0.5	5.320
-0.6	5.510
-0.7	5.785
-0.8	6.230

3. Model description and computational approach

Simulations in this study are performed with the VAC (Versatile Advection Code) [13, 14], in which several numerical methods for solving compressible hydrodynamics and magnetohydrodynamics (MHD) equations are implemented. In particular, this code can be used to solve the isothermal resistive MHD model, which is the one used in this work. Defining a length scale a (the cylinder radius), a density scale ρ_0 and a magnetic field scale B_0, we can write the equations in dimensionless units,

$$\frac{\partial \rho}{\partial t} + \nabla\cdot(\rho\mathbf{u}) = 0 \tag{3}$$

$$\frac{\partial \mathbf{u}}{\partial t} + \mathbf{u}\cdot\nabla\mathbf{u} + \left(\frac{c_s^2}{c_A^2}\right)\frac{1}{\rho}\nabla\rho = \frac{1}{\rho}(\nabla\times\mathbf{B})\times\mathbf{B} + \left(\frac{\nu}{a c_A}\right)\nabla\cdot\hat{\Pi} \tag{4}$$

$$\frac{\partial \mathbf{B}}{\partial t} + \nabla\times(-\mathbf{u}\times\mathbf{B}) = -\left(\frac{\eta}{\mu_0 a c_A}\right)\nabla\times(\nabla\times\mathbf{B}), \tag{5}$$

where $c_A = B_0/\sqrt{\mu_0\rho_0}$ is the Alfvén speed, $\hat{\Pi} = (\nabla\mathbf{u} + \nabla\mathbf{u}^{\mathrm{T}}) - \frac{2}{3}(\nabla\cdot\mathbf{u})$, and the pressure was eliminated using $p = c_s^2\rho$, where c_s is the sound speed. The three dimensionless parameters appearing in eq. (3)-(5) can be identified in the following way. The parameter $\beta = (c_s/c_A)^2 = \mu_0 p_0/B_0^2$, in the left hand side of eq.(4), quantifies the relative importance of hydrodynamics forces (coming from pressure gradients) and magnetic forces (Lorenz's force). Spheromaks are low β plasmas and thus to first order approximation, forces due to hydrodynamic pressure gradients may be ignored. For this reason, we do not advance the continuity equation, eq. (3), and we set a small value for c_s, compared to c_A. Since the initial condition has uniform density (and consequently pressure), the pressure forces vanish in our simulations. This simplification is usually referred to as the zero-β (or zero-pressure) approximation and is widely used when modelling low-β plasmas [10, 11]. The second dimensionless parameter is the dimensionless kinematic viscosity $\tilde{\nu} = \nu/(ac_A)$, the last one is the dimensionless resistivity, $\tilde{\eta} = \eta/(\mu_0 a c_A)$. The parameter $\tilde{\eta}$ may also be interpreted as the ratio of two time scales: $\tau_A = a/c_A$, the Alfvén time (representative of MHD propagation) and $\tau_r = \mu_0 a^2/\eta$, the resistive time. The inverse of this quantity is the Lundquist number $S = \tau_r/\tau_A$. In this work we use $\beta = 0$, $P_m = \tilde{\nu}/\tilde{\eta} = 1$ (magnetic Prandtl), and $\tilde{\eta} = 0.5 \times 10^{-4}$.

The system of partial differential equations that composes the physics model is solved in a uniform cartesian grid, with $N_x \times N_y \times N_z = 100 \times 100 \times 50$. The cylindrical flux conserver is constructed using appropriate values at ghost cells, i.e. knowing the solution inside the flux conserver ($r < a$) we set the values of external grid points ($r > a$) in such a way that the boundary conditions are satisfied (at $r = a$), up to the interpolation error. The perfectly conducting wall conditions employed are $(\mathbf{B} \cdot \hat{\mathbf{n}})|_{\partial\Omega} = 0$ and $(\mathbf{J} \times \hat{\mathbf{n}})|_{\partial\Omega} = 0$, and the condition $\mathbf{u}|_{\partial\Omega} = 0$. Note that this choice of boundary conditions is not compatible with the force-free equilibria that we use as initial conditions. The reason is that this boundary conditions imposes $\lambda|_{\partial\Omega} = 0$, while the initial condition has $\lambda|_{\partial\Omega} = \lambda(\psi = 0) = \bar{\lambda}(1 - \alpha) \neq 0$, from eq. (5) and the condition $\alpha < 0$. In general, an ideal force-free equilibrium may have a non vanishing tangential current at the flux conserver boundary. Nevertheless, we still use the boundary conditions mentioned above for compatibility with previous works [9, 10, 11] and because they are familiar Dirichlet and Neumann condition for $\mathbf{B}$. Conditions on the tangential derivative of $\mathbf{B}$ at the wall are not easy to apply in our representation, but can be applied in the context of spectral methods [15]. We will see, in the following section, that this boundary condition only affects the outer flux surfaces, i.e. the flux surfaces that are close to the wall. We note that in real laboratory plasmas, the current always drops off near the wall because the temperature there is lower and the resistivity is higher.

4. Results

The evolution of the poloidal flux for four different initial conditions is shown in Fig. 2 (ψ is normalized with its initial value and t is in Alfvén time units). Poloidal flux amplification, which is intimately related with toroidal current drive via magnetic relaxation, is observed in all the unstable cases (cases having $\alpha < -0.5$). We can see, in Fig. 2, that the activity that produces the flux amplification develops at earlier times in the most unstable cases and that this amplification is faster in those cases (has a shorter rise time).

In Fig. 3 the decay of the total magnetic energy and the energy of the $n = 0$ component of the Fourier decomposition in the toroidal direction are shown for the cases $\alpha = -0.6$ (a) and $\alpha = -0.8$ (b). The energies are normalized with their initial values. The case $\alpha = -0.6$, in Fig. 3(a), which presents a small amount of flux generation (Fig. 2), has an almost axisymmetric evolution (the evolution of W_{Tot} and $W_{n=0}$ can not be distinguished). We can also see that the energy decay of the configuration due to resistivity that results after the appearance of a small amount of higher mode activity (Fig. 3(c)), at times greater than 150, is faster than the decay of a Taylor state (in the presence of resistivity, the Taylor state decays exponentially, $\sim\exp(-2\eta\lambda_0^2 t)$, in dimensionless units [1], as shown with a black line in Fig. 3(a) and (b)).

The evolution of the total and $n = 0$ magnetic energies for the case $\alpha = -0.8$ is shown in Fig.

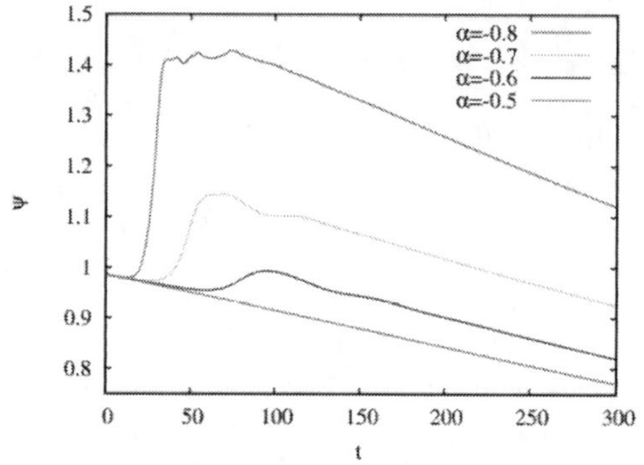

Figure 2. Poloidal flux vs time (in Alfvén time units) for four different cases. The initially unstable states (those with $\alpha < -0.5$) exhibit MHD activity resulting in poloidal flux amplification, which is directly related with toroidal current drive.

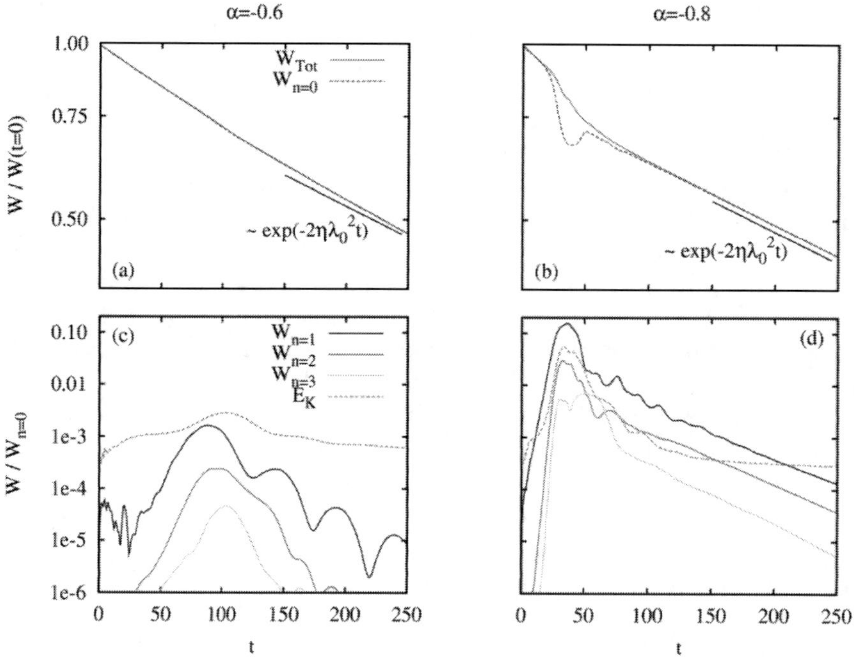

Figure 3. Evolution of the total (W) and the $n = 0$ ($W_{n=0}$) magnetic energies for the cases $\alpha = -0.6$(a) and $\alpha = -0.8$(b), relative to their initial value. Higher modes magnetic energies ($W_{n=1,2,3}$) and kinetic energy (E_K) relative to $W_{n=0}$ for the same cases ((c) and (d), respectively).

3(b). In this case, the system undergoes a stepwise relaxation event with a significant deviation from the axisymmetric situation. During this strong mode activity the total magnetic energy decays faster until axisymmetry is recovered and finally the magnetic energy of the system decays with the same slope than the Taylor state.

The higher mode activity can be seen in Fig. 3 (c) and (d), along with the kinetic energy, for the cases $\alpha = -0.6$ and $\alpha = -0.8$ respectively. The energies in this case are divided by

$W_{n=0}$ (at each time). In both cases, the initial perturbation in the velocity is translated almost instantaneously to $W_{n=1}$. In the $\alpha = -0.6$ case, this perturbation is not enough to trigger the instability, and the system resistively decays without activity until approximately $t = 25\tau_A$, when the instability begins to develop. This time delay is not observed in the more unstable case having $\alpha = -0.8$, where the magnetic energy of the $n = 1$ component grows exponentially from the beginning of the simulation (Fig. 3 (d)). In both cases the dominant mode is the $n = 1$ and the saturation of this mode is in temporal coincidence with the poloidal flux amplification.

The initial and final B_z and B_θ radial profiles at $z = h/2$ are shown in Fig. 4(a) for the $\alpha = -0.6$ case, and in 4(b) for the $\alpha = -0.8$ case. The profiles for the corresponding Taylor states are also shown. While in both cases the final state is more similar to the Taylor solution

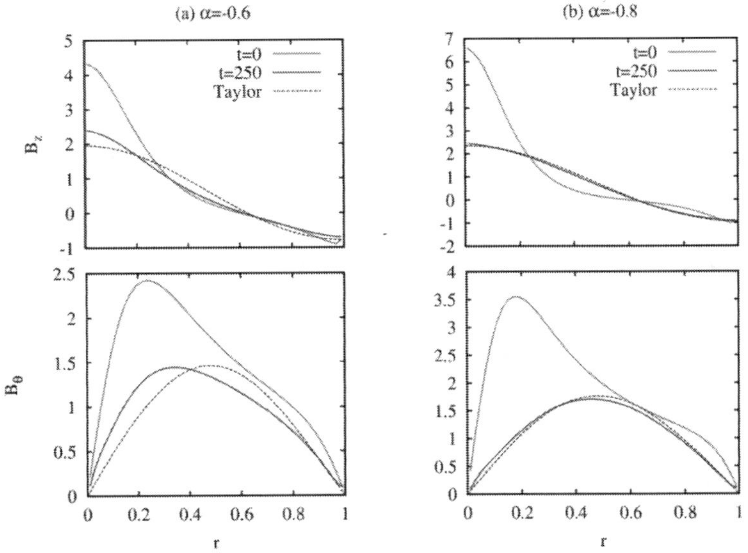

Figure 4. Initial, final and Taylor magnetic field radial profiles at $z = h/2$. Axial and toroidal components are shown for the cases $\alpha = -0.6$ and $\alpha = -0.8$. Note that the y-axis scale is different in each column.

than the initial state, the final profiles in case (b), with $\alpha = -0.8$, are much closer to the fully relaxed configuration described by the Taylor state than the final profiles in case (a).

The effect of the magnetic relaxation on the λ profiles can be observed in Fig. 5, for the case $\alpha = 0$ (a), and for the cases $\alpha = -0.6$ (b) and $\alpha = -0.8$ (c). As in the case of Fig. 4, these radial profiles correspond to $z = h/2$. The $\alpha = 0$ case shown in Fig. 5(a) does not present MHD activity because it is indeed the relaxed state (it is initially the Taylor state). However, the $\lambda(r)$ profile for this case is shown in order to see the effect of the boundary condition applied. Note that the boundary condition $(\mathbf{J} \times \mathbf{n})|_{\partial\Omega} = 0$ imposed on the flux conserver makes λ vanish at $r = 1$, so it is inconsistent with the uniform λ of the relaxed state. This perturbation introduced on the outer flux surfaces near the cylinder wall propagates, as time evolves, along these flux surfaces producing deviations from the uniform λ profile near the symmetry axis of the cylinder. Nevertheless, internal flux surfaces remain unaffected by this perturbation. A similar effect of this perturbation on the cases $\alpha = -0.6$ and $\alpha = -0.8$ can be observed in Fig. 5 (b) and (c), respectively.

The cases showed in Fig. 5(b) and 5(c) begin with a hollow λ profile indicating that the current is concentrated in the outer flux surfaces. After the magnetic relaxation, the current

Journal of Physics: Conference Series **166** (2009) 012010 doi:10.1088/1742-6596/166/1/012010

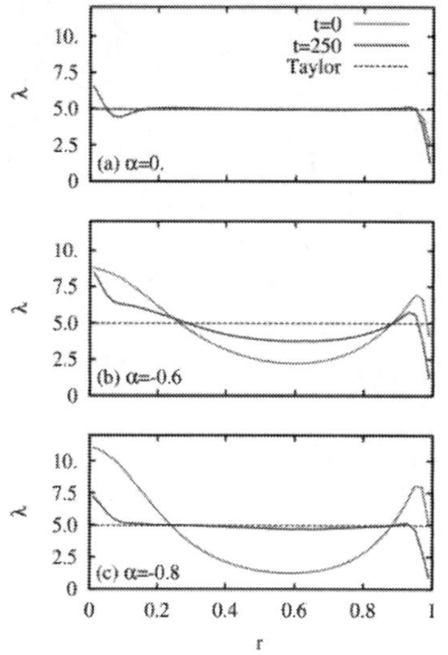

Figure 5. Radial λ profiles at $z = h/2$. The boundary conditions introduces a small perturbation that affects the outer flux surfaces in the three cases. The most unstable case $\alpha = -0.8$ (c) exhibits a complete λ uniformization.

is redistributed and the λ profile becomes more uniform. In particular, at the magnetic axis position ($r \sim 0.65$), the value of λ grows in these cases implying that some toroidal current has been driven along the magnetic axis. The uniformization of the λ profile in the case $\alpha = -0.6$ is incomplete, in contrast with the more unstable case $\alpha = -0.8$ which exhibits an almost uniform final λ profile.

In Fig. 6 we show the evolution of the ratio $2\mathrm{W}_{n=0}/\mathrm{K}_{n=0}$ for four cases, ranging from the initial Taylor state $\alpha = 0$, to the most unstable case $\alpha = -0.8$. The case $\alpha = 0$ evolves close to

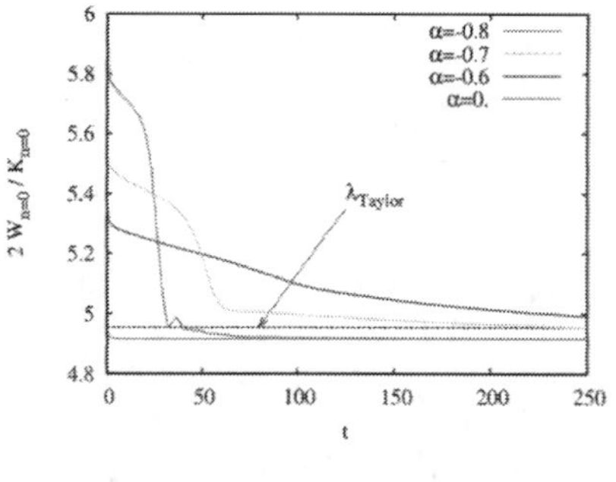

Figure 6. Selective decay of magnetic energy ($\mathrm{W}_{n=0}$) relative to the magnetic helicity ($\mathrm{K}_{n=0}$), in four cases. As the initial condition is more unstable, Taylor's hypothesis is better fulfilled during relaxation.

the theoretical value for the Taylor state (indicated as $\lambda_{\mathrm{Taylor}}$ in Fig. 6). The small discrepancy is due to the physical boundary condition imposed at the flux conserver walls ($(\mathbf{J} \times \mathbf{n})|_{\partial\Omega} = 0$), and possibly to numerical inaccuracies introduced by the discretization. In the cases with $\alpha < 0$ the ratio $2\mathrm{W}_{n=0}/\mathrm{K}_{n=0}$ decays. During the relaxation event itself, when MHD activity arises, the decay rate grows in the three cases, indicating that the energy is dissipated faster than helicity specially at that moment. After the activity, the decay rate of the energy becomes more similar to that of the magnetic helicity, specially for the initially more unstable cases.

From the three cases presented in Fig. 6 with $\alpha < 0$, only the case $\alpha = -0.8$ reduces its magnetic energy to the minimum value allowed by the helicity present in the system, satisfying the Taylor hypothesis. In the other cases the relaxation is incomplete, in the sense that there is still some magnetic energy above the minimum compatible with the helicity present.

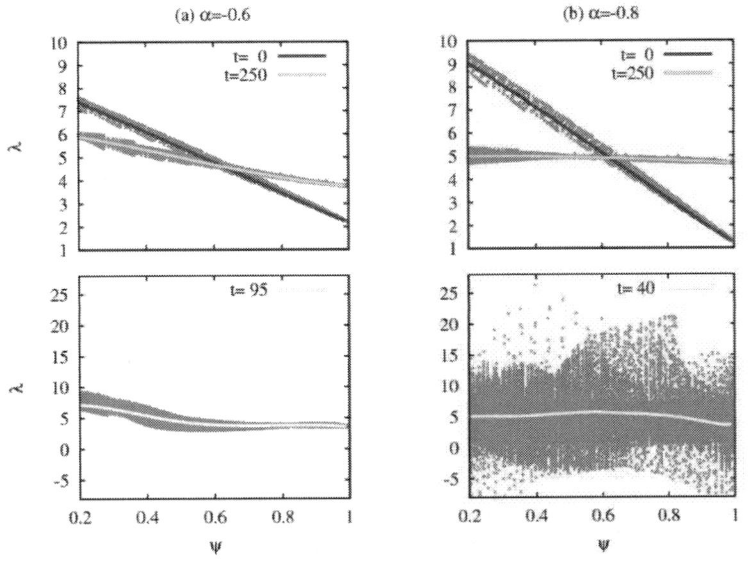

Figure 7. The values of λ and ψ are computed in each grid point and plotted in λ vs ψ space (orange points). The lines are the λ average at each ψ.

Another way to evaluate how close is a solution to the Taylor state is plotting the value of λ as a function of the poloidal flux ψ. This is done computing these two values at each grid point for the cases $\alpha = -0.6$ and $\alpha = -0.8$, for several times (Fig. 7). Each point on the graphs of Fig. 7 corresponds to one grid point and the λ average for each ψ is indicated by a line. It is observed that the net effect of the relaxation event is to reduce the slope of the $\lambda(\psi)$ profile. In the more unstable cases the MHD activity that arises produces the complete relaxation of the configuration, leading to a solution with an almost uniform λ. On the other hand, when the system is just above the stability threshold the relaxation is incomplete, and the final state after relaxation has a linear $\lambda(\psi)$ profile with a lower slope, that is below the linear stability threshold. During the relaxation event, at the time of maximum activity (when the axisymmetry is broken), the $\alpha = -0.6$ case ($t = 95$) presents a little dispersion on the λ values, concentrated at low values of ψ (outer flux surfaces). The $\alpha = -0.8$ case has a large dispersion in λ, with a flat distribution along ψ, and it presents regions with negative λ, which means that the current and the magnetic field are antiparallel. These regions may be associated with magnetic reconnection events occurring during the non-linear activity that causes the full relaxation of the system.

5. Conclusions

We have computed the non-linear evolution of several force-free equilibria representative of a helicity-injected spheromak. The cases below the linear stability threshold did not show significant MHD activity to provide poloidal flux generation. For unstable cases poloidal flux generation was observed, as a result of the appearance of MHD activity characterized by non-axisymmetric modes. This activity is gentle in cases just above the stability threshold, leads to little flux amplification and produces an incomplete relaxation of the configuration. More unstable cases exhibit much more activity, that introduces important deviations to the axisymmetry of the configuration, produces substantial flux amplification and makes the system relax completely, as expected if the Taylor hypothesis were valid.

Measurements made in experimental spheromak equilibria indicate that during sustainment the configurations differ from the minimum-energy state and are well described by a linear $\lambda(\psi)$ profile with negative slope [4]. Although the conditions in the simulations presented here are far from the actual conditions in real experiments (because of the limited resolution, the lack of magnetic flux intercepting the boundaries, etc.), the results presented in this work suggest that the process that occurs during spheromak sustainment is an incomplete relaxation event, which leads a marginally unstable configuration to a marginally stable one. That is to say, as the power source drives poloidal current in the outer flux surfaces and the toroidal current along the magnetic axis resistively decays, α (the slope of the $\lambda(\psi)$ profile) becomes more negative, until it crosses the stability threshold. Then, some MHD activity arises producing an incomplete relaxation of the plasma, like in the case $\alpha = -0.6$ presented above. This intermittent evolution around the stability boundary has been already proposed, analizing results from linear theory [16] and experiments [17]. Nevertheless, it has not been observed, up to our knowledge, that the non-linear evolution of a marginally unstable spheromak without sustainment ends in a marginally stable configuration, instead of fully relaxing towards a Taylor state.

References

[1] Bellan P M 2000 *Spheromaks: A Practical Application of Magnetohydrodynamic Dynamos and Plasma Self-Organization* (London: Imperial College Press)

[2] Jarboe T R 1994 *Plasma Phys. Control. Fusion* **36** 945

[3] Taylor J B 1986 *Rev. Mod. Phys.* **58** 741

[4] Knox S O, Barnes C W, Marklin G J, Jarboe T R, Henins I, Hoida H W and Wright B L 1986 *Phys. Rev. Lett.* **56** 842

[5] Browning P K, Cunningham G, Gee S J, Gibson K J, al-Karkhy A, Kitson D A, Martin R and Rushbridge M G 1992 *Phys. Rev. Lett.* **68** 1718

[6] Hooper E B, Pearlstein L D and Bulmer R H 1999 *Nucl. Fusion* **39** 863

[7] Matthaeus W H and Montgomery D 1980 *Ann. N. Y. Acad. Sci.* **357** 203

[8] Ting A C, Matthaeus W H and Montgomery D 1986 *Phys Fluids* **29** 3261

[9] Katayama K and Katsurai Makoto 1986 *Phys. Fluids* **29** 1939

[10] Izzo V A and Jarboe T R 2003 *Phys. Plasmas* **10** 2903

[11] Sovinec C R, Finn J M and del-Castillo-Negrete D 2001 *Phys. Plasmas* **8** 475

[12] Shumlak U and Jarboe T R 2000 *Phys. Plasmas* **7** 2959

[13] Tóth G 1996 *Astrophys. Lett. Comm.* **34** 245

[14] VAC code, `www.phys.uu.nl/~toth/`

[15] Mininni P D and Montgomery D C 2006 *Phys. Fluids* **18** 116602

[16] Brennan D P, Browning P K and Van der Linden R A M 2002 *Phys. Plasmas* **9** 3526

[17] Brennan D P, Browning P K, Van der Linden R A M, Hood A W and Woodruff S 1999 *Phys. Plasmas* **6** 4248

Self–similar asymptotics in convergent viscous gravity currents of non–Newtonian liquids

Julio Gratton[1] and Carlos Alberto Perazzo[2]

[1] INFIP–CONICET, Dpto. de Física, FCEyN, UBA, Ciudad Universitaria, Pab. I, 1428 Buenos Aires, Argentina.

[2] Universidad Favaloro and CONICET, Solís 453, 1078 Buenos Aires, Argentina.

E-mail: jgratton@tinfip.lfp.uba.ar, perazzo@favaloro.edu.ar

Abstract. We investigate the evolution of the ridge produced by the convergent motion of two substrates, on which a layer of a non–Newtonian power–law liquid rests. We focus on the self–similar regimes that occur in this process. For short times, within the linear regime, the height and the width increase as $t^{1/2}$, independently of the rheology of the liquid. In the self–similar regime for large time, the height and the width of the ridge follow power laws whose exponents depend on the rheological index.

1. Introduction

Mountains belts arise due to the shortening of the crust that occurs when two lithospheric plates collide, or when a plate is subducted beneath an other. However there is also abundant geological evidence showing that extension occurs in the central part of many mountain belts, notwithstanding shortening is taking place. This apparently paradoxical fact suggested one of us [1] the idea that mountain building occurs as a consequence of a dynamic balance between the shortening of the plate and the spreading flow, that occurs because an isostatically compensated range is not in hydrostatic equilibrium, and tends to spread and collapse unless restrained by appropriate stresses. Indeed, it can be easily estimated that the characteristic time for the collapse of a mountain range resulting from root spreading is of the same order of magnitude as the times involved in orogeny, which means that both processes (shortening and lateral spreading) occur simultaneously. Starting from these considerations the scaling laws for the evolution of orogenic belts were derived, based on simple physical hypotheses about the viscous flow caused by the shortening of the Earth's crust [1], which is assumed to have a velocity that depends on the depth.

Various physical models have been used to simulate the build–up and the evolution of mountain belts. All are extensions and variations of the thin viscous sheet model of England and McKenzie [2; 3]. A discussion and a classification of these models has been made by Medvedev and Podladchikov [4; 5], to which the reader is referred for more details. All these numerical calculations attempt to describe specific orogenies as realistically as possible. No effort has been made to gain a deeper physical insight of the process. In particular no scaling laws have been derived from these models. Unlike in [1] most thin viscous sheet models assume that the velocity field does not depend on the vertical coordinate, so that the main gradient is horizontal. It was shown in [1] that this assumption leads to scaling laws that are slightly different from those derived under the assumption that the main velocity gradient is vertical.

To achieve a better understanding about these kind of flows we investigated a simple model that consists of an initially uniform fluid layer that rests over an horizontal substrate. This substrate is divided in two parts, that for $t > 0$ are pushed one against the other [6]. This convergent motion drags the liquid, and produces a ridge (see figure 1). We studied the evolution of the flow of a Newtonian liquid, and we showed that there are two self–similar regimes that occur in different space–time domains. These regimes and their corresponding scaling laws were obtained analytically. Here we extend these results for liquids with a power–law rheology, because this behavior is more appropriate to describe the behavior of the lithosphere.

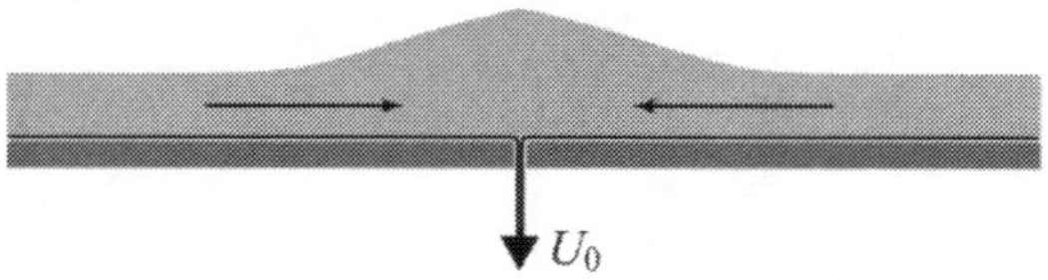

Figure 1. Formation of a ridge due to the convergent motion of the substrates.

2. Basic equations

The geometry of our model is shown in figure 1: an initially uniform liquid layer rests on two substrates, that start to move suddenly with velocities $U_0 > 0$ (the substrate for $X < 0$) and $-U_0$ (the substrate for $X > 0$). Thanks to the symmetry of the problem it is enough to consider the half–plane $X > 0$. Then we assume that a uniform semiinfinite liquid layer, whose thickness is H_0, rests on a rigid horizontal surface, and that at $X = 0$ there is a vertical wall over which the liquid slips. At $T = 0$ the substrate starts to move with the constant speed $-U_0$ so that the liquid accumulates against the wall. We assume a power–law rheology of the form

$$\tau_{ij} = AE^{(1-n)/n}\dot{\varepsilon}_{ij}, \quad E = (\dot{\varepsilon}_{ij}\dot{\varepsilon}_{ij})^{1/2}. \tag{1}$$

Here τ_{ij} and $\dot{\varepsilon}_{ij} = \frac{1}{2}(\partial_j v_i + \partial_i v_j)$ are the components of the stress and the strain rate tensors, and A, n are positive constants. A Newtonian liquid corresponds to $n = 1$, in which case the viscosity is $A/2$. Reasonable values for the lithosphere are $n = 3$ and $A = 10^{13}$ (c.g.s) [7]. We assume that the flow is slow and dominated by shear stresses, so we can employ the lubrication approximation [8] (we neglect surface effects). Let $H \equiv H(X, T)$ be the thickness of the layer, and $U \equiv U(X, T)$ its vertically averaged horizontal velocity. The boundary conditions are $U(0, T) = 0$ and $H(\infty, T) = H_0$. We define the dimensionless variables u, h, x, t by means of

$$U = U_0 u, \quad H = H_0 h, \quad X = X_0 x, \quad T = \frac{X_0}{U_0} t,$$

where

$$X_0 = \frac{\rho g}{A}\left[\frac{2^{\frac{1+n}{2}} H_0^{2n+1}}{(n+2)U_0}\right]^{\frac{1}{n}},$$

g is the acceleration of gravity and ρ is the density of the liquid. The evolution equation is then

$$h_t = h_x + s\left(h^{n+2}|h_x|^n\right)_x, \quad s \equiv \mathrm{sign}(h_x). \tag{2}$$

This equation can be written as the continuity equation

$$h_t = -(uh)_x, \tag{3}$$

where

$$u = -1 - s\, h^{n+1} |h_x|^n, \tag{4}$$

is the dimensionless vertically averaged horizontal velocity. In the case we are considering $s = -1$, then the initial and boundary conditions are

$$h(x,0) = 1, \quad h(\infty, t) = 1, \quad h^{1+\frac{1}{n}} h_x \Big|_{x=0} = -1. \tag{5}$$

The second of these conditions implies that for large x, where $h_x \to 0$, the velocity is essentially equal to that of the substrate. The third condition implies that $u(0,t) = 0$, which means that at $x = 0$ the nonlinear diffusion term in (4) exactly cancels the advective term. On the other hand, from mass conservation one has $\int_0^\infty (h-1)dx = t$. Then we see that the accumulation of mass near $x = 0$ due to advection is counteracted by the diffusion produced by gravity. Notice that the boundary condition at $x = 0$ depends on the rheology, and that as n increases the ridge tends to have a steeper peak.

3. Numerical solutions

The problem (2–5) does not have a closed solution. We show some numerical results in figures 2 and 3. Figure 2 displays the evolution of the ridge for $n = 3$. It can be observed that h has an inflexion point. It is located near $x = 0$ at the beginning, and migrates towards large x as t increases, tending to reach the leading part of the ridge, where h approaches 1. It can be also noticed that the aspect ratio (heigth/width) of the ridge decreases with time. Qualitatively similar results are found for other values of n. In figure 3 we compare the profiles for $n = 1$, 3 and 10 and for $t = 0.3$ and $t = 8$. It can be observed that the larger is n, the narrower and higher is the ridge.

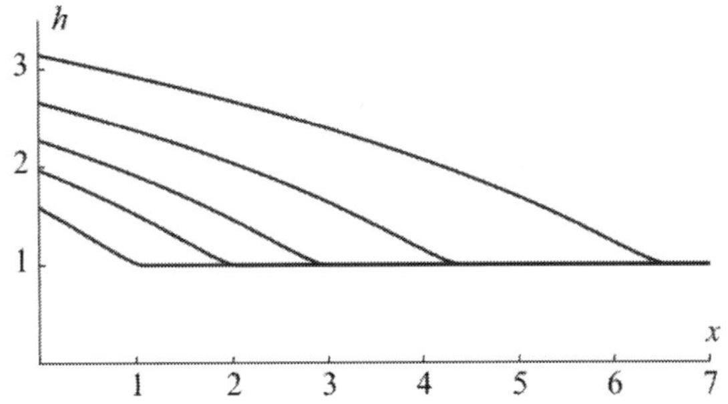

Figure 2. Numerical profiles for $n = 3$ and $t = 0.3$, 1, 2, 4 y 8.

4. Self–similar behavior for $t \ll 1$

For small t, when $h \approx 1$, equation (2) can be linearized. To this end we seek a solution of the form

$$h(x,t) = 1 + t^{1/2} f(\psi), \quad \psi = \frac{x}{2\sqrt{t}}, \tag{6}$$

in which the elevation difference $t^{1/2} f(\psi) \ll 1$. Here the exponents of t in front of f and in the definition of ψ are both equal to $1/2$. This choice ensures that the linearized boundary condition at $x = 0$ and the mass conservation do not depend on t, and implies that in this regime the height and the width of the ridge increase as $t^{1/2}$ regardless of n. Notice however that n appears in the equation for f:

$$0 = f - \psi f' - 2^{-n} n (-f')^{n-1} f'', \tag{7}$$

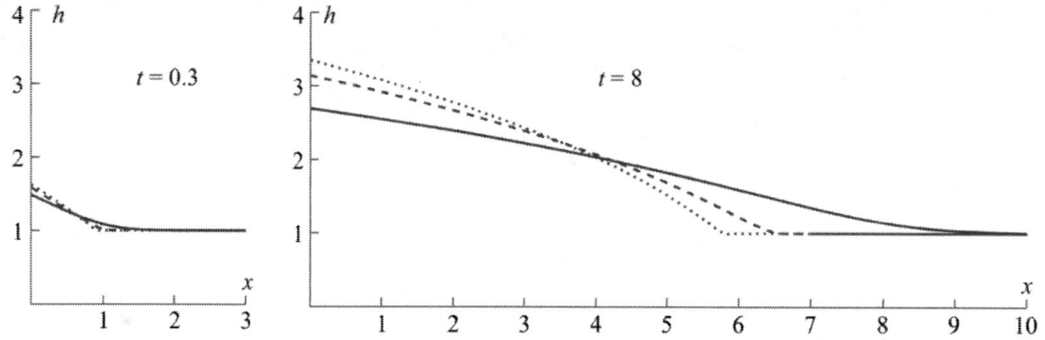

Figure 3. Numerical profiles at $t = 0.3$ (left) and $t = 8$ (right) for $n = 1$ (full line), 3 (dashes) and 10 (points).

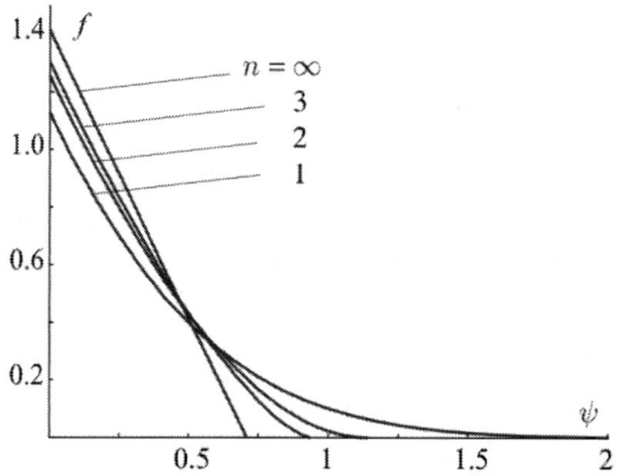

Figure 4. Profiles of f for $t \ll 1$ with $n = 1$, 2, 3, and ∞).

so that the shape of the ridge depends on n.

The boundary conditions are $f'(0) = -2$ and $f(\infty) = 0$. From the second of these conditions we obtain

$$f(0) = \frac{\sqrt{2}\,n}{n - 1 + \sqrt{\frac{\pi}{2}}}.$$

In the limit $n \to \infty$ the solution of (7) is $f_\infty = \sqrt{2} - 2\psi$. The solution for $n = 1$ was obtained in [6] and is

$$f_1 = \frac{2e^{-\psi^2}}{\sqrt{\pi}} - 2\psi \operatorname{erfc}(\psi),$$

where $\operatorname{erfc}(\psi) = 1 - \operatorname{erf}(\psi)$ is the complementary error function. For other values of n the function f is obtained by numerical integration. In figure 4 we show $f(\psi)$ for different values of the rheological index. The maximum height of the ridge in the linear regime is given by

$$h_w = 1 + \frac{\sqrt{2}\,n}{n - 1 + \sqrt{\frac{\pi}{2}}}\, t^{1/2}. \tag{8}$$

In figure 5 we show the evolution of the maximum elevation difference $h_w - 1$. It can be seen that for small t it tends to the self-similar behavior given by equation (8).

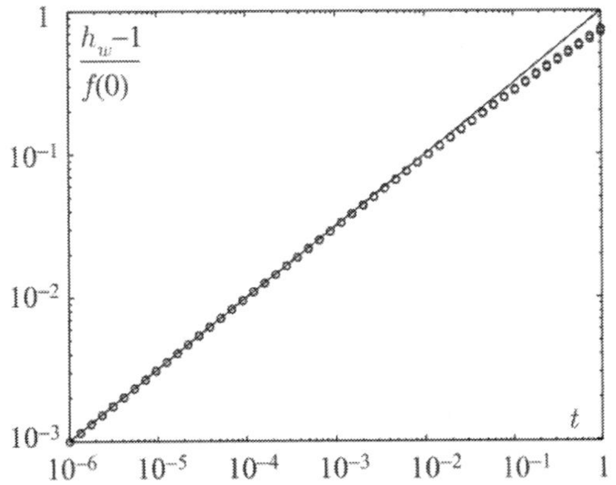

Figure 5. Evolution of the maximum height of the ridge. The full line corresponds to the scaling law $t^{1/2}$ and the circles to the numerical solutions with n =1, 2 and 3.

Summarizing, in this regime the height and the width increase as $t^{1/2}$ regardless of n, but h_w and the shape of the ridge depend on n.

5. Self–similar behavior for $t \gg 1$

For very large time the solution tends to another self–similar regime. To obtain it we seek a solution of the form $h = t^\alpha F(\xi)$, with $\xi = x/t^\beta$. Imposing the boundary condition at $x = 0$ and assuming mass conservation, we find that $\alpha = n/n_3$ and $\beta = n_2/n_3$, where $n_2 = 1 + 2n$ and $n_3 = 1 + 3n$. Replacing in the differential equation for h and integrating once, we find the following differential equation for F:

$$F - F^{n_2}\left(-F'\right)^n = A, \quad A = \text{const.} \tag{9}$$

By comparing this with the boundary condition at $\xi = 0$ we find that $A = 0$. Then the solution of (9) is

$$F = \left(B - \frac{n_2}{n}\xi\right)^{n/n_2}, \quad \xi = \frac{x}{t^{n_2/n_3}}.$$

In consequence

$$h = t^{n/n_3}\left(B - \frac{n_2}{n}\frac{x}{t^{n_2/n_3}}\right)^{n/n_2}.$$

To find B we define x_f as the place where $h = 0$, and we obtain $x_f = (n\,B/n_2)\,t^{n_2/n_3}$. Since $h \gg 1$ in the present regime, x_f can be called the "front" of the current. Next we impose $\int_0^{x_f} h_t\,dx = 1$, and we find $B = (n_3/n)^{n_2/n_3}$, so finally we obtain

$$h = h_w\left(1 - \frac{x}{x_f}\right)^{n/n_2}, \tag{10}$$

with

$$h_w \equiv h(0) = \left(\frac{n_3}{n}t\right)^{n/n_3}, \quad x_f = \frac{n}{n_2}h_w^{n_2/n}. \tag{11}$$

In figure 6 we show this solution for different values of n at a given time.

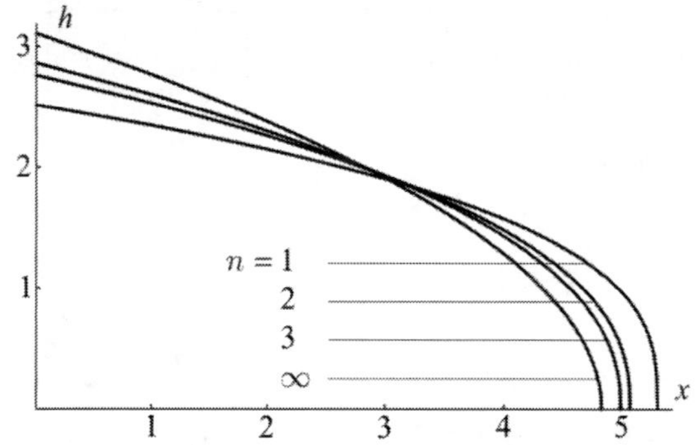

Figure 6. Self–similar solution (10–11) for $t = 10$ with $n = 1$, 2 y 3, and in red for $n \to \infty$.

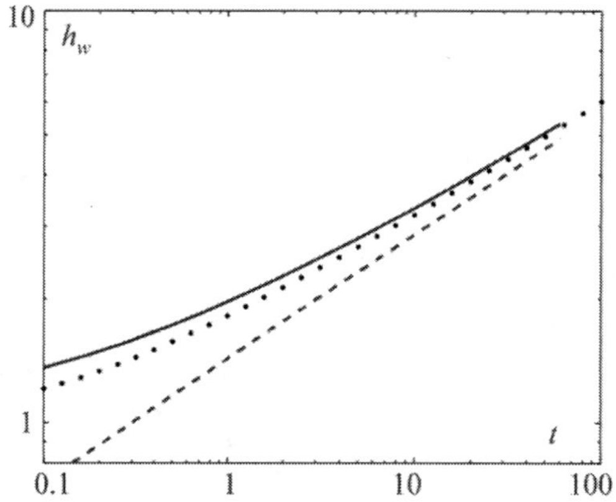

Figure 7. Evolution of maximum height of the ridge for $n = 3$ and large t. The full line corresponds to the numerical solutions, the dashed line corresponds to equation (11) and the dotted line correspond to the quasi–self–similar solution.

A better approximation can be achieved by means of the quasi–self–similar approach [9; 10]. For this purpose we define the position of the front as the place where $h = 1$, we proceed as before, and B is obtained as the solution of the following equation:

$$B = \left(\frac{B}{t^{n/n_3}} + \frac{n_3}{n} \right)^{n_2/n_3} .$$

From this equation we find the following equations for h_w and x_f in place of (11):

$$h_w^{n_3/n} = h_w^{n_2/n} + \frac{n_3}{n} t, \quad x_f = \frac{n}{n_2} \left(h_w^{n_2/n} - 1 \right) . \tag{12}$$

From these results one sees that for very large times the quasi–self–similar solution tends to agree with the self–similar one. Closed form expressions of h_w and x_f can only be obtained for $n = 1$ and ∞, but they are very cumbersome and we shall not reproduce them here. In figure 7 we compare the values of h_w for $n = 3$ from the numerical solution with those given by the self–similar and the quasi–self–similar solutions.

We can improve the description far from the crest if we follow an approach similar to that described in [6]. To this end, we look for a solution that satisfies $F(\infty) = t^{-n/n_3}$ (to ensure that

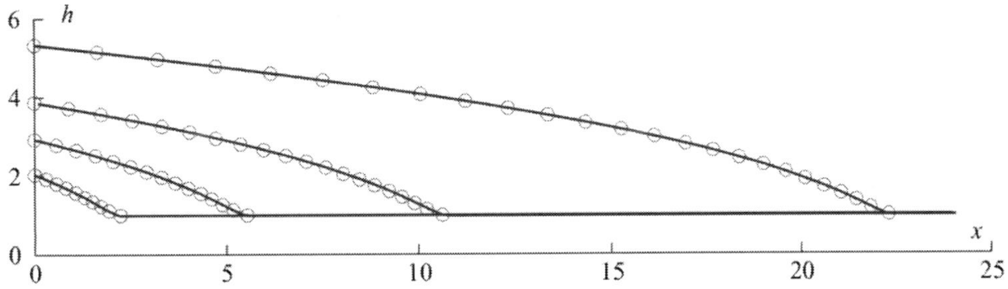

Figure 8. Comparison of the numerical profiles (lines) at $t = 1.2$, 6, 18 and 60 with the profiles of the rescaled solution (circles) from (15) in which h_w is given by the numerical solutions and $n = 3$.

$h(\infty) = 1$), sacrificing the condition at $\xi = 0$. Next we set $\alpha = n/n_3$, $\beta = n_2/n_3$ and integrate the leading term of the differential equation for f. The resulting equation is again (9), but now $A \neq 0$, and the solution is given in implicit form by

$$\xi = \frac{n}{2(1+n)} \left(-\frac{F^{1+n_2}}{A} \right)^{1/n} {}_2F_1 \left[\frac{1+n_2}{n}, \frac{1}{n}; \frac{1+n_3}{n}; \frac{F}{A} \right] + c. \tag{13}$$

Here $c =$ const. and ${}_2F_1[a, b; c; w]$ is the hypergeometric function. Now we allow the integration constants A and c to depend weakly on t. To satisfy the condition at infinity we set $A = t^{-n/n_3}$, and to find c we require that $h = h_w$ when $x = 0$. The solution is then

$$x = \int_h^{h_w} \left(\frac{s^{2+n}}{s-1} \right)^{1/n} ds. \tag{14}$$

Notice that this integral can be expressed in terms of the incomplete Beta function, although there is no advantage in doing that. Of course the solution (14) does not satisfy the boundary condition at $x = 0$ given in (5). However it is possible to modify (14) in order to fulfill this condition by rescaling x. This rescaling is needed to ensure that $h'(0)$ has the appropriate value, and can be achieved multiplying the r.h.s. of (14) by the factor $[(h_w - 1)/h_w]^{1/n}$. We then find

$$x = \left(\frac{h_w - 1}{h_w} \right)^{1/n} \int_h^{h_w} \left(\frac{s^{2+n}}{s-1} \right)^{1/n} ds. \tag{15}$$

So far we have not yet specified h_w. If we employ the value of h_w obtained from the numerical solutions the agreement is excellent, even for $t \approx 1$, as can be appreciated in figure 8. For t not very large the h_w from (12) corresponding to the quasi–self—similar solution differs only by a few percent (the difference tends to increase with n) from the exact value (see figure 7). Then if high precision is not needed, one can use the rescaled solution (15) with the value of h_w given by (12).

6. Conclusions

As in the Newtonian case the evolution of the ridge takes place in two stages. In the beginning the profile $h(x, t)$ looks like the self–similar solution (6–7). Later on $h(x, t)$ tends to the solution (10–11). The transition between these stages occurs around $t \approx 1$, and corresponds to the migration of the inflexion point of $h(x, t)$, that initially is close to $x = 0$, and approaches x_f

as $t \to \infty$. During the first stage the width and the relief increase as $t^{1/2}$. But as the height becomes large, the mass that is added mostly goes to increase the width of the ridge, which then grows faster than the height.

Summarizing, in the formation of a ridge there are two self–similar regimes. In the linear regime, when $t \lesssim 0.1$, the profile is given by (6–7) and the horizontal and vertical scales increase as $t^{1/2}$. As n varies the shape of the ridge varies between f_1 for $n = 1$ and a triangle for $n \to \infty$. In the large time regime (when $t \gtrsim 10$) the profile is given by (10–11) and the height scales as $t^{n/(1+3n)}$, while the width increases as $t^{(1+2n)/(1+3n)}$. These scaling laws coincide with those previously derived in [1] by means of dimensional analysis. For fixed t, as n increases, the peak of the ridge becomes more pronounced and its average slope becomes larger.

The analogy between the present results and those derived for a Newtonian liquid [6] can be traced to the fact that the constitutive relation (1) introduces a single-dimensional parameter (A) into the problem, as in the case of a Newtonian liquid (the viscosity coefficient). In both instances this dimensional parameter can be scaled out by an appropriate definition of the dependent variable, and thus does not appear in the final governing equations. For this reason one finds similarity solutions whenever the analogous Newtonian problem is self–similar. The dimensionality of A depends on the rheological index n and, as a consequence, the large time scaling laws have rheology–dependent exponents. It is interesting to observe that these dependences are weak. However, the differences between Newtonian and non–Newtonian flows are quantitatively significant.

Acknowledgments
We acknowledge grants PIP 5377 from CONICET, X031 from Universidad de Buenos Aires and PICTR 2002-00094 and PICTO 21360 from ANPCYT.

References

[1] Gratton J 1989 *J. Geophys. Res.* **94** 15627–15634
[2] England P and McKenzie D 1982 *Geophysical Journal International* **70** 295–321
[3] England P and McKenzie D 1983 *Geophysical Journal International* **73** 523–532
[4] Medvedev S E and Podladchikov Y Y 1999 *Geophysical Journal International* **136** 567–585
[5] Medvedev S E and Podladchikov Y Y 1999 *Geophysical Journal International* **136** 586–608
[6] Perazzo C A and Gratton J 2008 *Physics of Fluids* **20** 043103
[7] Ranalli G and Murphy D 1987 *Tectonophysics* **132** 281–295
[8] Gratton J, Minotti F and Mahajan S M 1999 *Phys. Rev. E* **60** 6960–6967
[9] Diez J A, Gratton R, Thomas L P and Marino B 1994 *J. Colloid Interface Sci.* **168** 15–20
[10] Gratton R, Diez J A, Thomas L P, Marino B and Betelú S 1996 *Phys. Rev. E* **53** 3563–3572

Self–similar asymptotics in non–symmetrical convergent viscous gravity currents

Carlos Alberto Perazzo[1] **and Julio Gratton**[2]

[1] Universidad Favaloro and CONICET, Solís 453, 1078 Buenos Aires, Argentina.
[2] INFIP–CONICET, Dpto. de Física, FCEyN, UBA, Ciudad Universitaria, Pab. I, 1428 Buenos Aires, Argentina.

E-mail: perazzo@favaloro.edu.ar, jgratton@tinfip.lfp.uba.ar

Abstract. We investigate the evolution of the ridge produced by the non–symmetrical convergent motion of two substrates over which an initially uniform layer of a Newtonian liquid rests. The lack of symmetry of the flow arises because the substrates move with different velocities. We focus on the self–similar regimes that occur in this process. For short times, within the linear regime, the height and the width increase as $t^{1/2}$ and the profile is symmetric, independently of degree of asymmetry of the motion of the substrates. In the self–similar regime for large time, the height and the width of the ridge follow the same power laws as in the symmetric case, but the profiles are asymmetric.

1. Introduction

As in the companion paper [1] we investigate here the physics involved in the scaling laws for the evolution of the orogenic belts previously obtained by means of dimensional analysis [2]. The basic aspects of the process of mountain building and of the models developed to study it are discussed in the introduction of [1] and we shall not repeat them here. Here we extend our recent work [3] to include the effects of an asymmetric motion of the substrate. To this purpose we investigate a simple model that consists of a uniform layer of a Newtonian liquid resting over a horizontal substrate divided in two parts, that for $t > 0$ are pushed one against the other with different velocities. These asymmetric motions drag the liquid and produce a ridge as shown schematically in figure 1. Here we give the details of the evolution of the current that, as in the symmetric case, has two self–similar regimes that occur in different space–time domains and whose scaling laws we investigate.

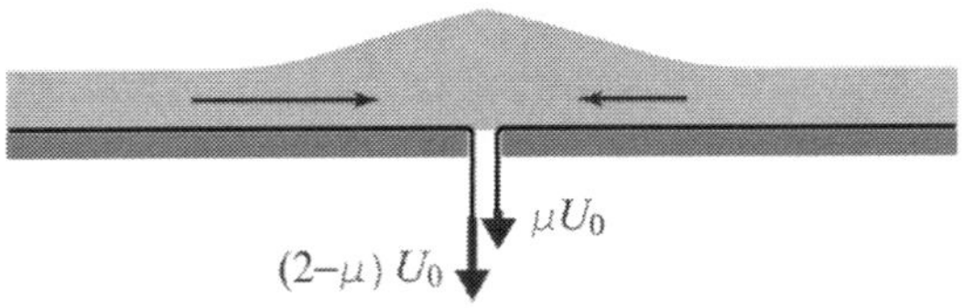

Figure 1. Formation of a ridge due to the asymmetric convergent motion of the substrates.

2. Basic equations

We consider a uniform liquid layer with thickness H_0 that rests on a rigid horizontal surface. At $T = 0$ the left half of the substrate starts to move with velocity $(2-\mu)U_0 > 0$ and the right half with velocity $-\mu U_0 < 0$, where $0 \le \mu \le 1$ (so that the left substrate has a velocity that is larger or equal than that of the right substrate). Then $\mu = 0$ corresponds to the fully asymmetric case where the right substrate remains at rest, and $\mu = 1$ correspond to the symmetric case[1] when both substrates move with equal speed U_0. The total mass inflow is $2U_0H_0$ for any μ.

Since we consider only $0 \le \mu \le 1$ the mass flux coming from the left is larger then the mass flux coming from the right. We assume that the flow is slow and dominated by viscosity so that we can apply the lubrication approximation [4] and we neglect surface effects. Let $H \equiv H(X,T)$ be the thickness of the liquid layer and $U \equiv U(X,T)$ the vertically averaged horizontal velocity. We define the dimensionless variables u, h, x, t by means of

$$U = U_0 u, \quad H = H_0 h,$$

$$X = \frac{g}{3\nu}\frac{H_0^3}{U_0}x, \quad T = \frac{g}{3\nu}\frac{H_0^3}{U_0^2}t,$$

where g is the acceleration of gravity and ν is the kinematic viscosity. Then the evolution is governed by the following equations:

$$\begin{aligned} h_t &= (\mu-2)h_x + \left(h^3 h_x\right)_x, \qquad &\text{for } x<0 \\ h_t &= \mu h_x + \left(h^3 h_x\right)_x, \qquad &\text{for } x>0 \end{aligned} \tag{1}$$

The boundary condition at $x \to \pm\infty$ is $h(\pm\infty, t) = 1$. At $x = 0$ the thickness h is continuous, but $\partial h/\partial x$ has a discontinuity given by

$$\left.\frac{\partial h}{\partial x}\right|_{x\to 0^-} - \left.\frac{\partial h}{\partial x}\right|_{x\to 0^+} = \frac{2}{h(0,t)^2}. \tag{2}$$

This condition implies that at $x = 0$ there is a mass flux form left to right, which did not happen in the symmetric case. Then for a given μ one must solve the equations (1) with the initial condition $h(x,0) = 1$, and subject to the matching condition (2) in $x = 0$. Notice that that now the conservation of mass takes the form

$$\int_{-\infty}^{\infty} h_t dx = 2.$$

It should be mentioned that Buck and Sokoutis [5] performed sandbox experiment to produce in the laboratory a convergent current of this kind, with only the left substrate moving ($\mu = 0$). They also derived an evolution equation for the totally asymmetric case that is equivalent to equation(1) and presented a solution for the linear regime (see below) that however is incorrect.

3. Numerical solutions

The problem (1–2) does not have closed form solutions so that it must be solved numerically. Some results are shown in figure 2. Although the current is not symmetric the peak of the ridge remains at $x = 0$ for all t. In addition the profile is initially symmetric for all μ, but the asymmetry appears later and grows with t. As in the symmetric case, $h(x)$ has an inflexion point in each side, that moves towards larger $|x|$ as t increases, tending to approach the leading part of the ridge, where h is close to 1. It can be also noticed that the aspect ratio (heigth/width) of the ridge diminishes with time. Qualitatively similar results are found for all values of μ.

[1] This is the case considered in [3]; in this reference some preliminary results for the asymmetric case were also reported but notice that a different notation was employed. The formulae given there can be obtained from those given in the present paper by the substitution $U_0 \to U_0/(2-\mu)$ and $\mu \to (2-\mu)/\mu$.

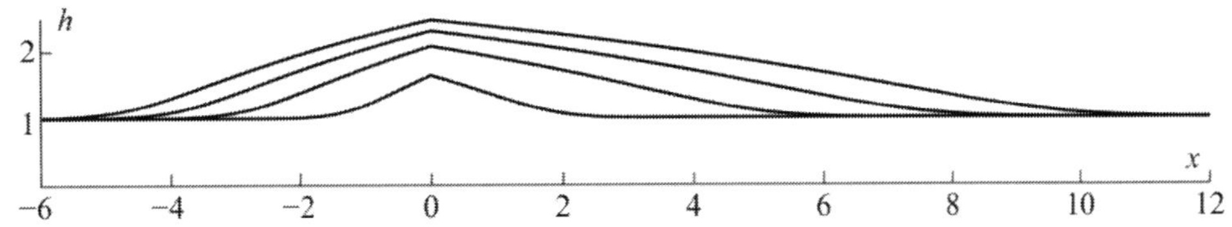

Figure 2. Numerical profiles with $\mu = 0.1$ and for $t = 0.6, 2.4, 4.2, 6$.

4. Behaviour for $t \ll 1$

For small t, when $h \approx 1$, we can write $h = 1 + z$ with $z \ll 1$ and linearize equations (1), that reduce to

$$z_t = -(2-\mu)z_x + z_{xx}, \qquad \text{for } x < 0$$
$$z_t = \mu z_x + z_{xx}, \qquad \text{for } x > 0$$

At $x = 0$ the continuity of z must be satisfied and the matching condition takes the form

$$\left.\frac{\partial z}{\partial x}\right|_{x\to 0^-} - \left.\frac{\partial z}{\partial x}\right|_{x\to 0^+} = 2.$$

We propose now a self–similar solution with the form

$$z = t^{1/2} f_1(\psi) \quad \text{for } \psi < 0$$

$$z = t^{1/2} f_2(\psi) \quad \text{for } \psi > 0,$$

where $\psi = x/2t^{1/2}$. Here the exponents of t in front of $f_{1,2}$ and in the definition of ψ are all equal to $1/2$ to ensure that the linearized matching condition at $x = 0$ and the mass conservation do not depend on t. This implies that in this regime the height and the width of the ridge increase as $t^{1/2}$ regardless of μ. However μ appears in the equations for $f_{1,2}$:

$$f_1'' + 2\left[\psi - t^{1/2}(2-\mu)\right] f_1' - 2f_1 = 0 \quad \text{for } \psi < 0,$$

$$f_2'' + 2\left[\psi + t^{1/2}\mu\right] f_2' - 2f_2 = 0 \quad \text{for } \psi > 0.$$

For t very small ($t \ll \min[x/2\mu, x/2(2-\mu)]$) we neglect the second terms in both brackets to obtain

$$f_1'' + 2\psi f_1' - 2f_1 = 0 \quad \text{for } \psi < 0,$$

$$f_2'' + 2\psi f_2' - 2f_2 = 0 \quad \text{for } \psi > 0.$$

The f functions must vanish at infinity. At $\psi = 0$ we require $f_1(0) = f_2(0)$ and $f_1'(0) - f_2'(0) = 4$. We finally obtain the following solution

$$f_1 = 2\frac{e^{-\psi^2}}{\sqrt{\pi}} + 2\psi\left[\operatorname{erf}(\psi) + 1\right], \qquad \text{if } \psi < 0,$$

$$f_2 = 2\frac{e^{-\psi^2}}{\sqrt{\pi}} + 2\psi\left[\operatorname{erf}(\psi) - 1\right], \qquad \text{if } \psi > 0,$$

where $\operatorname{erf}(\psi)$ is the error function. From this it can be observed that in this regime μ does not appear in the solution, that is symmetric despite that the motion of the substrates is not symmetric.

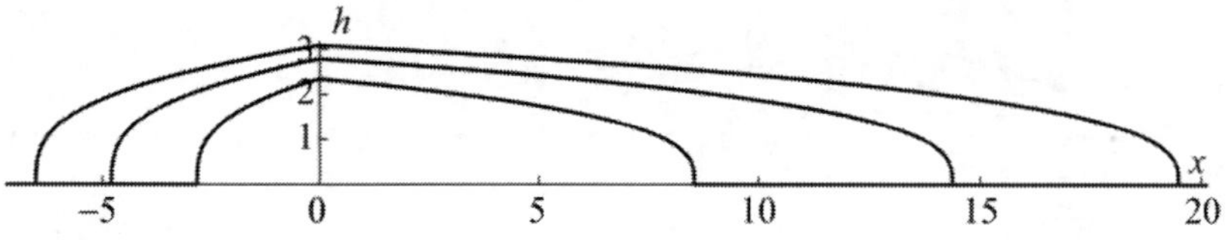

Figure 3. Self–similar solution (3–4) with μ =0.5 and t =10, 20 and 30.

5. Behavior for $t \gg 1$

For very large times the solution approaches a different self–similar regime. In this limit we shall assume first that $\mu \neq 0$, because the totally asymmetric case $\mu = 0$ must be studied separately. For $\mu \neq 0$ we seek a solution of the form

$$h = t^{1/4} F_1(\xi) \quad \text{for } \xi < 0,$$

$$h = t^{1/4} F_2(\xi) \quad \text{for } \xi > 0,$$

where $\xi = x/t^{3/4}$. This choice of the exponents of t is necessary in order to ensure that the matching condition as well as the conservation of mass are independent of t. With these hypothesis, assuming $t \gg 1$ and consequently retaining only the leading terms, and then integrating once the resulting equations we obtain the following equations for F_1 and F_2

$$-(2-\mu)F_1 + F_1^3 F_1' = A_l \quad \text{if} \quad \xi < 0,$$

$$\mu F_2 + F_2^3 F_2' = A_r \quad \text{if} \quad \xi > 0,$$

where A_l and A_r are integration constants. The conditions at $\xi = 0$ are now

$$F_1(0) = F_2(0) \equiv F_w, \quad F_1'(0) - F_2'(0) = \frac{2}{F_w^2}.$$

To satisfy the last condition it is necessary that $A_l = A_r \equiv A$. Notice however that it can be shown that the solution thus obtained does not satisfy the conditions at $\xi \to \pm\infty$. Following our previous work [3] we assume $A = 0$, and then the self–similar solution for $t \gg 1$ that satisfies mass conservation is

$$h = \begin{cases} 0 & \text{for} \quad -\infty < x < x_l, \\ h_w(1 - \frac{x}{x_l})^{1/3} & \text{for} \quad x_l < x < 0, \\ h_w(1 - \frac{x}{x_r})^{1/3} & \text{for} \quad 0 < x < x_r, \\ 0 & \text{for} \quad x_r < x < \infty, \end{cases} \tag{3}$$

where

$$h_w = \left[4\mu\,(2-\mu)\,t\right]^{1/4}, \quad x_l = -\frac{h_w^3}{3(2-\mu)}, \quad x_r = \frac{h_w^3}{3\mu}. \tag{4}$$

Since $h \gg 1$ in the present regime, $x_{l,r}$ can be called the "fronts" of the current at the left and the right, respectively. From (3–4) we see that the right front is farther from $x = 0$ than the left front, as can be expected since the flow from the left is larger than that from the right (see figure 3). From (4) we have $|x_r/x_l| = (2-\mu)/\mu$, the fraction of the total mass that accumulates for $x > 0$ is constant. The width of the ridge $x_r - x_l$ decreases with μ, while its height as well as its aspect ratio increase (see figure 4). The powers of t in the expressions of the width and the height of the profile agree with those of the scaling laws obtained in [2].

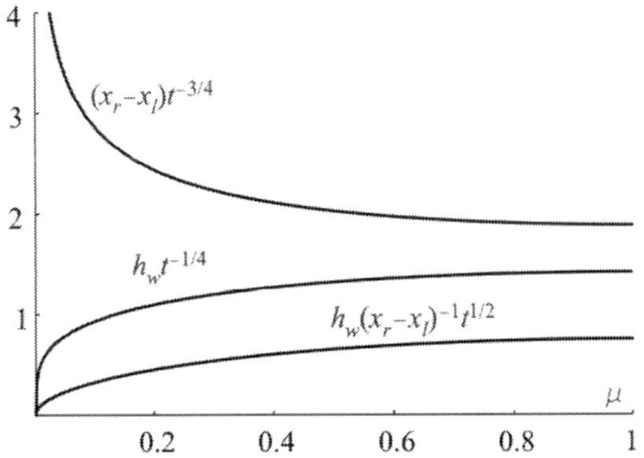

Figure 4. Dependence of the width, height and aspect ratio with $\mu > 0$.

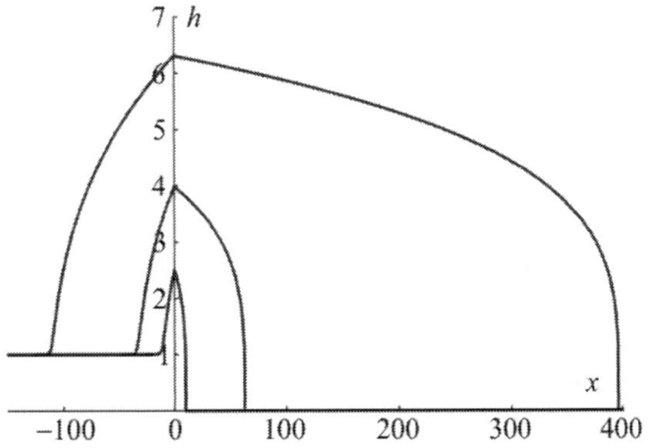

Figure 5. Solution (5–6) in the totaly asymmetric case for t =10, 100 and 1000.

Clearly the previous treatment fails for $\mu = 0$, so that this case must be considered separately. We proceed as before, but now we cannot set $A = 0$ because this choice leads to a solution independent of x for $x > 0$. Then we take $A = -t^{-1/4}$ so that $h(x \to -\infty) = 1$. It can be shown that this choice allows to obtain an approximate solution for $\mu = 0$ and $t \gg 1$ of the form

$$\begin{aligned} x &= \tfrac{(h^3-h_w^3)}{3} + \tfrac{(h^2-h_w^2)}{2} + h - h_w - \log\left(\tfrac{h_w-1}{h-1}\right) \quad \text{for} \quad -\infty < x < 0, \\ h &= \begin{cases} h_w(1-\frac{x}{x_r})^{1/4} & \text{for} \quad 0 < x < x_r, \\ 0 & \text{for} \quad x_r < x < \infty, \end{cases} \end{aligned} \tag{5}$$

where

$$h_w = (10t)^{1/5}, \quad x_r = \frac{h_w^4}{4}. \tag{6}$$

Notice that in the totaly asymmetric case the height of the ridge growths as $t^{1/5}$ and that $x_r \propto t^{4/5}$. On the other hand if we define $x_l < 0$ as the place where $h = \text{const.} \approx 1$, it can be shown that $x_l \propto t^{3/5}$. Then the fraction of the total mass of the ridge that remains at $x < 0$ decreases as $t^{-1/5}$. In the figure 5 we show this solution at different times. This totally asymmetric approximate solution is not self–similar. However the part of the solution corresponding to $x > 0$ is self–similar. The part corresponding to $x < 0$ is actually quasi–self–

similar [3; 6; 7] and for very large t tends to self–similarity, but of a different kind as that for $x > 0$, since $x_l \propto t^{3/5}$ and $x_r \propto t^{4/5}$.

6. Conclusions

The main results we have obtained are that the lack of symmetry of the flow does not modify the scaling laws that govern the growth of the ridge, that are the same as those obtained in the symmetric case [3]. The maximum height of the ridge remains at $x = 0$ for all t, regardless of the asymmetry. In the linear self–similar regime (small t) the ridge is symmetric around $x = 0$, its shape does not depend on μ, and its height and its width scale as $t^{1/2}$. This solution is given by f_1 and f_2. For larger t the ridge becomes asymmetric and the profile for $x < 0$ is narrower and steeper than for $x > 0$. For very large t and $0 < \mu \leq 1$ a self–similar regime occurs given by (3) and (4). The profile of the ridge is asymmetric, but its height and its width scale as $t^{1/4}$ and $t^{3/4}$, respectively, as in the symmetric case. The case $\mu = 0$ is different. As t increases, the portion of the ridge in $x > 0$ becomes more dominant and its height and its width scale as $t^{1/5}$ and $t^{4/5}$, respectively.

Acknowledgments

We acknowledge grants PIP 5377 from CONICET, X031 form Universidad de Buenos Aires, PICTR 2002-00094 and PICTO 21360 from ANPCYT.

References

[1] Gratton J and Perazzo C *Journal of Physics: Conference Series* this volume

[2] Gratton J 1989 *J. Geophys. Res.* **94** 15627–15634

[3] Perazzo C A and Gratton J 2008 *Physics of Fluids* **20** 043103

[4] Buckmaster J D 1977 *J. Fluid Mech.* **81** 735–756

[5] Buck W R and Sokoutis D 1994 *Nature* **369** 737–740

[6] Diez J A, Gratton R, Thomas L P and Marino B 1994 *J. Colloid Interface Sci.* **168** 15–20

[7] Gratton R, Diez J A, Thomas L P, Marino B and Betelú S 1996 *Phys. Rev. E* **53** 3563–3572

X Meeting on Recent Advances in the Physics of Fluids and their Applications IOP Publishing
Journal of Physics: Conference Series **166** (2009) 012013 doi:10.1088/1742-6596/166/1/012013

Experimental and numerical study of patterns in laryngeal flow

N. E. Chisari[1], **G. Artana**[1] **and D. Sciamarella**[2]

[1]LFD, Facultad de Ingeniería, Universidad de Buenos Aires, Av. Paseo Colón 850, C1063ACV Buenos Aires, Argentina

[2]CNRS, Bâtiment 508, Université Paris-Sud, 91403 Orsay, France

E-mail: `Denisse.Sciamarella@limsi.fr`

Abstract. Unsteady airflow is investigated in a channel with a geometry approximating that of the human larynx. The laryngeal flow is simulated by solving the Navier-Stokes equations for an incompressible two-dimensional viscous fluid, and visualized using the Schlieren technique in an experimental setup consisting of a rigid replica of the larynx, with and without ventricular bands. This study shows the spontaneous formation of vortex couples in several regions of the laryngeal profile, and at different stages of the evolution of the starting glottal jet.

1. Introduction

The interest in unsteady flow through the larynx has been renewed with the evidence that certain unsteady effects may be relevant to model the phonation process correctly [1, 2]. The dynamics of an unsteady flow through a channel with constrictions is a very general problem in fluid mechanics. The human larynx forms an airway with two consecutive constrictions. The first constriction is defined by the vocal folds, which are a pair of mucous membranes stretched horizontally across the larynx (see Fig. 1). The second constriction is determined by the so called false vocal folds or ventricular bands, which are a pair of thick mucous structures located slightly above the true folds. During voice production, pressure from the lungs supplies the energy for sustaining vocal fold oscillations. The false vocal folds are poor vibrators, due to their more viscous and less stiff properties which prevent their vibration in normal speech. Independently of whether they vibrate or not, the literature indicates the tendency of the false vocal folds to adduct toward the midline during phonation. Once adducted, both true and false vocal folds, define a region between them called the laryngeal ventricle (or Morgagni's ventricle).

In 1955, Lee studied a steady laminar flow through a double constriction, varying the size of the second constriction, the proximity of the two constrictions and the flow rate. He found that the farther the two constrictions were from each other, the less complex was the effect on the flow near the first constriction. It is therefore reasonable to speculate that the presence of the laryngeal ventricle is not innocuous when it comes to characterizing the flow through the glottis. It is an open question whether airflow complexity certainly affects vocal output by modifying glottal flow, altering airflow resistance and affecting the source acoustics. In any case, as indicated by both van den Berg and Lee [3, 4], the effects of supraglottal structures on the flow will ultimately depend on their shape and size.

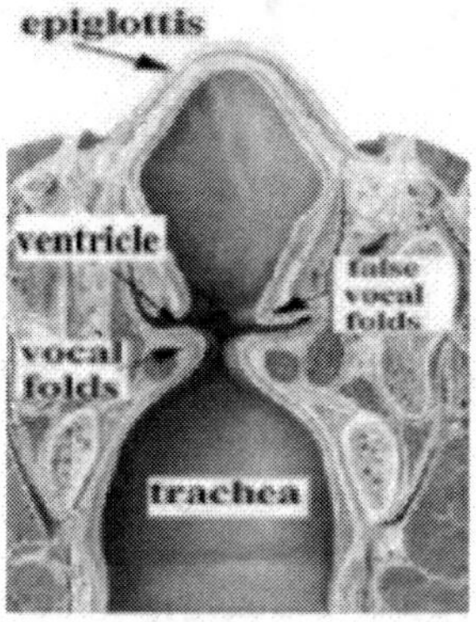

Figure 1. Geometry of the laryngeal channel.

In order to investigate the influence of the geometry on laryngeal flow, the authors have considered the aerodynamic impact of the ventricular bands of both in vitro and in silico (numerical) experiments [5]. These experiments are performed in a static but adjustable geometry, in which the false vocal folds can be either removed or inserted at different locations in the laryngeal channel. The first constriction is modeled as a slitted nozzle in a duct. Previous experiments have shown that the essentially planar jet that exits the vocal folds has the shape of a sliced mushroom-like current [1]. This mushroom structure corresponds in fact to a dipolar vortex front, i.e. to a couple of closely packed counter-rotating circulations. Indeed, plane jets and vortex dipoles are supposed to be united by the same mechanism of generation: the action of a localized source of momentum in a viscous fluid [6].

In this work, we present experimental and numerical evidence that these flow patterns are characteristic of the first stages in the evolution of the starting jet that traverses the double constriction defined by the folds in the larynx. A qualitative description of the relevant phenomena involving this type of structures in a schematic laryngeal geometry for an impulsively starting jet is provided.

2. Methods

2.1. Experimental setup

Our experimental setup to analyze an impulsively starting flow can be divided into three main parts. One of them is the Schlieren system, an optical system that allows us to observe index of refraction gradients in a transparent fluid [7]. Changes of fluid density are the usual source of variations in the refraction index. The second part of the setup consists of a pneumatic circuit that feeds a replica of the larynx, the third part of the setup. The laryngeal replica is composed of a metallic base including rigid $25mm$-thick vocal folds, with a rounded geometry. This base is coupled to a supraglottal structure completely designed in acrylic glass (PMMA) with unmountable false vocal folds, also having a rounded shape (radius of curvature of $5mm$). The parameters that can be varied in this geometry are the distance h_{vf} between the vocal folds (1 or $2mm$), the separation h_{vb} between the ventricular bands ($2, 4$ or $6mm$) and the distance L_v between the vocal folds and the ventricular bands ($10, 20$ or $30mm$) that define the laryngeal ventricle.

The Schlieren system (Fig. 2) built in our laboratory is a typical symmetrical $Z-$type system with two $0.5m$ diameter spherical mirrors with focal lengths of $3.0m$. A rapid CMOS camera served as acquisition device for the Schlieren images, allowing recording frequencies of $50, 100$ or

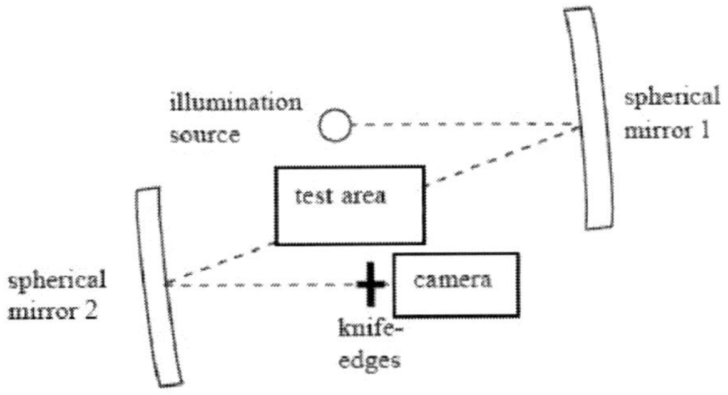

Figure 2. Diagram of the $Z-$type Schlieren system. Mirrors are spherical with a diameter of approximately $0.5m$ and a focal distance of $3m$.

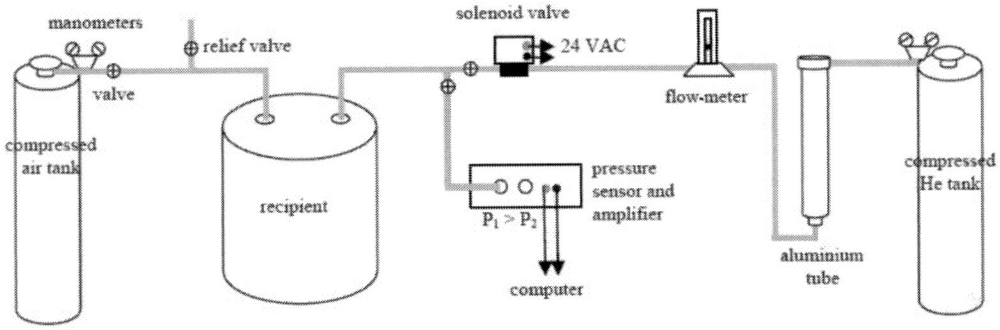

Figure 3. Diagram of the pneumatic circuit.

200 frames per second. When the frequency doubles, the spatial resolution of the image halves in the vertical direction. Both the illumination source and the knife-edges are placed at the mirror's focal distance. A pair of mutually perpendicular knife-edges placed in the focus of the second mirror help to improve the contrast in the Schlieren images.

Helium gas is introduced in the flow as a tracer through a lateral orifice immediately upstream the folds. A Helium compressed tank is used to feed the system with this gas. A pressure regulation device connected to the tank enables to control the mass flowrate of this gas, which tends to result quite reduced compared to the air mass flowing between the folds. Because the device is operated with a small amount of tracer introduced in the flow, the system enables a very good contrast during the transients.

The air pneumatic circuit (Fig. 3) consists mainly of a $400dm^3$ recipient fed by a compressed air tank. The amplified signal from a calibrated differential pressure sensor is transmitted to a computer, which monitors the pressure build-up inside the recipient when it is filled with dry air. Once the desired pressure is attained, the tank is isolated from the circuit by means of a two way valve. A system of hoses connects the recipient in series with a solenoid two way valve, a flow-meter and finally to the lower extreme of an $80cm$ length and a $2.5cm$ (inner) diameter aluminium tube. The response time of the solenoid valve (normally closed) at the opening is approximately $5ms$.

The larynx replica is mounted on the tube's superior extreme, where pressure equals the

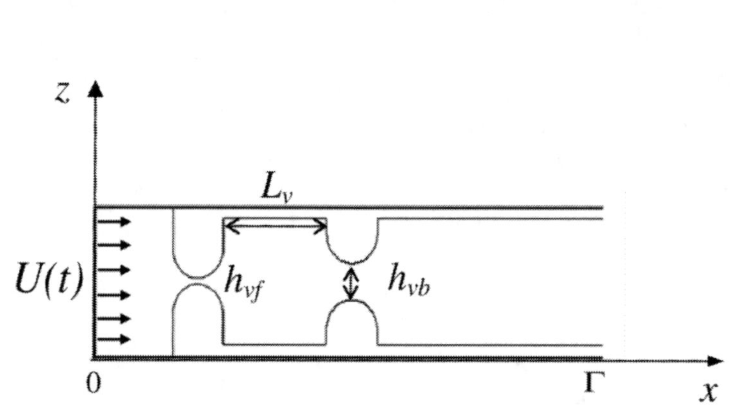

Figure 4. Laryngeal channel geometry and boundary conditions for the direct numerical simulations. The incoming velocity is set to increase following an initial linear ramp of 5 ms.

atmospheric pressure. Thus, when the solenoid valve is energised, the air flows along the ducts from the recipient to the replica, passing through the vocal folds. The flow rate, which is conserved along the circuit, is measured in steady regime by the flow-meter. This value is chosen to be representative of a suitable airflow velocity between the vocal folds in accordance with typical values for the human larynx (15 m/s).

2.2. Numerical technique

In order to simulate the airflow through a schematic larynx corresponding to the experimental setup, we use a numerical code which has been specifically conceived to simulate unsteady airflow through a two-dimensional glottal model [2]. It uses a multigrid finite-difference method and a fixed Cartesian grid that guarantees that the discretization scheme is not degraded by non orthogonal frames. The equations are integrated with a time marching algorithm based on a prediction projection method. The spatial discretization uses a staggered grid. The immersed boundaries are taken into account through a phase function set to 0 or 1 for fluid or structure cells respectively.

Solutions are computed in a rectangle of unit height and adjustable length, shown in Fig. 4 The design of the flow boundaries corresponds to the geometry of the channel and replica used in the experimental setup. The rectangle has $N_x \times N_z$ rectangular cells. The Reynolds number is built on the channel's maximum height and the incoming velocity. The computations in this work are performed on grids with $N_x = 512$ or 1024 and $N_z = 128$ or 256. The dimensionless time step used in the calculation is of order 10^{-4} to 10^{-5}. For the lower spatial resolution, the coarse mesh is refined around the axis and next to the boundaries.

Boundary conditions are also set in accordance with the experiment. The value of the incoming velocity can be set either constant or time-dependent. We have performed numerical simulations with an initial linear ramp in order to match the response time of the solenoid valve. After this transient, the value of the flow rate is kept constant in time, both in numerical and physical experiments. The maximum value of the flow rate is established in correspondence with the value measured with the flow-meter of the experimental set up at very large times, a value which is used to choose the Reynolds number for the numerical experience. Velocities are set to zero in the structure cells. The inlet flow enters a channel where the fluid is initially still. On the solid walls we impose the classical no-slip and no-injection condition. At the

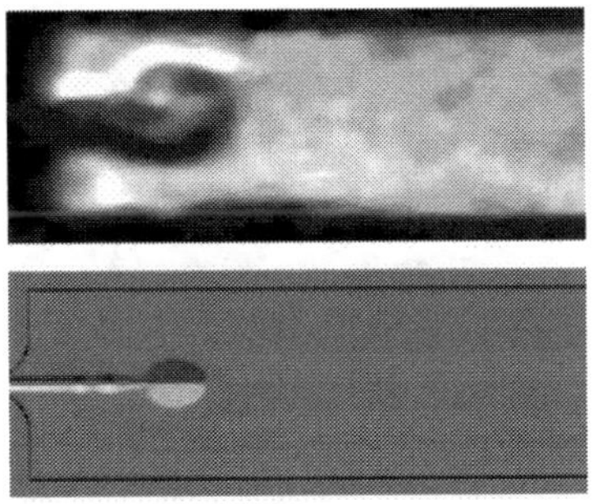

Figure 5. Jets emerging from the vocal folds at the initial stages of the flow onset with $Re = 900$ without ventricular bands. Above: Schlieren image, below: vorticity in numerical simulation.

downstream boundary, we impose the so-called advective-type boundary condition, the most natural boundary condition for a transparent artificial boundary. This condition reduces to a homogeneous Neuman condition if the flow reaches a steady state. No symmetry condition is forced along the axis, to capture possible flow asymmetries.

3. Results

Photographs of the different stages in the evolution of the starting jet in laboratory and numerical experiments are shown in Fig.5-9. At the exit of both constrictions, laboratory results clearly show the spontaneous formation of flow patterns which can be associated to a jet with a vortex dipole at its head. Smaller dipoles seem to form spontaneously within the ventricular cavity defined by the false vocal folds.

Numerical simulations show, during the jet formation stage, a starting jet that emerges from the vocal folds with a front composed by symmetrical opposite-sign vortices forming the mushroom-like cap which is typical of a pulsed injection of fluid (Fig. 5). This mushroom structure is easy to identify in the Schlieren images. The head of the jet progresses along the duct with a convection speed which amounts to about $9m/s$ for $Re = 900$ and a $1mm$ aperture for the vocal folds. This value is not significantly altered by the presence of the ventricular bands (Fig. 6). Along the boundary layers of opposite vorticity that delimit the evolving jet, it is possible to observe, after a certain time, the subsequent formation of the series of smaller dipolar structures, which bear a strong resemblance with the structures observed in certain oceanographic situations [8].

When the ventricular bands are present, a second jet is born at the exit of the constriction that they define (Fig. 7). This second jet is born simultaneously with the first jet if both structures (folds and bands) are equally constricted. If the aperture at the bands is wider than the glottal aperture, there is a delay in the birth of the second jet which is proportional to the ratio between the band and fold apertures. The jet emerges from the ventricular bands at a speed which depends on the ventricular aperture: the less constricted the bands are, the slower the final jet progresses. The size of the jet past the ventricular bands (which is also determinant of the size of the dipolar head of the final supralaryngeal jet) is also proportional to the ventricular aperture.

The flow patterns observed in the ventricle after the instant in which the first jet (the glottal jet) impinges on the ventricular bands are quite complex in all the tested geometrical configurations. Numerical simulations show that small-sized vortex dipoles are shed symmetrically in the upstream side of the walls of both ventricular bands. These small patterns initiate a sinuous trajectory firstly away from the jet and then back towards the jet axis,

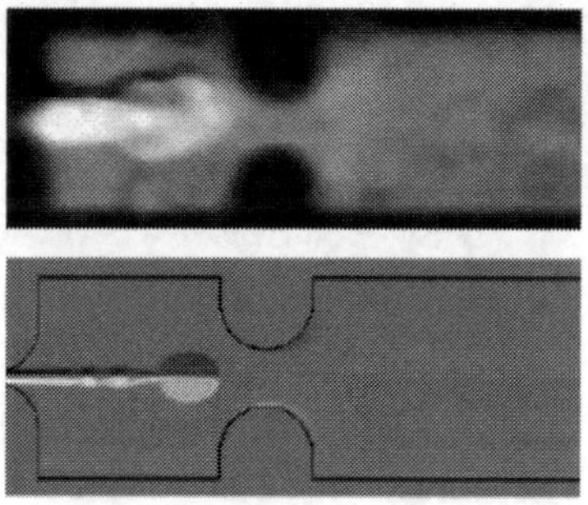

Figure 6. Jets emerging from the vocal folds at the initial stages of the flow onset with $Re = 900$ with ventricular bands ($L = 25mm$, $h_{vb} = 6mm$). Above: Schlieren image, below: vorticity in numerical simulation.

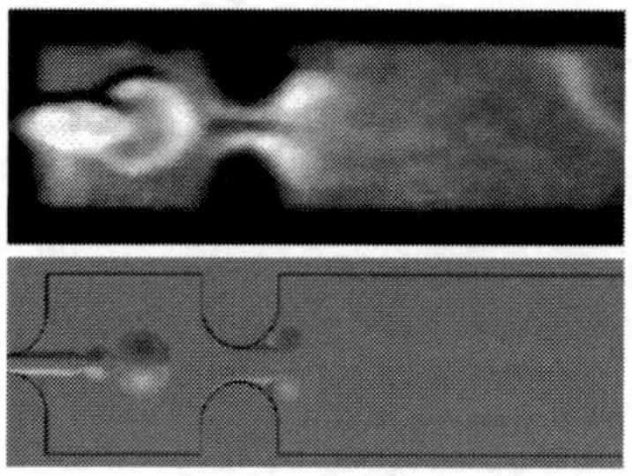

Figure 7. Jet emerging from the ventricular bands with $Re = 900$ for a glottal opening of $2mm$ and a band opening of $4mm$. $h_{vb}/h_{vf} = 2$. Above: Schlieren image, below: vorticity in numerical Simulation.

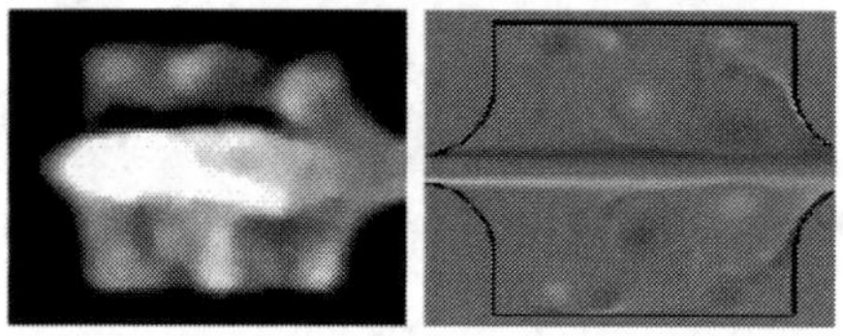

Figure 8. Persistent structure of symmetrical vortices of opposite sign within the ventricle for $Re = 900$ in two different geometrical configurations. Left: Schlieren image, right: vorticity in numerical simulation.

contributing to form, after the initial phase, two relatively large vortices of opposite sign in either side of the jet within the ventricle (Fig. 8) which tend to persist in all the numerical experiments performed with ventricular bands. These large vortices are also visible and persistent in the Schlieren visualizations. The time needed for these vortical structures to settle depends on $U(t)$ and h_{vb}.

At larger times, a three dimensional flow pattern is observed for the laboratory flow past the ventricular bands (Fig. 9). At some distance from the ventricular bands a slug flow type is established that occupies the whole section. At the exit of the ventricular aperture, the jet is still visible, surrounded by vortical structures which are similar to the ones observed within the ventricle in the previous stage of the flow. The two-dimensional numerical experiments do

(a)
(b)
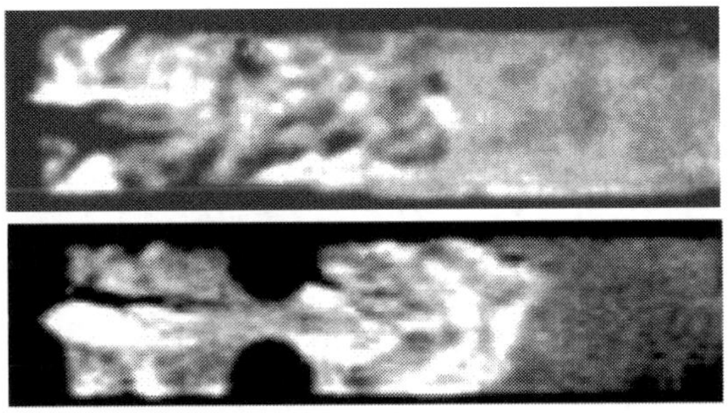

Figure 9. Loss of stability of the supralaryngeal jet Above: through vocal folds, below: through ventricular bands.

not reproduce this 3D instability. They are not meant to correctly estimate the transition to turbulence, but to study the flow patterns that prevail in the initial stages, which they predict correctly.

4. Conclusions

In the present study, we have performed numerical and laboratory experiments on the basis of two-dimensional and rigid models of the larynx, with and without false vocal folds.

Summarizing the results presented above, the dynamics of an impulsively starting flow through a double constriction is characterized by a double jet (one per constriction). The jets exiting the slitted constrictions have a head which is structured as a vortex dipole. These dipoles are known to have a fast translating motion and a relatively long lifetime, unless they interact with other structures or collide against a wall. This is exactly what happens when the ventricular bands or false vocal folds are present. As a result, the complexity of the flow is clearly affected by the presence of the laryngeal ventricle, in which resident vortical patterns seem to settle after a collective interaction of a sequence of small scale dipoles shed by the upstream side of the ventricular bands.

The next natural step in our study is to quantify the properties of the observed patterns and, if possible, to give a theoretical model of the dynamics of these structures that helps unveiling their possible role in the larynx.

Acknowledgments

This research was supported in part by the Regional Programme for Scientific Cooperation STIC-AmSud Project 06/07.

References

[1] Pelorson X, Hirschberg A, Hasselt R V, Wijnands A and Auregan Y 1994 *J. Acoust Soc Am.* **96** (6) 3416–3431
[2] Sciamarella D and Le Quéré P 2008 *European Journal of Mechanics - B/Fluids* **27** (1) 42–53
[3] van der Berg J W 1955 *Folia Phoniatrica* **7** 57–69
[4] Lee T S 1994 *J. Fluids Mech.* **116** 66–71
[5] Sciamarella D, Chisari E, Artana G, Bailly L and Pelorson X 2008 *J. Acoust Soc Am.* **123** (5) 3577
[6] Voropayev S I, Afanasyev Y D and Filippov I A 1991 *J. Fluid Mech.* **227** 543
[7] Setlles G 2001 *Schlieren and Shadowgraph Techniques, Visualizing Phenomena in Transparent Media* (Springer-Verlag)
[8] Wells M G and van Heijst G J F 2003 *Dynamics of Atmospheres and Oceans* **37** 223–244

Reduced order models for EHD controlled wake flow

Juan D'Adamo†, Roberto Sosa†, Ada Cammilleri‡ and Guillermo Artana†

† CONICET, Laboratorio de Fluidodinámica, ‡ Departamento de Matemática, Facultad de Ingeniería, Universidad de Buenos Aires.
CP 1063 Avda. Paseo Colón 850, Buenos Aires, Argentina

E-mail: jdadamo@fi.uba.ar, rsosa@fi.uba.ar

Abstract. We present in this work a way to construct reduced order model (ROM) for wake flows at low Reynolds numbers, from experimental data of particle image velocimetry (PIV). The proper orthogonal decomposition method (POD) allows the extraction of a small number of functions or modes that describes the flow velocity field. Galerkin projection of Navier Stokes equations onto these modes leads to obtain a system of ordinary equations, the reduced order model.

We take interest particularly in the study of the modifications introduced by an electrohydrodynamic (EHD) actuator on the flow. EHD action is obtained from a gaseous discherge of dielectric barrier type (DBD), produced by electrodes flush mounted on a cylinder surface. The reduced order model presents as an effective tool in order to analyse the coupled electrical and hydrodynamic phenomena. We focus on the flow changes considering the voltage between the electrodes as a control parameter.

1. Introduction

Reduced order models are widely used in fluid mechanics in order to represent the main dynamical behaviour of a flow. Within this framework, a few number of modes can be enough to perform analysis, understanding and predictions for the different states of a given flow. Wake flows are particularly interesting in terms of academic and many industrial purposes and we will focus our study to this kind of flows. The flow around a cylinder has long been a benchmark problem concerning wake flows. In order to obtain such models, we focus on the POD technique which main aspects are described next.

Also known as Karhunen-Loève Decomposition, Singular Value Decomposition, Principal Components Analysis, POD has been first introduced in the context of turbulence by Lumley (1967)[1]. It has been used by different authors (see for instance a review by Holmes *et al.*(1996)[2]) as a method to obtain approximate descriptions of the large scale or coherent structures in laminar and turbulent flows. Without any a priori hypothesis on the flow, the POD method provides a flow representation in terms of a linear combination of basis functions, or modes, ordered decreasingly by their kinetic energy content. Indeed, once the basis is obtained, a Galerkin projection is performed to reduce the order of Navier-Stokes equations. It results a system of ordinary differential equations that should describe approximately the flow. However, the solution of such system is not straightforward. After the Galerkin projection, iIt cannot be assured that the error of the approximation is bounded. Additional hypothesis on the model

are required.
Previous works on the flow around a cylinder started with Deane *et al.*(1991)[3] and since then, many efforts have been performed in order to construct a control system of the POD low order model. Some drawbacks of this technique are the lack of stability and the lack of robustness of the model away from the flow observations.
In order to improve these aspects, we had proposed in D'Adamo et al (2007)[4] a scheme based on data assimilation theory. Another method was presented by Kalb & Deane(2007)[5] which performs a linear correction on the dynamical systems coefficients, and it was tested on direct numerical simulation data and on the Lorenz system. On the other hand, Noack *et al.*(2003)[6] had pointed out that the error sources start solving the system directly obtained by the Galerkin projection. It is not possible to assure, a priori, the system structural stability. Little perturbations on the equation system coefficients can produce qualitatively different solutions. For this reason, the authors proposed the POD-Galerkin reduction to an invariant, *inertial*, manifold towards a normal form, a non linear equation which has a prototypical structure.
On the other hand, in the context of wake flows study, POD modes can be interpreted as fluctuations that superpose to a stationary solution, a *base flow*, unstable, of the Navier Stokes equations. Noack *et al.*(2003)[6] constructed their model from this hypothesis. However, in recent articles (Thiria et al.(2007) [7], Barkley (2006) [8]) it has been emphasized the importance of the time averaged flow as basic state that can account for stability properties of wakes under forcing conditions. On these bases, we developed a POD-ROM constructed from experimental (PIV) data.
Electro-hydrodynamic (EHD) actuators for flow control have been receiving special attention in the last years Artana et al.(2003)[9]. Through the ionization of flowing air close to the surface of the body, EHD devices produce a modification of the condition of the flow at the wall. We propose in this work, to identify such modifications in the flow around a cylinder by means of the POD based ROM.

2. POD-Galerkin model

2.1. Introduction

The POD-Galerkin method has been extensively presented in many articles. The method foundations can be studied in Holmes *et al.*(1996)[2]. Let us review its main characteristics.
A flow field $u(x,t) \in L^2(\Omega)$, where Ω is a physical domain, can be represented in terms of basis functions $\phi_i(x)$ and their temporal coefficients $a_i(t)$ such that:

$$u(x,t) = \sum_{i=0} a_i(t)\phi_i(x) \tag{1}$$

It is necessary to define an inner product as

$$(u,v) = \int_\Omega u \cdot v dx$$

The modes $\phi(x)$, global and mutually orthogonal, are extracted from a set of observations $u(x,t_i)$ obtained experimentally or numerically, and they are optimal with respect to the average kinetic energy representation of the flow contained in Ω. The spatial modes inherit by construction properties of the flow such as its boundary conditions and, when we work with incompressible flows, they are also divergence-free.
When working on finite dimension, the modes identification reduces to solve the singular value decomposition (SVD) of a V matrix that represents in its columns the different snapshots of the velocity fields $u(x,t_i)$.

The Navier Stokes equations, and the incompressibility condition,

$$\frac{\partial u}{\partial t} + (u \cdot \nabla)u = -\nabla p + \frac{1}{Re}\nabla^2 u \tag{2}$$

$$\nabla \cdot u = 0 \tag{3}$$

are then reduced by means of a Galerkin projection onto the subspace generated by the first s vectors of the POD basis, $\{\phi_1 \ldots \phi_s\}$. The ROM has the form:

$$\frac{da}{dt} = \dot{a} = \mathbb{M}(a) \tag{4}$$

where $\mathbb{M}$ represents a nonlinear operator which is a polynomial that contains the projected Navier-Stokes terms and its boundary conditions. The projection produces the following terms:

$$\left(\frac{\partial u}{\partial t}, \phi_k\right) = \frac{da_k}{dt}$$

$$\left(\frac{1}{Re}\nabla^2 u, \phi_k\right) = \frac{1}{Re}\int_\Omega \nabla^2 u \phi_k dx$$

$$= \frac{1}{Re}\int_\Omega \sum_{i=1}^{s} a_i \nabla^2 \phi_i \cdot \phi_k dx = \frac{1}{Re}\sum_{i=1}^{s} a_i \int_\Omega \nabla^2 \phi_i \cdot \phi_k dx \tag{5}$$

$$(\nabla p, \phi_k) = \int_\Omega \nabla p \phi_k dx = \int_{\delta\Omega} (p\phi_k) \cdot dl \tag{6}$$

$$((u \cdot \nabla)u, \phi_k) = \sum_{i=1}^{s}\sum_{j=1}^{s} a_i a_j \int_\Omega (\phi_i \cdot \nabla)\phi_j \cdot \phi_k dx \tag{7}$$

These can be included in a polynomical form on the temporal coefficients $a_1 \ldots a_s$ for $\mathbb{M}$.

$$\dot{a} = F(a) = \sum_{k=1}^{N_m} c_k P_k \tag{8}$$

where $P_k \quad k = 1 \ldots N_m$ is a vector of all the monomials of different degrees in variables $a = (a_1, \ldots a_s)$, and c_k are their corresponding coefficients. The number N_m depends on the number of degrees and on the number of modes considered, the resolved modes.

2.2. System identification

The solution of (8) is not straightforward. Equation (8) determines in general a dynamical system that in many cases may converge to erroneous states after a relative short time of integration. Indeed, different authors have observed that the predictions at long time may fail and that, for instance, models describing a cylinder wake flow, after a few cycles of vortex shedding may exhibit differences of phase and amplitudes with respect to empirical values, as pointed out by Kalb & Deane(2007) [5]. Holmes *et al.*(1996)[2] had mentioned as possible responsible for this behaviour the following points: Neglect of the incidence of boundary or pressure terms in the computation of the dynamical system coefficients; inaccurate estimation of derivatives of spatial modes that determine the dynamical system coefficients; and that the low-dimensional truncation produces loss of dissipation and energy transfer between the different scales of the flow and it only reproduces velocity fields which are close to the spatial structures averaged in the decomposition.

Concerning the pressure term, several authors have neglected the influence of the pressure term in wake flows arguing that this effect could be disregarded when large wake domains were considered. However, depending on the flow geometry and truncation level, neglecting this term could add uncertainty in the ability of the computed dynamical system to faithfully represent the actual flow dynamics. Noack *et al.*(2005)[10] proposed an additional quadratic term which is estimated solving Poisson equation. The Poisson equation, determined applying the divergence operator on the Navier-Stokes equations, can be written as:

$$\nabla^2 p = -\nabla \cdot (u \cdot \nabla u) \qquad in \quad \Omega \tag{9}$$

and allows to estimate pressure from the velocity field described as in Eq. (1).
Another way to correct the model (8) is to apply the intrinsic stabilization algorithm developed by Kalb & Deane(2007)[5], which corrects the linear terms and is effective in situations where the pressure is adequately expressed by a linear model, i.e. channel flow.

Using PIV data, the system coefficients may be poorly determined. To avoid the direct estimation of these coefficients, Perret *et al.*(2006)[11] proposed a polynomial identification procedure which consists in writing *a priori* the ROM. Provided the temporal coefficients a_j and their time derivatives $\dot{a}_j$, we can derive from (8) a linear system, which may be solved by least mean square or singular value decomposition. This is why we had not stated, intentionally, the degree of (8). The unresolved modes can be represented as functions, polynomials, of the resolved ones and therefore increase the system degree. We remark that the pressure field (9) is implicitly included in this formulation.
It must be pointed out that estimating $\dot{a}(t_j)$ from traditional schemes of finite differences leads to important numerical errors, further amplified when working with experimental data. To deal with this problem, it is possible to perform a polynomial aproximation for each a_j for every instant t_i. Consecutive time intervals allow to calculate $a_j(t_i) \simeq p_j(t)$, where p_j is a polynomial of degree n, and then, the derivatives are obtained analytically deriving $p_j(t)$. This idea was introduced by Maquet *et al.*(2004)[12] in a work on dynamical systems identification.
We had proposed in D'Adamo et al. (2007)[4] a method to work with experimental data improving the polynomial identification with an assimilation technique. Considering a noise $\delta(t)$ as the difference between the model prediction and an observations set, $\dot{a} = F(a) + \delta(t)$, it is minimized iteratively. The scheme does not deal with the system structural instability but it provides a way to work with noisy data.
A dynamical system is structural unstable when any little perturbation on its coefficients leads to solution qualitatively different. In our problem, the methods listed above don't assure structural stability. In order to construct a robust system, once its coefficients are determined, we analyse it by means of dynamical system theory: particularly the center manifold and normal form theory. Indeed, a first reduction can be achieved by studying the evolution of the variables contained in the center manifold of (8) and considering that the remaining variables are *slaved* to them. Noack et al.(2003)[6] studied the particular problem of POD models for wakes by means of the center manifold theory. On the other hand, once the problem dimension is reduced, we may simplify the nonlinear terms of the resulting equation, according to simmetry considerations. This nonlinear change of coordinates is called the normal form theory and it is applied particularly to this problem in D'Adamo et al.(2008a)(2008b)[13][14].

2.3. Dynamical system analysis

We know that the physical model for the wake flow problem is represented by the Stuart-Landau equation (see, for instance [15]):

$$\frac{dA}{dt} = (\gamma + i\omega)A + (c_r + ic_i)|A|^2 A \tag{10}$$

Journal of Physics: Conference Series **166** (2009) 012014 doi:10.1088/1742-6596/166/1/012014

and the vortex onset is described as a Poincaré-Andronov-Hopf bifurcation (see [16]§3.1B for a definition) in agreement with experiments Provansal et al.(1987)[15]. These reasons sustain the hypothesis of an equation of degree three for the modes amplitudes that resumes the coherent structures behaviour.

On the other hand, in agreement with Barkley's hypothesis, the fluctuations $a(t)$ are superposed to the mean flow, and the solution $a = 0$, a fixed point, corresponds to the base flow. A center manifold reduction on this point can simplify our problem.

The center manifold theory is a rigorous mathematical technique used to reduce dynamical systems dimension. As we pointed out, a Poincaré-Andronov-Hopf bifurcation must be described by our model for $a = 0$. We consider a bifurcation parameter $\mu = Re - Re_c$, where Reynolds critical number Re_c corresponds to the value of Re for the vortex shedding onset.

A linear change of coordinates (Real Jordan form) transforms the $s-$dimensional Eq. (8) into the form:

$$\begin{cases} \dot{x} = Ax + f(x, y, \mu) \\ \dot{y} = By + g(x, y, \mu) \end{cases} \tag{11}$$

where $x \in \mathbb{R}^c$ and $y \in \mathbb{R}^{u+e}$, for our problem the center manifold dimension $c = 2$ and $s = c + u + e$, being u and e the dimension of the unstable and stable manifold respectivly. We consider then

$$A = \begin{pmatrix} Real(\lambda) & -Imag(\lambda) \\ Imag(\lambda) & Real(\lambda) \end{pmatrix}$$

where $\lambda = \lambda(\mu) = \alpha(\mu) + i\omega(\mu)$ represents the eigenvalues of A and we suppose that B does not contain eigenvalues of zero real part.

In agreement with the Center Manifold theory (see i.e. Wiggins(2003)[16] §2.1) we know that the center manifold can be represented locally by:

$$\{(x, y)/y = h(x), h(0) = 0, Dh(0) = 0\} \tag{12}$$

and that, in such a case, the dynamics of Eq. (11) is determined by:

$$\dot{x} = Ax + f(x, h(x), \mu) \tag{13}$$

for $\|x\| < \delta$, sufficiently small. In our applications we consider $h(x) = p(x)$, a polynomial of degree 3 in $x \in \mathbb{R}^c$. This is a practical choice: higher order polynomials do not assure a better reconstruction and, on the other hand, they amplify the coefficients' noise.

However, we cannot assure yet that the solution of Eq. (11) has the characteristic behaviour of a stable limit cycle, convergent towards the trayectory defined by the set $\{a_j(t_i)\}$. The polynomial $f(x)$ can be simplified depending on A eigenvalues. For this problem, when $\mu = 0$, these are a pair of conjugate complex of zero real part. When $\mu > 0$, the A eigenvalues become of positive real part (i.e. α) and, based on this fact, the resulting normal form(see i.e. Wiggins(2003)[16]) is

$$\begin{cases} \dot{x}_1 = \alpha x_1 - \omega x_2 + (a x_1 - b x_2)(x_1^2 + x_2^2) \\ \dot{x}_2 = \omega x_1 + \alpha x_2 + (b x_1 + a x_2)(x_1^2 + x_2^2) \end{cases} \tag{14}$$

without considering terms of degree equal or greater than 5. We verify that this equation is equivalent to Stuart-Landau (10), but we have obtained it from the nonlinear analysis of POD-Galerkin system. In polar coordinates, Eq. (14) has the form:

$$\begin{cases} \dot{r} = \alpha r + a r^3 \\ \dot{\theta} = \omega + b r^2 \end{cases} \tag{15}$$

The analysis of the dynamics of these equations, for the values of α and ω of our applications, confirms that the original system has in the origin an unstable focus and an asymptotically stable

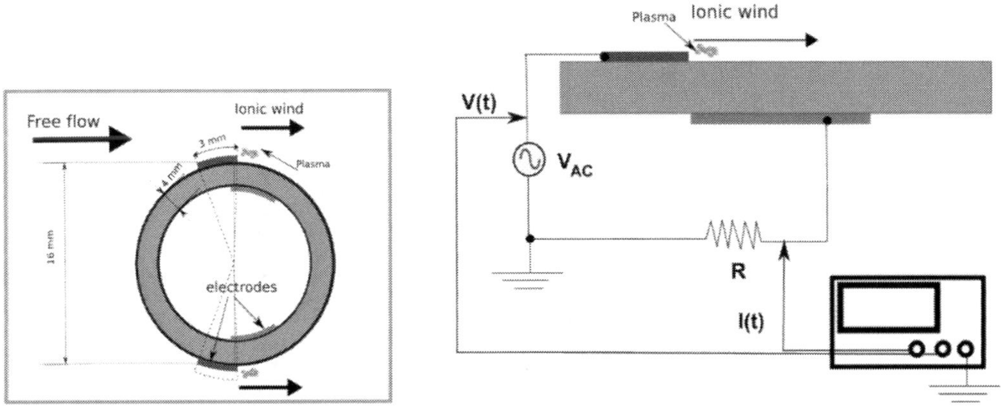

Figure 1. Schematic of the EHD actuator. *Left:* Electrodes disposed flush-mounted. *Right:* Electric circuit.

orbit. Detailed descriptions of these methods can be found in a general context in Guckenheimer & Holmes (1986)[17], Wiggins(2003)[16].
It results for the cylinder wake flow in this regime, a third order system that describes the dynamics of a two dimensional manifold. The remaining variables have dynamics that *collapse* into the center manifold. For this reason, we restrict our analysis of the wake flow modification to the study of just 2 variables, that are constructed from POD modes, that describes the center manifold.

3. Experimental Setup

3.1. Wind tunnel and cylinder

The experimental configuration we settled for this work consists of a flow around a circular cylinder at low Reynolds number, $Re = 100$. The velocity field measurements were undertaken with the cylinder placed in a closed loop wind tunnel with a test section of 18x18 cm^2. In our study the EHD actuators were mounted on the surface of a polymethyl methacrylate cylinder that were 20 mm in diameter D. The cylinder was a hollow tube with a 4 mm thick wall.

3.2. EHD excitation

The DBD plasma actuator used in our configuration consists in set of couples of flat aluminium foils electrodes disposed symmetrically in the spanwise direction of the cylinder surface, in an arrangement as shown in Fig. 1. For each pair, one of the electrodes is exposed to the air while the other is fully covered by the dielectric material. In this case, the cylinder thickness is used as the dielectric layer. The covered electrode is grounded, and the a.c. voltage is applied to the exposed electrode. Both electrodes are slightly overlapped (of the order of 0.5 to 1 mm), in order to ensure a uniform plasma in the full spanwise direction. The a.c. voltage amplitude ranges from 3 to 5 kV, and the operating frequency of the a.c. voltage supplied to the electrodes is 7.5 kHz. The plasma actuator is operated in "steady" manner, i.e. the input driving a.c. frequency is well above the fluid response frequency (taking into account that the Strouhal number $St \sim 0.2$ and the characerisitc length D=0.16 m, the vortex shedding frequency is $\sim 1Hz$), and therefore the flow senses a constant body force.

3.3. Velocity field measurements

The velocity field was measured using a PIV technique. These experiments were conducted using a Pixelfly PCO VGA camera, with a resolution of 640x480 pixels, 1/50 s of time between images.

Figure 2. Flow Visualization for the flow around a circular cylinder. Von Karman vortex street for $Re = 100$.

A green laser Intelite GM32-150IH, 150 mW, combined with a rotating polyhedric mirror, was used to provide illumination on the test section.
The algorithms used in this work belongs to GPIV software [18] which is under GNU General Public License. We have considered for our images a two step grid refinement, so the final interrogation size is 16x16 pixels with a 50% overlapping.
The flow domain covers about 6×8 diameters and it allows to describe the large structures of the flow. The time resolution was suitable to recover the flow dynamics, in a way that the vortex shedding frequencies are lower than the acquisition frequency.

4. Results

Typical visualizations of this flow may be found i.e. in the early experiments by Homman (1936) presented in Zdravkovich (1997)[19]. The Fig. 2 shows for this regime the characteristic vortex shedding.
The modifications introduced by the plasma actuator can be described firstly by the mean flow field. We observe that the near wake region behind the cylinder enlarges under increasing voltage. The resulting mean flow resembles more to the stationary solution, characterized by a long recirculation zone.
The modal decomposition allow us firstly to construct a ROM for each case. In Fig. 3, the first POD spatial mode shows the structure of the large vortex associated to the von Karman street. Indeed, superposing the second mode and modulating these modes with their temporal coefficients, the main characteristics of the flow are recovered. For this kind of actuation, these global structures are conserved under EHD actuation.
The temporal coefficients are plotted in Fig. 4. We distinguish no changes on their frequency, the Strouhal number $St = fD/U$ remains practically the same (for this flow, $St \sim 0.17$). From the observations, we note that temporal modes amplitudes decrease with increasing EHD voltage. The actuator achieves a stabilizing effect on the wake as it dumps progressively the fluctuations as it can be verified from Fig. 4a) to c). This result is in agreement to the considerations remarked on mean flow modifications.

On the other hand, the ROM solution captures the flow attractor for each case, a limit cycle. The model is obtained after applying the polynomial identification technique, the center manifold and the normal forms reduction. Fig. 4d) shows the phase diagram for $AC = 5kV$, where the solution of any initial condition (i.e. points 1 to 4) is conducted, by means of the model, into the

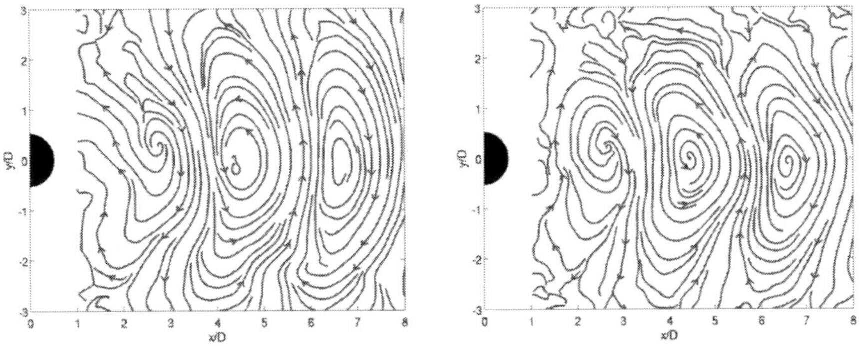

Figure 3. First POD spatial mode for the flow field. Left: Free of EHD actuation. Right: $AC = 5kV$ Geometrical description by streamlines.

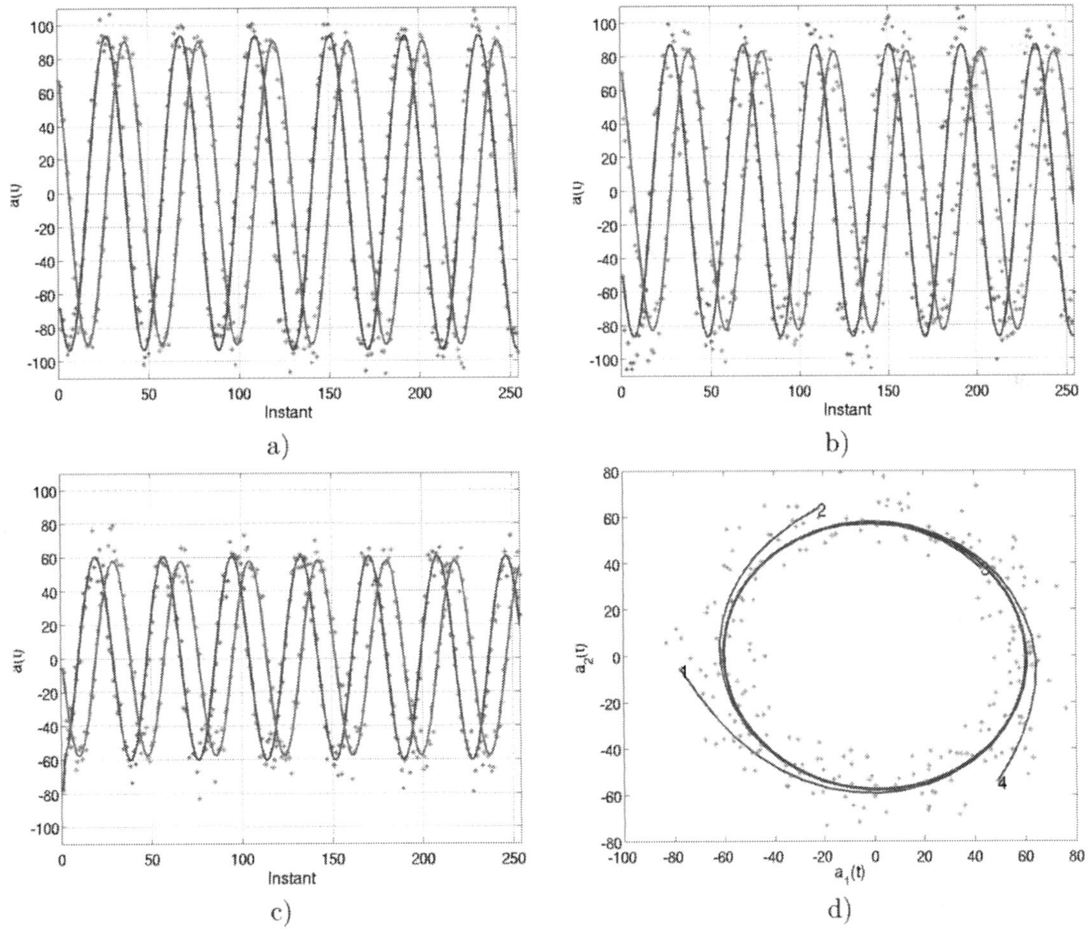

Figure 4. Temporal coefficients evolution. Symbols: measurements. Lines: Reduced order model solution. a) No actuation. b) $AC = 3.5kV$. c) $AC = 5kV$. Phase diagram for $AC = 5kV$.

limit cycle. Structural stability of the dynamical system is therefore assured:solutions starting from different initial conditions are always attracted to the limit cycle. Hence, the model can reproduce transient states and it results an attractive tool in the study of transient excitations and to solve optimization problems.

5. Conclusions

From the benchmark of the flow around a cylinder problem, we proposed a tool in order to compare flow modifications introduced by an EHD actuator. A reduced order model is constructed from experimental data, PIV measures, and it allows to capture the essential dynamics of the flow. We applied the POD-ROM to a set of images corresponding to the flow around a cylinder at a low *Re* number, and the resulting flow under EHD actuation. As experimental data is subjected to many error sources, we performed methods of dynamical system identification and reduction to guarantee the model reliability and stability.

The mean flow is modified as the recirculation bubble length is increased by the EHD action. The actuator has a stabilizing effect on the flow, confirmed by the decrease of the fluctuations amplitudes, represented, in the ROM by the temporal coefficients. The POD-ROM scheme results a convenient tool to identify the dynamical changes introduced by the EHD actuator. The system is robust enough to solve solutions starting far from the limit cycle, and it presents as an interesting tool to study transient states and optimization.

We remark that our study was exposed for the center manifold reduction, scheme valid for Reynolds numbers close to the threshold (for the cylinder wake $Re_c \sim 50$. Nevertheless, this can be extended to the so called *inertial manifold*(see i.e. Noack et al.(2003)[6]), which is constructed by identifying the real positive eigenvalues of the jacobian $\partial F_i/\partial a_j$ in (8), and referring to the corresponding eigenmodes the reamaining ones.

Further studies may be conducted after this work, as the introduction of a control parameter to produce only one dynamical model to represent each actuated case to allow close loop control schemes.

Acknowledgement This work has been undertaken with ANPCYT grant PICT 38070 and UBACYT grant number I017

References

[1] Lumley J 1967 *Atmospheric Turbulence and Radio Wave Propagation* 166–178
[2] Holmes P, Berkooz G and Lumley J 1996 *Turbulence, Coherent Structures, Dynamical Systems and Simmetry* (Cambridge University Press)
[3] Deane A, Karniadakis G and Orszag S 1991 *Physics of Fluids A* **3** 2337–2354
[4] D'Adamo J, Papadakis N, Mémin E and Artana G 2007 *Journal of Turbulence* **8** 1–22
[5] Kalb V and Deane A 2007 *Physics of Fluids* **19** 18
[6] Noack B, Afanasiev K, Morzynski M, Tadmor G and Thiele F 2003 *Journal of Fluid Mechanics* **497** 335–363
[7] Thiria B and Wesfreid J 2007 *Journal of Fluid Mechanics* **579** 137–161
[8] Barkley D 2006 *EPL (Europhysics Letters)* **75** 750–756
[9] Artana G, Sosa R, Moreau E and Touchard G 2003 *Experiments in Fluids* **36** 580–588
[10] Noack B, Papas P and Monkewitz P A 2005 *Journal of Fluid Mechanics* **523** 339–365
[11] Perret L, Collin E and Delville J 2006 *Journal of Turbulence* **7** 1–15
[12] Maquet J, Letellier C and Aguirre L 2004 *Journal of Theoretical Biology* **228** 421–430
[13] D'Adamo J, Cammilleri A and Artana G 2008 *INMAT* (Buenos Aires, Argentina: Facultad de Ingeniería (UBA))
[14] D'Adamo J, Cammilleri A, Mémin E and Artana G 2008 *Escola de transicao e turbulencia* (San Carlos, Brasil: ABCM)
[15] Provansal M, Mathis C and Boyer L 2006 *Journal of Fluid Mechanics Digital Archive* **182** 1–22
[16] Wiggins S 2003 *Introduction to Applied Nonlinear Dynamical Systems and Chaos* (New York: Springer) ISBN ISBN: 978-0-387-00177-7
[17] Guckenheimer J and Holmes P 1986 *Nonlinear Oscillations, Dynamical Systems, and Bifurcation of Vector Fields* (Springer)

[18] van der Graaf G 2008 Gpiv, open source software for particle image velocimetry http://gpiv.sourceforge.net
[19] Zdravkovich M M 1997 *Flow Around Circular cylinders.* vol 1 (Oxford University Press)

Circular cylinder drag reduction by three-electrode plasma actuators

R Sosa, J D'Adamo and G Artana

Laboratorio de Fluidodinámica, Universidad de Buenos Aires, Av. Paseo Colon 850, C1063ACV, Buenos Aires, Argentina. CONICET

rsosa@fi.uba.ar

Abstract. The drag reduction in a circular cylinder was explored by means of a novel three electrode plasma actuator (*DBDE*). The *DBDE* actuator can reduce the drag coefficient up to a ~25% respect to the base flow drag coefficient. It has been demonstrated that, within the present experimental conditions, the *DBDE* actuator, for a fixed value of the power coefficient, adds a higher momentum to the flow and, consequently, produces a higher drag reduction than the *DBD* actuator with the same power coefficient. The actuator efficiency was analysed in terms of the momentum added to the flow revealing similar behaviour for both kind of actuators. However to produce similar levels of actuation both kind of actuators require different values of V_{AC} voltages that resulted always lower for *DBDE*. The reduction in this high voltage value is highly beneficial as is directly related to: a lower HV AC source power requirement, a reduction in the dielectric breakdown probability of the device and a reduction in leakage currents.

1. Introduction

Flow over bluff bodies, for Reynolds number exceeding a critical value, presents vortex shedding in the wake hence resulting in a significant pressure drop on the rear surface of the body. This vortex shedding occurs over a wide range of Reynolds numbers causing structural vibrations, acoustic noise and resonance, enhanced mixing, and significant increases in the aerodynamic drag.

Active flow control techniques have been studied to improve aerodynamic performance of bluff bodies. Most of the flow actuation techniques are concentrated on the reattachment of separated flow which results in enhancing lift and/or reducing drag forces. Flow reattachment can be achieved using surface blowing to construct a pseudo streamline-shaped surface [1-2], surface suction to obtain a thinner boundary layer [3], oscillations of the body (rotary [4-5], streamwise [6] and transverse [7-8]) or synthetic jets [9-10] to utilise the effects of momentum injection into a boundary layer. However, these techniques add different kind of complexity to the system and offer in general a limited controllability in eliminating or delaying flow separation.

Recently, the use of plasma actuators has been proposed as an alternative method to control flow separation (see for instance the comprehensive review [11]). These plasma actuators use a non-thermal surface plasmas and the flowing air close to the surface of the body is ionized. As a consequence of the elastic collisions between the migrating charged particles and the neutral species of the gas, the neutrals increase their momentum giving rise to an 'electric wind' that takes place in the close vicinity of the wall surface. This additional body force resulting from the strong directional ion flow can suppress flow separation as it can be used to overcome the destabilising adverse pressure gradient. Also other mechanisms like alterations of the physical properties of the gas (density, viscosity, etc), may eventually contribute to enhance the electromechanical coupling of discharge-airflow [12-13]. The plasma actuators are technologically attractive because of their simplicity (they have no moving parts) and their very short response time (typically of the order of tens of milliseconds, see for instance [11]).

Plasma actuators can be classified considering the number of electrodes comprised and their arrangement. A first group involves devices employing two electrodes arrangements. At the same time this actuators group may be divided if we consider devices in which the electrodes are separated by a dielectric barrier, usually known as dielectric barrier discharge actuators (*DBD*), or devices in which the two electrodes are bared and flush mounted on a dielectric surface (*BED*).

The *DBD* actuators use periodically excited electrodes and the dielectric barrier interposed between them plays an important role in the stabilisation of the discharge. A large number of aerodynamic studies of flows actuated with *DBD* have been carried out (see for instance [14-19]). Typically, the *DBD* actuator can generate an electric wind with a velocity up to about $8ms^{-1}$. Nevertheless the plasma area extension is limited as the ionization may extend up to ~2 cm from the electrode edge. This might be a crucial drawback for large scale applications.

The *BED* actuators may produce distinctive gaseous discharges depending on the applied potential and the flow conditions [20], these are: unipolar coronas discharge (*UC*), bipolar coronas (*BC*) and plasma sheet discharge (*PS*). In any case the electrodes excitation is produced either with DC or with a periodic voltage supply. The aerodynamic performances of *BED* plasma actuators have been reported in previous works for different flow conditions [21-27].

The main drawback of *BED* plasma actuators is that the electrical discharge may become rather unstable under certain atmospheric conditions. For instance, corona-to-arc transition have been detected in some cases when the relative air humidity exceeded a threshold of about 55% [28].

More recently [29], a new group of plasma actuators consisting in three electrodes devices has been developed (*TED*). Figure 1 shows the arrangement where two electrodes are flush mounted on the air exposed surface (upper electrodes (1) and (3)), and a third one is placed on the opposite side of the insulating surface (lower electrode, (2)). If the electrode (3) and the electrode (2) are grounded biased while the upper electrode (1) is at an AC voltage a typical surface dielectric barrier discharge is formed between electrodes (1) and (2), as is schematically showed in figure 1a. In this case the *TED* actuator works like a conventional *DBD* actuator.

On the other hand when the electrode (1) is at an AC voltage and a high enough DC voltage is applied to the electrode (3) one can produce different discharges, depending on the polarity of the DC voltage (V_{DC}). Thus for $V_{DC} < 0$ a luminescent plasma sheet (named sliding discharge, *SD*) that occupies the whole electrode gap is produced (see figure 1c), and if $V_{DC} > 0$ a discharge that visually looks like a *DBD* (see figure 1b), but with different electrical and mechanical characteristics, is established. Both discharge regimes are as stable as the *DBD* discharge and have the advantage that they allow large-scale applications because the discharge extension may be increased up to the gap between electrodes (1) and (3) [30-32].

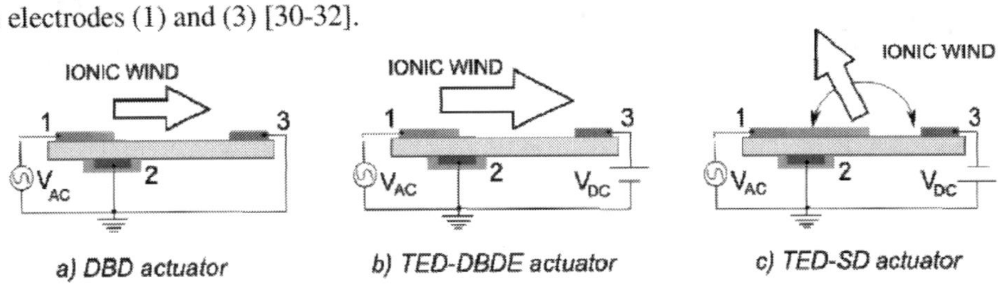

Figure 1. Different plasma actuators. a) Dielectric Barrier Discharge actuator (*DBD*), b) Three Electrode Device with $V_{DC} > 0$ (*TED-DBDE*), c) Three Electrode Device with $V_{DC} < 0$ (*TED-SD*)

In the *TED* actuators the induced electric wind morphology is strongly dependent on the polarity of the DC voltage applied to electrode (3). When $V_{DC} < 0$, and the sliding discharge (*SD*) takes place, two induced jet flows, with origin at the electrodes proximity, result directed towards the inter-electrode space. In this case the global induced flow direction also dependent on the applied AC voltage. By varying the AC voltage value the global plume like flow direction and span angles almost lying in the range 0-180º referred to the dielectric surface (see figure 1c). More detailed experiments concerning induced flow by the *SD* actuator can be found in [32].

On the other hand if we consider a *TED* actuator with a positive DC component applied to electrode (3) the induced electric wind direction (figure 1b) is coincident with that observed in a single *DBD* actuator (figure 1a) but the positive DC voltage allowed to increase the induced electric wind, respect to the case $V_{DC} = 0$, without a noticeable increase in the electrical power consumption [31]. This last configuration will be referred as enhanced dielectric barrier discharge or *DBDE* actuator.

The objective of this work is to determine the capability of *DBDE* plasma actuators to reduce the aerodynamic drag experienced by a circular cylinder in uniform translation. We are especially interested in comparing, for this flow configuration, the performance of the *DBDE* actuation with the more traditional *DBD* ones [33-35].

To undertake this analysis we performed two different set of experiments. One of them has as objective to characterise the induced electric wind when the plasma actuator is applied in quiescent air. We focus our attention in the difference between the cases $V_{DC} = 0$ (*DBD*) and $V_{DC} > 0$ (*DBDE*). This set of experiments were done mounting the plasma actuator on a flat plate in order to simplify the measurement of the induced velocity. Electrical measurements complete this study. In a second set of experiments we analyse the modifications produced in the aerodynamic drag exerted on a circular cylinder (within the Re range of $6.10^3 < Re < 13.10^3$) by either *DBD* or *DBDE* actuators.

2. Experimental set-up

2.1. Induced flow in quiescent air and electrical measurements

A scheme of the experimental set up is shown in figure 2. In order to analyse the induced electric wind in a quiescent air the plasma actuator was mounted in a dielectric flat plate.

In this case the electrode arrangement consisted in two flat aluminium foils (electrodes (1) and (3)) flush mounted on the surface of a polymethyl methacrylate flat plate (150 mm x 350 mm, thickness of 3 mm) and a third one disposed at the opposite side (electrode (2)). Electrode (2) was buried in order to inhibit discharges on this dielectric plate side. The electrodes dimensions are 50 μm thickness, length 300 mm (in the z direction, see figure 2) and width 5 mm. The inter-electrode gap between electrodes (1) and (3) was 28 mm.

A variable DC power supply (V_{DC} in the range +10 ÷ +20 kV, 10 W) biased positively electrode (3). Another variable AC sine voltage power supply (frequency: $f = 8$ kHz, zero mean value; peak to peak voltage V_{AC} up to 20 kV) was applied to electrode (1). The AC power supply consisted in a function generator coupled to an audio amplifier (of 150 watt) that fed a high voltage transformer coil [36]. The electric current flowing in the system, $I_2(t)$ and $I_3(t)$, were measured with shunt resistances ($R = 50\Omega$) connected to an oscilloscope with 60 MHz of analogical bandwidth and 1Gs/s of sampling rate. The AC voltage applied to electrode (1), and the DC voltage applied to electrode (3) were measured with a HV probe (1000 × / 3.0 pF / 100MΩ).

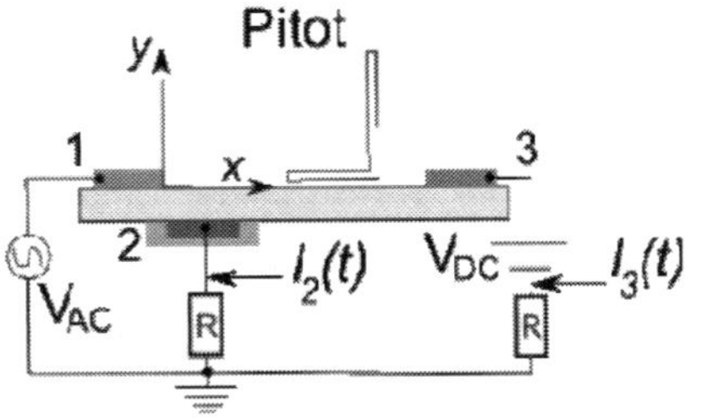

Figure 2. Experimental set-up for measure the induced electric wind in a quiescent air and the actuators electrical characteristics

The set-up was placed horizontally in the test section of a low speed wind tunnel with a square test section of 450 mm x 450 mm (850 mm length) in order to avoid external flow perturbations.

One pressure probe (Pitot tube in figure 2) was mounted on the dielectric surface, placed in front of electrode (1) at $x = 15$ mm, and y = 0.48 mm. This probe was made of electrically insulating glass tube (internal diameter: $ID = 0.97$ mm, outer diameter $OD = 1.3$ mm) and was connected by mean of a tygon tube to a low differential pressure transmitter (0 to 30 Pa, accuracy of 0.4% full scale). The reference pressure value was measured at the test section but far away from the electrodes.

2.2. Aerodynamic drag measurement

In this study the electrodes were mounted on a PMMA cylinder that was 40 mm in diameter D (length L = 300 mm). The cylinder was a hollow tube with a 3-mm thick wall.

The electrode arrangement consists in three electrodes flush-mounted on the cylinder surface (two couple electrodes (1)-(2) and one electrode (3), see figure 3a). The electrodes dimensions were 50 μm thickness, length 300 mm (in the z direction) and width 5 mm. All the electrodes were parallel to the cylinder axis (see figure 3a) and the inter-electrode gap between electrodes (1) and (3) was 28 mm.

Our experimental device was placed in an open low speed wind tunnel (0-5 m/s) with a square test section of 450 mm x 450 mm (850 mm length) with the cylinder axis placed perpendicular to the main flow (U_0) and oriented such that the electrode (3) occupied the rear stagnation point ($\theta = 0$) and the pairs of electrodes (1) and (2) occupied positions $\theta = 90^o$ and $\theta = -90^o$ respectively (see figure 3a).

We have measured the drag exerted on the cylinder by means of a balance that enable us to detect changes of 10 mN in the force. The cylinder was hanged from this balance by two supports (fixed on both cylinder ends through pieces of nylon entering in the hollows of the tube) and the force was detected by equilibrating the moment of the drag force with the moment of the weights of the balance. (an outline of the wind tunnel and aerodynamic drag balance is shown in figure 3b). The blockage ratio was 8.9% and no correction has been applied to the drag data for blockage effects.

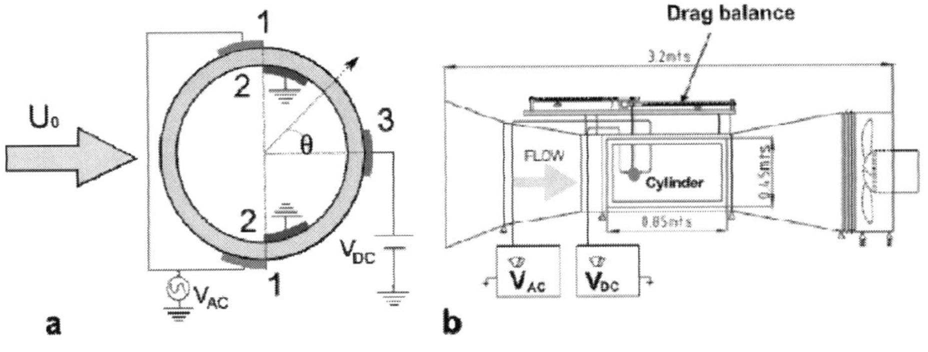

Figure 3. Experimental set-up for the aerodynamics measures. a) Detail of the *TED* actuator mounted on the cylinder, b) Scheme of the wind tunnel and aerodynamic balance

Journal of Physics: Conference Series **166** (2009) 012015 doi:10.1088/1742-6596/166/1/012015

3. Results.

3.1. Induced electric wind and power consumption

The morphology of the induced flow in the case of the *DBD* actuator has been widely studied [19,37-39]. The *DBD* induces a depression above the right edge of electrode (1), and then the fluid is accelerated tangentially to the wall towards electrode (3). In the case of the three-electrode device actuator with $V_{DC} > 0$, it seems that the depression occurs above the whole discharge region, (i.e. between electrodes (1) and (3)). It was previously demonstrated that when a positive DC component is applied to electrode (3) the induced momentum increase, respect to the momentum induced by a *DBD* actuator (i.e. $V_{DC} = 0$), whatever the x position [31]. This increase also occurred downstream of the visible discharge extension. This behaviour was associated to alterations of the asymmetric character of the induced flow in the *DBD* actuators (the negative *DBD* discharge, occurring during the negative half cycle, resulted in a faster velocity than the positive one, [38]). The velocity increase when $V_{DC} > 0$ due to the acceleration of the negative space charge created by the *DBD* in the negative half-cycle results higher than the velocity reduction during the *DBD* positive half-cycle, producing a higher time-averaged velocity.

In the present work a single *DBDE* actuator has been mounted on a flat plate in order to characterise the induced flow in terms of the applied voltages (V_{DC} and V_{AC}) and the consumed electrical power (P). The same geometrical parameters that those considered for the cylinder flow control case have been used (i.e. dielectric thickness, electrodes dimensions and inter-electrode gap between electrodes (1) and (3)). For the flat plate tests we have measured the induced flow velocity at fixed point in all the considered cases to characterise the induced flow (complete velocity profiles can be found in [31]). For the *DBD* and *DBDE* actuators we will adopt the following nomenclature: DBD_I and DBD_IE_J, where the subscripts I and J indicate the applied V_{DC} and V_{AC} voltages respectively.

Figure 4 shows the velocity (U_j) measured at $x = 15$ mm and $y = 0.48$ mm versus V_{AC}, without the DC component (*DBD* actuator case) and with $V_{DC} = +12$ kV (named DBD_IE_{12}) and +16 kV (named *DBD_IE16*). This figure enables to show that the induced electric wind velocity increases with the applied DC component value.

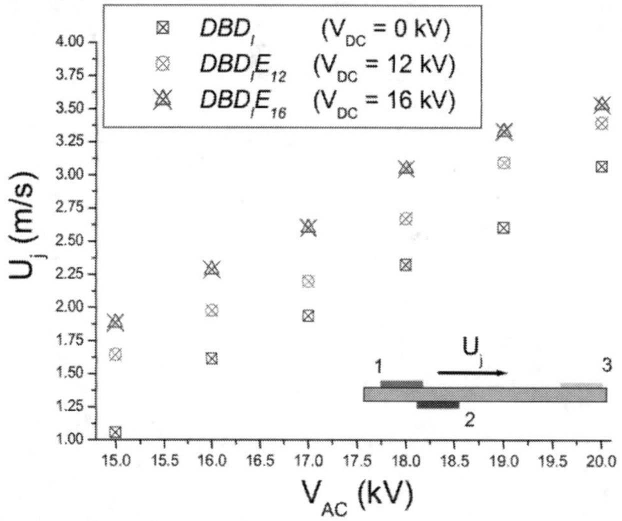

Figure 4. Induced flow velocity U_j vs. V_{AC} for different V_{DC} values

We also can observe that the same induced values of U_j observed for the *DBDE* actuator can be achieved with a *DBD* actuator but applying higher values of V_{AC}. For instance if we consider the *DBDE* actuator with $V_{DC} = 16$ kV and $V_{AC} = 15$ kV (named $DBD_{15}E_{16}$) the induced flow velocity results $U_j \approx 1.9$ m/s, and if we consider a single *DBD* actuator with $V_{AC} = 17$ kV (named DBD_{17}) the induced velocity has been resulted almost the same. Bellow we will discuss what does this higher V_{AC} value mean in terms of the actuator electrical power consumption.

It is worth nothing that the upper limit for the V_{DC} value was ~16 kV because under our experimental conditions (i.e. electrodes radii and inter-electrode gap) an increase of V_{DC} beyond this upper limit will produce sparking between electrodes (1) and (3).

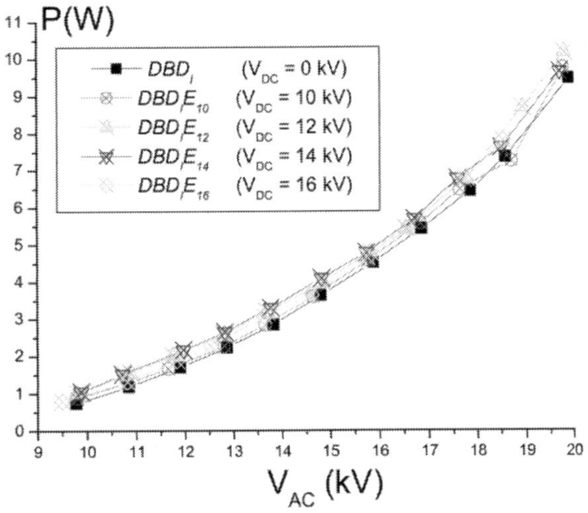

Figure 5. Electrical power consumption (*P*) as a function of V_{AC} with V_{DC} as a parameter for the *DBD* and *DBDE* actuators

Figure 5 shows the electrical power consumption (*P*) versus V_{AC}, for a single *DBD* actuator and for a single *DBDE* actuator considering several positive values of V_{DC}. It shows that for a fixed value of V_{AC} an increase in the applied V_{DC} value produce a slight increase in the electrical power consumption. For instance with V_{AC} = 15 kV, P ≈ 3.6 W for a single *DBD* (i.e. V_{DC} = 0, DBD_{15}) whereas $P \approx 4.1$ W for a single *DBDE* actuator with V_{AC} = 15 kV and V_{DC} = 16 kV ($DBD_{15}E_{16}$) resulting in a power consumption increase of ~12%. In general the increase in the electrical power consumption, with the increase in V_{DC}, was confined within the range ~5÷15%.

On the other hand we can observe that in all the cases the electrical power increases with V_{AC} according to a parabolic function $P = K(V_{AC} - V_0)^2$, been $K = 5.73\times10^{-8}\frac{W}{V^2}$ and $V_0 = 7.58\times10^3 V$ for the single *DBD* actuator under the present experimental conditions.

Now we can complete our comparison of performance of both kind of actuators, to induce a flow, considering energy consumption arguments. The analysis shows that although similar values of U_j can be achieved with both type of actuators the power consumption required by the *DBDE* actuator results always lower. For instance in the previously worked out example the *DBDE* actuator requires $P \approx 4.1$ W to induce a velocity of $U_j \approx 1.9$ m/s ($DBD_{15}E_{16}$) whereas to induce the same velocity the *DBD* actuator consumption is $P \approx 5.4$ W (DBD_{17}). This example shows that the *DBDE* requires only the ~76 % of the electrical power consumed by the *DBD* actuator to induce the same flow velocity. To reinforce this experimental observation Table 1 shows another examples comparing the consumed power of the *DBD* and *DBDE* actuators (P^{DBD} and P^{DBDE} respectively) to induce the same velocity U_j.

Table 1. Comparisons between the different actuators consumed power to induce the same velocity

U_j(m/s)	*DBD*		*DBDE* (V_{DC} = 16 kV)		P^{DBDE}/P^{DBD} (%)
	V_{AC} (kV)	P^{DBD} (W)	V_{AC} (kV)	P^{DBDE} (W)	
1.9	17	5.4	15	4.1	76
2.3	18	6.4	16	4.8	75
2.6	19	7.4	17	5.7	77
3.1	20	9.5	18	6.8	72

In conclusion the *DBDE* actuators requires less power consumption to produce the same induced flow velocity. In order to highlight this behaviour we will compare the actuators performance in terms of non-dimensional parameters. One of them is the non-dimensional momentum coefficient (C_μ), defined as:

$$C_\mu = \frac{U_j^{\,2} b}{U_0^{\,2} D} \tag{1}$$

Where *b* is the characteristic height of the jet induced by the plasma actuator. The momentum coefficient evaluates the ratio between plasma actuator added momentum ($\propto \rho U_j^{\,2} bL$) and a characteristic momentum flow in the free stream ($\propto \rho U_0^{\,2} DL$). Another important parameter will be the power coefficient (C_W) defined as:

$$C_W = \frac{P}{\frac{1}{2}\rho U_0^3 LD} \tag{2}$$

The power coefficient evaluates the ratio between the plasma actuator consumed electric power (P) and a characteristic flow of kinetic energy in the free stream flow.

Figure 6 shows the DBD_I and DBD_IE_{16} (*DBDE* with V_{DC} = 16 kV) momentum coefficient (C_μ) as a function of the power coefficient (C_W) for different *Re* numbers ($\mathrm{Re} = U_0 D/\nu$).

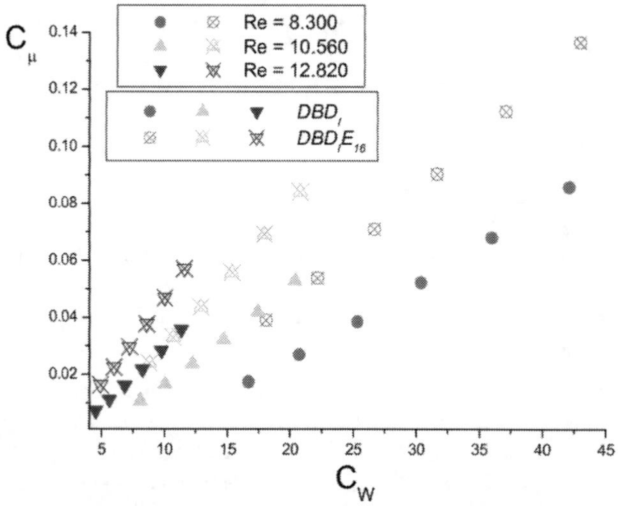

Figure 6. DBD_I and DBD_IE_{16} (*DBDE* with V_{DC} = 16 kV) momentum coefficient (C_μ) as a function of the power coefficient (C_W) for different *Re* numbers

We can observe that for the same non-dimensional power coefficient the momentum addition to the flow is higher if we consider a *DBDE* actuators rather than a *DBD* ones. This last behaviour was observed for all the considered *Re* numbers.

3.2. Aerodynamic drag force

Figure 7 shows the drag force (F_D) as a function of the free stream velocity (U_0) for the base flow (i.e. without discharge) and for the flow modified by means of two single *DBD* actuators considering different V_{AC} values. It can be observed that a significant drag reduction can be achieved by means of the *DBD* actuation. Moreover, if the V_{AC} value increase the drag reduction is higher. Typically the *DBD* actuation can reduce the drag force within ~15÷25% for all the free stream velocity tested values.

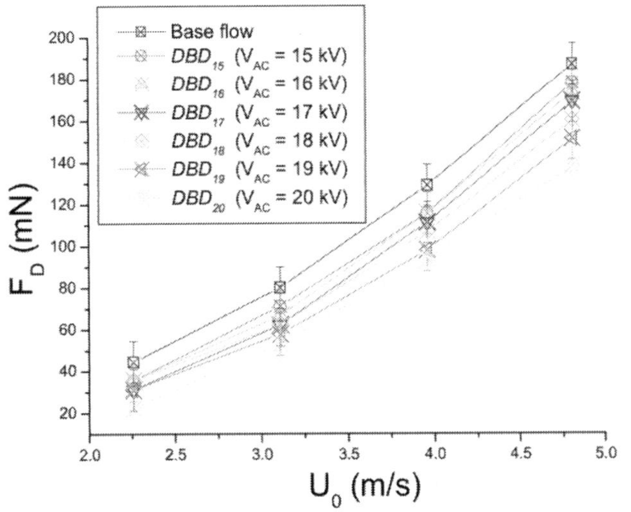

Figure 7. Drag force (F_D) as a function of the free stream velocity (U_0) for the base flow and for the flow modified by means of *DBD* actuators

In order to compare the drag reduction achieved with the different kind of actuators Figure 8 shows the drag force (F_D) as a function of the free stream velocity (U_0) for the base flow (i.e. without discharge) and for the flow modified either by two single *DBD* or *DBDE* actuators. The values of V_{DC} and V_{AC} were chosen such as the induced wind velocity (U_j) in quiescent air result almost the same for both kind of actuators in each figure. So figure 8a compare the case of *DBD* actuators with V_{AC} = 17 kV (DBD_{17}) with the case of *DBDE* actuators with V_{AC} = 15 kV and V_{DC} = 16 kV ($DBD_{15}E_{16}$), and figure 8b compare the case of *DBD* actuators with V_{AC} = 18 kV (DBD_{18}) with the case of *DBDE* actuators with V_{AC} = 16 kV and V_{DC} = 16 kV ($DBD_{16}E_{16}$).

It is clear from figures 8a y 8b that if we adjust the actuators parameters (i.e. V_{AC} and V_{DC} values) in order to induce the same flow velocity we produce almost the same drag reduction on the cylinder whatever the actuator used.

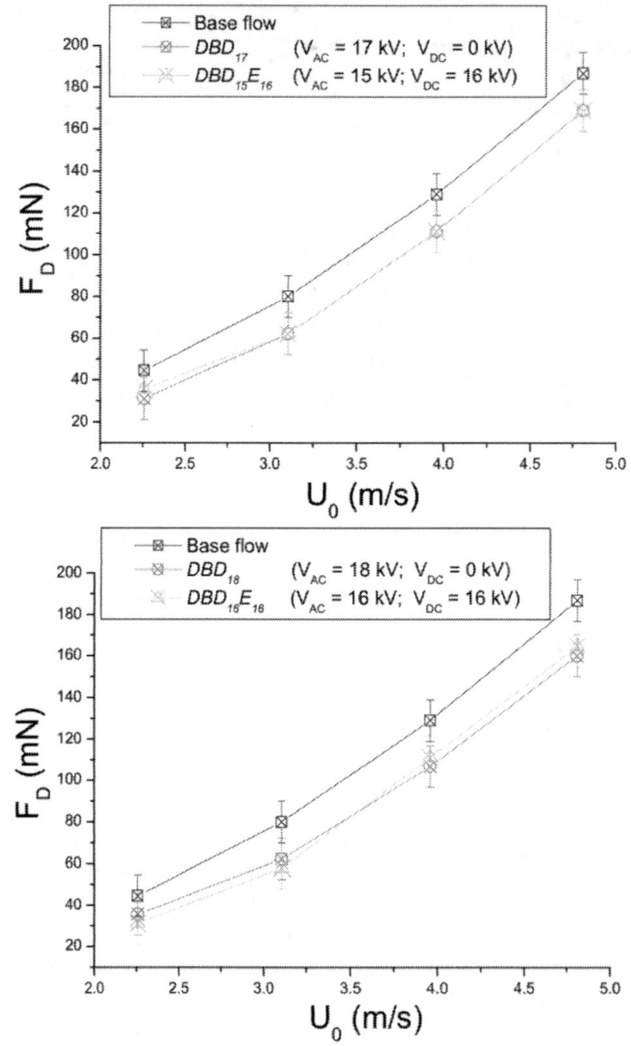

Figure 8. Drag force (F_D) as a function of the free stream velocity (U_0) for the base flow and for the flow modified by means of: upper) *DBD* actuators (V_{AC} = 17 kV and V_{DC} = 0 kV) and *DBDE* actuators (V_{AC} = 15 kV and V_{DC} = 16 kV); lower) *DBD* actuators (V_{AC} = 18 kV and V_{DC} = 0 kV) and *DBDE* actuators (V_{AC} = 16 kV and V_{DC} = 16 kV);

It is useful to compare both kind of actuators considering the drag force in its non dimensional form (i.e. the drag coefficient, C_D):

$$C_D = \frac{F_D}{\frac{1}{2}\rho U_0^{\,2} D L} \tag{3}$$

Figure 9 shows the drag coefficient (C_D) as a function of the power coefficient (C_W) for the *DBD* and *DBDE16* actuators considering two different Reynolds numbers.

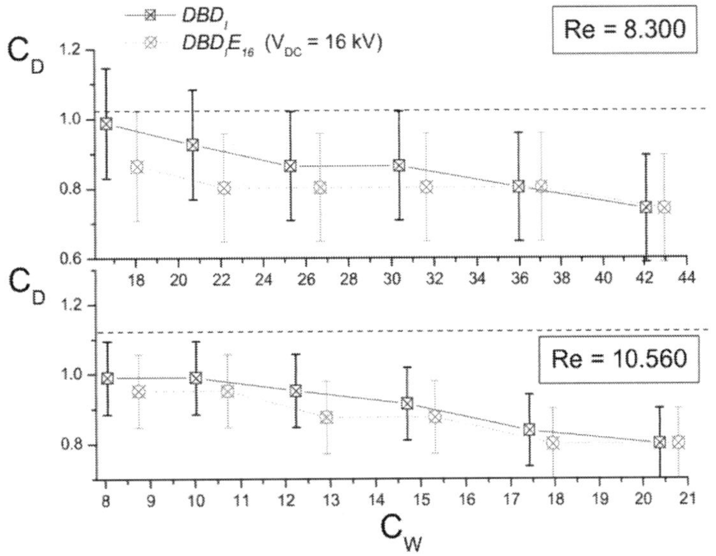

Figure 9. Drag coefficient (C_D) as a function of the power coefficient (C_W) for the *DBD* and *DBDE* actuators (with V_{DC} = 16 kV). (upper) Re = 8300 and (lower) Re = 10.560.

The horizontal dash line ($C_D = 1.1$) represent the drag coefficient in the base flow (i.e. without plasma actuation). We observe that a considerable drag reduction can be achieved with both types of actuators. The plotted curves $C_D(C_W)$ has a very small negative slope and the curves corresponding to the *DBDE* actuation (dot lines) always appear below (or almost equal to) the curves associated to the *DBD* actuation (solid lines). This trend was also observed within all the C_W and Re studied ranges.

Another important parameter to compare the actuators performances is the actuator efficiency (η) defined as:

$$\eta = \frac{\left(F_D^{\,OFF} - F_D^{\,ON}\right)U_0}{P} \tag{4}$$

Where $F_D^{\,ON}$ and $F_D^{\,OFF}$ are the cylinder drag force with and without plasma actuator respectively. The actuator efficiency evaluates the ratio between the power saved as a consequence of the drag reduction ($\left(F_D^{\,OFF} - F_D^{\,ON}\right)U_0$) and the electrical power required to energise the actuator (P).

Figure 10 shows the actuator efficiency (η) as a function of the power coefficient (C_W) for different Reynolds numbers. In this case we have compared the efficiency of the *DBD* actuators with that of the *DBDE* actuators with V_{DC} = 16 kV (DBD_IE_{16}).

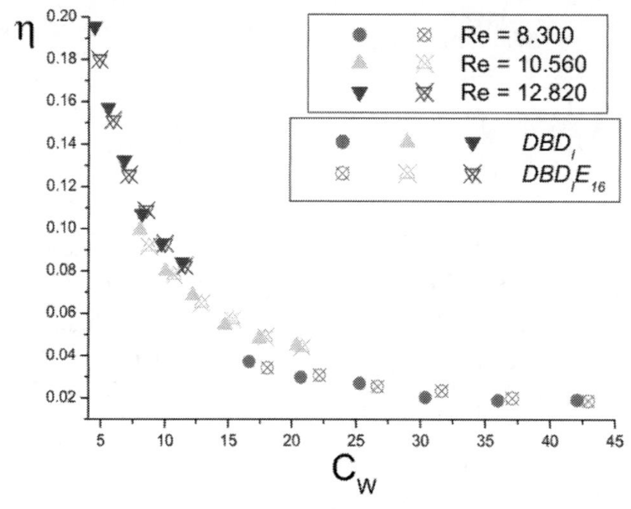

Figure 10 shows the DBD_I and DBD_IE_{16} (*DBDE* with V_{DC} = 16 kV) actuators efficiency (η) as a function of the power coefficient (C_W) for different *Re* numbers

This figure hides in some sense that the *DBDE* actuation enables to improve efficiency of *DBD* actuators as we expected from our previous experimental results. In fact for the same power coefficient (C_W) we have observed a higher momentum addition and higher drag reduction when a *DBDE* actuation was considered (see figure 6 and 9 respectively) and consequently the expected *DBDE* efficiency must be higher than the *DBD* ones. However the differences between the drag reduction achieved with the different kind of actuators (see figure 9) are not enough to revel this behaviour when these non-dimensional parameters are considered.

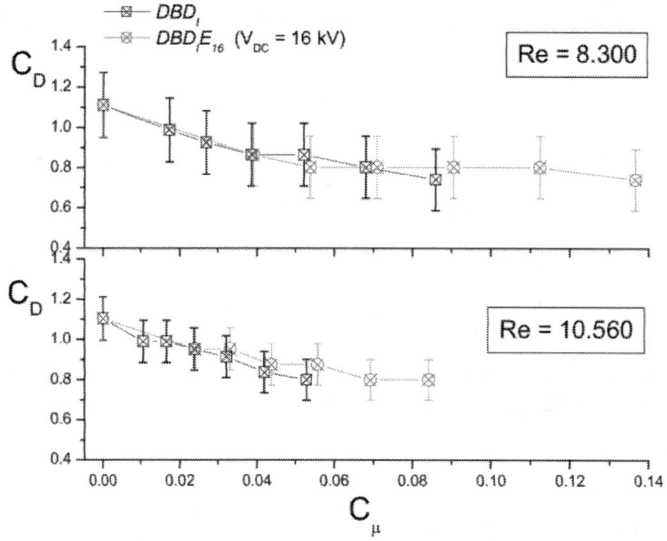

Figure 11. Drag coefficient (C_D) as a function of the momentum coefficient (C_μ) for the *DBD* and *DBDE* actuators (with V_{DC} = 16 kV). (upper) Re = 8300 and (lower) Re = 10.560.

In this figure it can also be observed that the actuator efficiency tends to diminish (for both kinds of actuators) when momentum coefficient increase. This efficiency reduction could be explained in terms of a saturation of the actuation with the increase in C_μ.

In Figure 11 we represent the drag coefficient (C_D) as a function of the momentum coefficient (C_μ) for the *DBD* and *DBDE* actuators for a typical case.

This figure enables to show that the increase in the momentum coefficient produce a similar decrease in the drag coefficient for both kinds of actuators. However for values of C_μ higher than $C_\mu \approx 0.06$ the drag coefficient seems to reach a saturation value. The actuator efficiency decreases for higher values of C_μ is in concordance with this saturation in the actuation (i.e. a saturation in the achievable increase of the $\left(F_D{}^{OFF} - F_D{}^{ON}\right)$ value for a fixed *Re* number). In other words for the higher C_μ values the consumed electrical power increase without a noticeable increase in the cylinder drag reduction. This plateau in the function $C_D(C_\mu)$ as been theoretically predicted by an analytical study of a tangential wall jet applied to the control of separated flow past a circular cylinder[40].It has been shown that the wall jet is very efficient in reducing drag by delaying the separation point in the range of small blowing strength because the suction force induced by the wall jet is negligible compared to the drag due to the external stream. However when the blowing strength is large this suction force finally produces an increase of the drag [40].

4. Conclusions

The drag reduction in a circular cylinder was explored by means of a novel three electrode plasma actuator (*DBDE*). The *DBDE* actuator can reduce the drag coefficient up to a ~25% respect to the base flow drag coefficient. It has been demonstrated that, within the present experimental conditions, the *DBDE* actuator, for a fixed value of the power coefficient, adds a higher momentum to the flow and, consequently, produces a higher drag reduction than the *DBD* actuator with the same power coefficient. For the flow configuration considered the characteristics of the actuation mechanism itself (steady momentum addition) is rather inefficient and it is very difficult to increase the drag reduction by increasing the amount of momentum due to the very small slope in the curves $C_D(C_\mu)$. For both kind of actuators the device efficiency has been analysed in terms of the power added to the flow. It has been found that the efficiency attained similar values of almost 20% but at high levels of actuation the efficiency reduced to a much lower level of 2 %. This saturation has been also observed with syntethic jets devices where it has been shown that the steady injection of momentum reduce the drag coefficient but when higher values of the momentum coefficient were considered the drag reduction reached a plateau and the efficiency actuator consequently decayed.

In the case under study the *DBDE* actuator does not increase appreciably the actuation efficiency respect to that achieved with *DBD*. However to produce similar levels of actuation both kind of actuators require different values of V_{AC} voltages (always lower for *DBDE*). The reduction in this high voltage value is highly beneficial as is directly related to: a lower instantaneous electrical power requirement (i.e. the electrical reactive power is lower), a reduction in the dielectric breakdown probability and a reduction in the leakage currents (which are produced by loss of insulation at high voltage and frequencies).

The work was done as a first proof of concept of *DBDE* actuators as efficient tools to govern the flow around bluff bodies. Presently we are working on unsteady actuation, which seems to hold even more promise than steady devices in matters of efficiency.

Acknowledgement
This work has been undertaken with ANPCYT grant PICT 38070 and UBACYT grant number I017

References

[1] Lin J, Towfighi J and Rockwell D 1995 Near-wake of a circular cylinder: control by steady and unsteady surface injection *J. Fluids Struct.* **9** 659–69
[2] Sevilla A and Martinez-Bazan C 2004 Vortex shedding in high Reynolds number axisymmetric bluff-body wakes: local linear instability and global bleed control *Phys. Fluids* **16** 3460–69
[3] Arcas D and Redekopp L 2004 Aspects of wake vortex control through base blowing/suction *Phys. Fluids* **16** 452–56
[4] Poncet P 2004 Topological aspects of three-dimensional wakes behind rotary oscillating cylinders *J. Fluid Mech.* **517** 27–53
[5] Choi S, Choi H and Kang S 2002 Characteristics of flow over a rotationally oscillating cylinder at low Reynolds number *Phys. Fluids* **14** 2767–77
[6] Cetiner O and Rockwell D 2001 Streamwise oscillations of a cylinder in a steady current. Part 1: locked-on states of vortex formation and loading *J. Fluid Mech.* **427** 1–28
[7] Blackburn H and Henderson R 1999 A study of two-dimensional flow past an oscillating cylinder *J. Fluid Mech.* **385** 255–86
[8] Carberry J, Sheridan J and Rockwell D 2003 Controlled oscillations of a cylinder: a new wake state *J. Fluids Struct.* **17** 337–43
[9] Amitay M, Smith B L and Glezer A 1998 Aerodynamic flow control using synthetic jet technology *36th AIAA Aerosp. Sci. Meet. Reno Nev. AIAA Pap. No.* 98–0208
[10] Glezer A and Amitay M 2002 Synthetic jets *Annu. Rev. Fluid Mech.* **34** 503–29
[11] Moreau E 2006 Airflow control by non-thermal plasma actuators *J. Phys. D: Appl. Phys.* **40** 605-636
[12] Sosa R Mecanismos de Acople en actuadores EHD PhD Thesis University of Buenos Aires Faculty of Engineering October 2007 (in Spanish)
[13] Scherbakov A et al 2000 Drag reduction by ac streamer corona discharges along a wing-like profile plate *AIAA Paper No* 2000-2670
[14] Roth J R, Sherman D M and Wilkinson S P 2000 Electrohydrodynamic flow control with a glow discharge surface plasma *AIAA J.* **38** 1172–9
[15] Jukes T N, Choi K S, Jonhson G A and Scott S J 2004 Turbulent boundary layer control for drag reduction using surface plasma *AIAA Meeting Portland USA paper* #2004-2216
[16] Corke T C, Mertz B and Patel M P 2006 Plasma flow control optimized airfoil *AIAA Meeting Reno USA paper* #2006-1208
[17] Roupassov D V, Zavyalov I N and Starikovskii A Y 2006 Boundary lauer separation plasma control using low-temperature non-equilibrium plasma of gas discharge *AIAA Meeting Reno USA paper* #2006-373
[18] Sung Y, Kim W, Mungal M G and Cappeli M A 2006 Aerodynamic modification of flow bluff objects by plasma actuation *Exp. Fluids* **41** 479–86
[19] Forte M, Jolibois J, Pons J, Moreau E, Touchard G and Cazalens M 2007 Optimization of a dielectric barrier discharge actuator by stationnary and non-stationnary measurements of the induced flow velocity: application to airflow control *Exp.Fluids* **43** 917-928
[20] Moreau E, Leger L and Touchard G 2006 Effect of a DC surface non-thermal plasma on a flat plate boundary layer for airflow velocity up to 25m/s *J. Electrostatics* **64** 215–25
[21] Artana G, D'Adamo J, Leger L, Moreau E and Touchard G 2002 Flow control with electrohydrodynamic actuators *AIAA J.* **40** 1773–9
[22] Artana G, Sosa R, Moreau E and Touchard G 2003 Control of the near wake flow around a circular cylinder with electrohydrodynamic actuators *Exp. Fluids* **36** 580–8
[23] Sosa R and Artana G 2006 Steady control of laminar separation over airfoils with plasma sheet

actuators *J. Electrostat.* **64** 604–10

[24] Sosa R, Moreau E, Touchard G and Artana G 2004 Stall control at high angle of attack with periodically excited EHD actuators *AIAA Oregon AIAA Paper No* 2004-2738

[25] Sosa R, Artana G, Moreau E and Touchard G 2007 Stall control at high angle of attack with plasma sheet actuators *Exp.Fluids* **42** 143-167

[26] Magnier P, Dunpin Hong, Leroy-Chesneau A, Pouvesle J M and Hureau J 2007 A DC corona discharge on a flat plate to induce air movement *J. Electrost.* **65** 655-659

[27] Magnier P, Dunpin Hong, Leroy-Chesneau A, Bauchire J M and Hureau J 2007 Control of separated flows with the ionic wind generated by a DC corona discharge *Exp. Fluids* **42** 815–825

[28] Léger L, Moreau E and Touchard G 2002 Effect of a DC corona electrical discharge on the airflow along a flat plate *IEEE Trans. Indust. Appl.* **38** 1478–85

[29] Louste C, Artana G, Moreau E and Touchard G 2005 Sliding discharge in air at atmospheric pressure: electrical properties *J. of Electrostat.* **63** 615-620

[30] Moreau E, Loustea C and Touchard G 2008 Electric wind induced by sliding discharge in air at atmospheric pressure *J. of Electrost.* **66** 107-14

[31] Moreau E, Sosa R and Artana G 2008 Electric wind produced by surface plasma actuators: a new dielectric barrier discharge based on a three-electrode geometry *J. Phys. D. Appl. Phys.* **41** 115204

[32] Sosa R, Arnaud E, Memin E and Artana G 2008 Study of the Flow Induced by a Sliding Discharge *IEEE Transactions on Dielectrics and Electrical Insulation* In press

[33] Asghar A and Jumper E J 2004 Phase synchronized of vortex shedding from two side-by-side circular cylinders using plasma actuators *AIAA Meeting Portland, USA paper* #2004-925

[34] McLaughlin T E, Munska M D, Vaeth J P, Dauwalter T E, Goode J R and Siegel S G 2004 Plasma-based actuators for cylinder wake vortex control *AIAA Meeting Portland USA paper* #2004-2129

[35] Thomas F O, Kozlov A and Corke T C 2006 Plasma actuators for bluff body flow control *AIAA Meeting San Francisco USA paper* #2006-2845

[36] Yadav, Manish Dai, Xin and Roth J R 2004 Measurement of the Boundary Layer Velocity Induced by Paraelectric Plasma Actuators using an Inexpensive High Voltage Power Supply *Proc. 31st IEEE Int. Conference on Plasma Science Baltimore Maryland Page 244, IEEE Catalog No.* 04CH37537, ISBN 0-7803-8334-6.

[37] Jacob J, Rivir R and Carter C 2004 Boundary layer flow control using AC discharge plasma actuators *AIAA 2nd Flow Control Meeting Portland Paper* 2004-2128

[38] Pons J, Moreau E and Touchard G 2005 Asymetric surface dielectric barrier discharge in air at atmospheric pressure:electrical properties and induced airflow characteristics *J. Phys. D: Appl. Phys.* **38** 3635-42

[39] Boucinha V, Magnier P, Leroy-Chesneau A, Weber R, Joussot R, Dong B and Hong D 2008 Characterization of the Ionic Wind Induced by a Sine DBD Actuator used for Laminar-to-Turbulent Transition Delay *AIAA Meeting Seatle USA paper* #2008-4210

[40] Oh S and Roberts L Control of separated flow past a cylinder using tangential wall jet blowing 1989 *JIAA TR-93*

Journal of Physics: Conference Series **166** (2009) 012016 doi:10.1088/1742-6596/166/1/012016

Reduced order models for wake control with a spinning cylinder

A Gronskis[1], J D'Adamo[1,2], A Cammilleri[1] and G Artana[1,2]

[1] Lab. Fluidodinámica - Facultad de Ingeniería - Universidad de Buenos Aires - Argentina
[2] CONICET - Argentina

E-mail: aleg@fi.uba.ar

Abstract. We study the formulation of reduced order models for a circular spinning cylinder in laminar vortex shedding regime. Proper orthogonal decomposition techniques based on the snapshots generated from velocity fields and a Galerkin projection are used to generate a low dimensional representation of the infinite dimensional system described by the Navier Stokes equation. This kind of problem has generally been studied numerically, without taking into consideration the constraints associated to physical experimental conditions or under the assumption that actuations do not alter significantly the time averaged velocity field. We propose a reduction technique that enables to obtain a control function based on the mean velocity field deviations from the unperturbed reference flow. This function is educed with a Gram Schmidt construction and gives rise to an extended base in which the velocity fields of the perturbed modes are expanded. The analysis allows for easy determination of a suitable model to describe a large range of the control parameter space, without needing to recalculate the coefficients that define the system for new operating conditions.

1. Introduction

Progress in controlling the wakes produced by bluff bodies can lead to significant benefits in different areas like energy or transport, that have close links to environmental protection.

When the geometric shape of an aerodynamic body is fixed and passive flow control devices have been fully exploited or are found to be inappropriate, the study of active flow control problems becomes essential as it may lead to additional improvements in aerodynamic performance.

An important area of flow control research is dedicated to the analysis of vortex shedding phenomena behind bluff bodies, as it has associated problems related to drag, noise, fluid induced vibration and mixing.

In laboratory environments, the wake of circular cylinders has been particularly studied, being considered a benchmark problem representative of the possibility to effectively control wakes produced by more complex bluff bodies.

Among the different devices that can be used to alter the characteristics of the wake dynamics, the mechanical devices that introduce either a steady rotation or rotary oscillations of the cylinder around its axis have probably concentrated the larger number of research efforts.

Other studies have also tried to imitate the effect of a rotating wall through localized momentum injection, using in this case either synthetic jets technology or the so called plasma actuators technology (see for instance [1]). Particularly this last kind of device seems to open in the future the possibility to develop more complex strategies for flow control as it does not involve any moving parts and flow excitation is possible in a large range of frequencies.

One of the drawbacks of these devices is that numerical simulations of the actuation they produce, remain in general a rather difficult task. As a consequence, in order to test tools for flow control strategies, it still seems more convenient to avoid these kinds of complexities and initially restrict to simple cases with uniform moving boundaries.

The problem can be further simplified restricting the analysis to cases where the actuation considered is not time dependent. In this case the characteristics of the wake flow are determined only by two non-dimensional parameters: the Reynolds number $\mathrm{Re} = \frac{U_\infty d}{\nu}$ and a rotation parameter that represents the circumferential speed at the cylinder surface normalized by the free-stream velocity $\gamma = \frac{\dot{\theta} d}{2 U_\infty}$ (d cylinder diameter, U_∞ incident flow velocity, $\dot{\theta}$ angular rotation velocity).

Contrary to its initial simple appearance, this actuation may largely increase the different possible flow configurations, producing at the same time alterations of the mean velocity fields and of the wake dynamics.

For instance at the laminar vortex shedding range ($50<Re<200$), different works have shown [2, 3] that even though at moderate values of the rotation parameter ($0<\gamma<1.95$) the non dimensional frequency of shedding (f_s) remains quite constant,

$$St = \frac{f_s d}{U_\infty} \approx 0.2$$

the mean values of the forces induced on the cylinder and their fluctuations are quite sensitive to the rotation parameter's value.

This behavior can be illustrated with typical polar graphs like that of figure 1 representing the instantaneous drag coefficient (C_D) versus the instantaneous lift coefficient (C_L). This kind of graphs also enables to show other phenomena: we can see that once rotation exceeds a certain threshold that depends on the Reynolds number (i.e. $\gamma \sim 1.95$ for $Re \sim 180$), vortex shedding fully disappears giving rise to a steady flow regime.

More recently different authors [2, 4] reported that when exceeding a higher second threshold of the rotation parameter ($\gamma \sim 4.2$ for $Re \sim 180$) a second vortex shedding regime may re-appear with a characteristic frequency which is quite different from that of the first regime. This regime, which corresponds to very high angular velocity of the spinning cylinders, is out of the scope of this article.

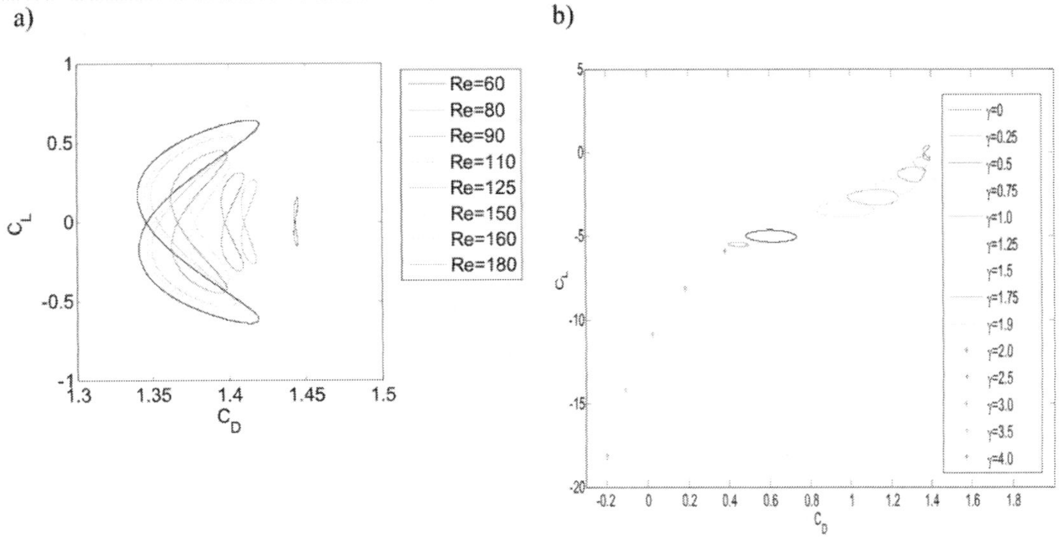

Figure 1. Polar graphs C_D vs C_L: a) Non-actuated flow, b) Spinning cylinder at Re=125.

Journal of Physics: Conference Series **166** (2009) 012016 doi:10.1088/1742-6596/166/1/012016

1.1. Reduced Order Modeling

It is possible to classify in groups the different kinds of flow control problems that one can consider, for instance *control of a final state, optimization, stabilization* or *estimation* [5].

The first kind of problem is how to determine the control action that will move the system from an initial state towards a final state. The second kind of problem studies how to find the optimal set of input parameters that maximizes an indicator of the system performance. In the study of the third kind of problem one tries to determine a control input that will keep the state of the system in a neighborhood of some solution. The last one concerns the reconstruction of the system given its incomplete or noisy data.

Usual tools of control theory can only be applied when it is possible to have at one's disposal a simple model expressed by a set of ordinary differential equations (reduced order model (ROM)), capable of giving accurate representations of the "essential" dynamical behaviour of the flow.

This set may be determined using snapshots of experimental results (numerical or physical experiments) and considering a representation of the flow fields expressed by a finite set of a n-orthonormal spatial base functions $\Phi_i(\mathbf{x})$, each with an associated time dependent coefficient $a_i(t)$. To obtain a consistent model with the observations, the velocity field expansion

$$\mathbf{u}(\mathbf{x},t) \cong \sum_{i=0}^{n} a_i(t)\Phi_i(\mathbf{x}) \tag{1}$$

has to satisfy the boundary conditions.

To determine the base it is possible to use Karhunen Loeve (KL) decomposition (or Proper orthogonal decomposition (POD)), where the spatial modes and temporal coefficients may be calculated from a singular value decomposition of the experimental data matrix by standard techniques [6]. This technique is optimal in the sense that it enables to obtain the set of functions that better approximates energy in comparison to any other base. The modes are ordered with decreasing energy and from the initial set of modes (equal to the set of snapshots) a number of modes may be finally retained fixing an energetic or spectral gap truncation criterion.

As the spatial modes are orthonormal, a Galerkin projection of the Navier Stokes equation under the hypothesis of incompressible flow reduces the system to an ODE system of the kind [6]

$$\frac{da_i}{dt} = i_i + \sum_{j=1}^{s} l_{ij} a_i + \sum_{j=1}^{s} \sum_{k=1}^{s} c_{ijk} a_j \, a_k \tag{2}$$

This system is a reduced order model (POD ROM) and the coefficients involved depend on the basis functions and pressure field. The influence of the pressure field is in general included with other contributions in the independent term. For the case of wake flows there is evidence that this term will have negligible effects.

The POD ROM largely simplifies the complexity of the NS equation but it presents different deficiencies.

One of them is to provide reliable predictions away from the reference orbit from which the empirical model was extracted. To successfully capture the flow dynamics of different states the basis functions used to build the model have to be broad enough to span a hyperplane (Karhunen Loeve space) that includes all relevant flow structures associated to other states of interest.

Even if this is satisfied the coefficients of the set of equations described by (2) may drastically change for the different states.

It is clear that is always possible to restrict the validity of the POD ROM to the neighborhood of the reference state and consider adaptive methods in which the bases are regularly updated when the effectiveness of the existing POD ROM fails to represent accurately the flow for the new control parameter values [7, 8, 9]. However, in order to define flow control strategies, it results more attractive to determine the POD functions, and consequently the coefficients of equation (2), once for all without the need of any recalculation.

Another drawback of POD ROM is the difficulty to account for changes in the boundary conditions suitably. This turns to be crucial when the interest is focused in developing flow control strategies based on boundary condition alterations. How to adequately include the effect of changes of a control parameter in the set of equations (2) may also represent a problem in this kind of formulation.

In this work we are interested in the study of these problems. In the next paragraphs we will give a brief summary of the background of our work.

1.2 *Basis generation*

In general, determining the coefficients of equation (2) using snapshots of a limit cycle associated to a stationary and periodic flow configuration (i.e. the one produced in a non actuated flow or when a continuous or regular periodic excitation is imposed to the cylinder), leads to a poor choice of the basis functions that may cancel out the benefits of the control.

To improve the ability of the model to represent different flow conditions, it is possible to generate the basis using additional snapshots. For instance, including snapshots of transient flows approaching a periodic flow configuration or limit cycle [7, 8].

Other authors [7, 9], have considered the basis generation with data sets associated to flow conditions with excitations producing rich transient configurations. The snapshots can be issued for instance from an ad-hoc time-dependent excitation following a varying-frequency sinusoidal function (chirp). By doing this, the long term and transient predictions of the POD ROM showed an improvement. However, a lack of robustness of this procedure was observed. The improvement not only depends on the nature of the excitation but also on the flow conditions which were applied when the forcing started [7, 9].

Another approach proposed in [10] or [11] adds to the original POD basis special modes called non-equilibrium modes. These non-equilibrium modes can be particular modes known to play a major role in the description of the flow dynamics not evidenced by the original decomposition (stability eigenmodes for example) or deviation modes (also named shift modes); they allow the representation of a missing dynamic mean field correction.

The continuous mode interpolation technique [12] proposes a scheme in which the basis of POD is issued from data corresponding to two close operating conditions. Between them, a POD basis is deduced considering an interpolated kernel of the Fredholm eigenproblem whose eigensolutions are associated to the spatial modes of the basis. This strategy requires in general an important number of reference states and it may have additional problems since these modes do not necessarily keep orthonormality [12].

The so called sequential proper orthogonal decomposition might also be used for generating a basis [13]. With this approach, the ensemble of snapshots correspond to different stationary and periodic flow states. Centroidal Voronoi Tessellations [14] may provide in this case a criterion to undertake the sampling of the different reference states. This kind of strategy has been applied to analyze instabilities and bifurcations in a driven cavity flow [13]. To our knowledge it has not been used to generate POD ROM with goals associated to flow control.

1.3 *Representation of actuation*

When a flow state is altered by some forcing, it is desired to have a POD ROM system able to describe the new physical situations by tuning some variables associated to the actuation. To incorporate the variable input control in the model, different approaches have been proposed: the "*interpolation method*", the "*penalty method*" and the "*control function method*" [9].

In the "*interpolation method*" the temporal coefficients of the system of equations (2) are linearly interpolated between two close operating conditions. This approach has been used in combination with the basis interpolation method in order to achieve flexible modes that can change at different flow conditions [13].

Journal of Physics: Conference Series **166** (2009) 012016 doi:10.1088/1742-6596/166/1/012016

In the "*penalty method*" approach, the non-actuated POD basis is considered and non-homogenous boundary conditions are imposed in a weak implicit sense. This method presents as a drawback a parameter for which appropriate values cannot be determined a priori. Also it leads to a formulation in which appear pressure terms that have to be evaluated at the boundaries of the flow domain. That evaluation may be difficult to perform in many cases.

In the "*control function method*" [7, 9] the boundary actuation is incorporated in the system adding in (2) a term of the form

$$\mathbf{u}(\mathbf{x},t) = \Lambda(t)\varphi_c(\mathbf{x}) + \sum_{i=0}^{s} a_i(t)\varphi_i(\mathbf{x}) \tag{3}$$

where $\varphi_c(\mathbf{x})$ is an arbitrary divergence free control function satisfying the homogenous boundary conditions.

There is no specific way to prescribe the control function. One usual way is to associate it to the flow field produced when the cylinder spins in a quiescent fluid (numerically this is achieved imposing at the cylinder surface a steady rotation with $\Lambda=1$ and assuming on the outer boundaries of the flow domain null velocity ($\mathbf{u}=\mathbf{0}$) [9]). Using this approach the POD ROM is generated in the usual way from a snapshot set of a modified velocity field:

$$\mathbf{u_r}(\mathbf{x},t^k) = \mathbf{u}(\mathbf{x},t^k) - \Lambda(t^k)\varphi_c(\mathbf{x})$$

The spatial modes of the $\varphi_i(\mathbf{x})$ thus generated have zero velocity on the cylinder surface and expansion (3) automatically satisfies the boundary condition of the problem. Thus, the system (2) can be rewritten as [7]

$$\begin{aligned}\frac{da_i}{dt} &= i_i + \sum_{j=1}^{s} l_{ij}a_j + \sum_{j=1}^{s}\sum_{k=1}^{s} c_{ijk}a_j a_k \\ &+ d_i\frac{\partial\Lambda}{\partial t} + \left(e_i + \sum_{j=1}^{N} f_{ij}a_i\right)\Lambda + g_i\Lambda^2\end{aligned} \tag{4}$$

Some problems may arise when the control function is obtained in this way.

First, the effect of the control mode results quite localized around the cylinder vanishing rapidly due to the absence of a convective free-stream flow in the formulation. As a consequence, changes produced by actuation in the far wake are more difficult to reproduce.

Besides, this kind of formulation of the control function has in general been restricted to cases where the data are issued from numerical simulations. In these cases the modes are generated with data that at close proximity of the boundary are relatively reliable. Unfortunately, when the experimental data are produced with real physical experiments the situation may depart from the "ideal" numerical case.

For instance, for experiments undertaken with the PIV techniques it results quite difficult to obtain information with high spatial definition in the wake and at regions close to the boundaries in the same snapshot. In addition, at close vicinity of the surface the error in the measurements may increase because of improper seeding of the flow in this region, or due to illumination problems like reflections or shadowing of the light sheet produced by the body itself.

Finally, most authors have used this approach in situations where the actuation did not significantly alter the mean flow field. The coefficients of the system of equations (4) depend not only on the characteristics of the basis function but also on the mean velocity field. A model that is not capable to adapt to these changes will result only well suited to describe cases in the neighborhood of the reference state.

An alternative to remediate this is to generate a control mode based on the difference of the mean values of non-actuated flow conditions ($\mathbf{u}_{m0}(\mathbf{x})$) and an actuated reference case ($\mathbf{u}_{mact}(\mathbf{x})$). The control function is then associated to a displacement vector defined as:

$$\varphi_c(\mathbf{x}) = \mathbf{u}_{\mathbf{mact}}(\mathbf{x}) - \mathbf{u}_{\mathbf{m0}}(\mathbf{x}) \tag{5}$$

This kind of approach has been proposed recently in some works [15] and explored for plasma actuated flows [16, 17].

It is in the scope of this article to develop accurate POD ROM of wake cylinders flows altered by the steady rotations of the cylinder axis. We focus our efforts on determining suitable strategies to generate the basis of the velocity field expansion and control functions.

Our study concerns the use of empirical models derived from spatio-temporal data issued from numerical simulations in which the limitations imposed by the constraints of "real" experimental conditions of laboratory tests are considered.

Next sections are organized as follows: in section 2.1 we describe the procedure used to obtain the velocity fields of the tests under study. In section 2.2 we describe the method to obtain the reduced order model and how we include in this model a control function that enables to describe the alterations of the wake produced by the rotation of the cylinder. The section 3 is dedicated to the discussion of results. A summary with the more important achievements of our work is proposed at the end of the manuscript.

2. Methods

Our POD ROM is fed with numerical experiments of the flow around a circular cylinder spinning at constant speed.

The Reynolds range under analysis ($50<Re<180$) corresponds to the laminar vortex shedding regime [18] where flow remains two dimensional and 3D effects can be disregarded [19]. The range of the control parameter of interest is $0<\gamma<2$.

2.1. Numerical technique

An incompressible code [20, 21] was used to perform the DNS studies. This code solves the continuity and momentum equations which are discretized using a finite difference method on a uniform Cartesian grid. Spatial derivatives are approximated with a sixth-order, compact-centered, finite difference scheme, while temporal derivatives are evaluated using a third-order low-storage Runge-Kutta method. The boundary condition on the cylinder surface is imposed using the virtual boundary technique.

Mass and momentum conservation principles are represented by the Navier-Stokes equations, which have the following form for an incompressible fluid

$$\nabla \cdot \mathbf{u} = 0 \tag{6}$$

$$\frac{\partial \mathbf{u}}{\partial t} = -\nabla p - \boldsymbol{\omega} \times \mathbf{u} + \upsilon \nabla^2 \mathbf{u} + \mathbf{f} \tag{7}$$

where υ is the kinematic viscosity, $p(\mathbf{x},t)$ the pressure field and $\boldsymbol{\omega}(\mathbf{x},t)$ the vorticity field. The external volumetric force field $\mathbf{f}(\mathbf{x},t)$ is used here to impose the boundary condition on the cylinder surface (Ω_0). Figure 2 shows a schematic view of the boundary conditions over the inflow, lateral and outflow regions of the domain.

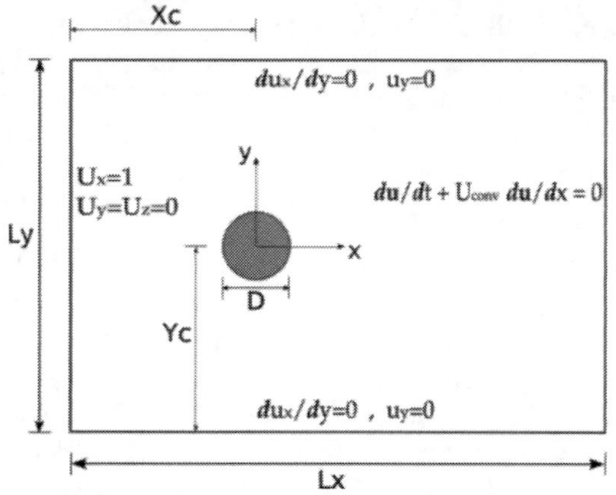

Figure 2. Computational domain and boundary conditions imposed.

2.1.1. Spatio Temporal Discretization. The simulations were performed using 1081×1081 grid points in the x and y directions respectively. The domain size was Lx=30D, Ly=30D, and the centre of the cylinder was located at Xc=15D, Yc=15D. The grid resolution was 36 points per diameter. We have selected these values taking into account the results obtained in previous simulations with this code [20, 21] and others [22].

Numerical simulations were performed until the flow reached a fully developed state, in which all the flow characteristics were analysed.

2.1.2. Simulation Results. A comparison between the results obtained with the code described in this work and those of other research efforts is shown on figures 3 and 4 [4, 19]. In figure 3 we illustrate the effect of varying the rotation parameter γ on the time averaged value of the pressure drag, total drag, pressure lift and total lift coefficients and compare with results obtained by Kang et al. [3]. The figure 4 shows how the code reproduces changes in Strouhal number as a function of Reynolds number and as a function of the rotation parameter. Results are compared with the works of Williamson [18] and Kang et al. [3], respectively.

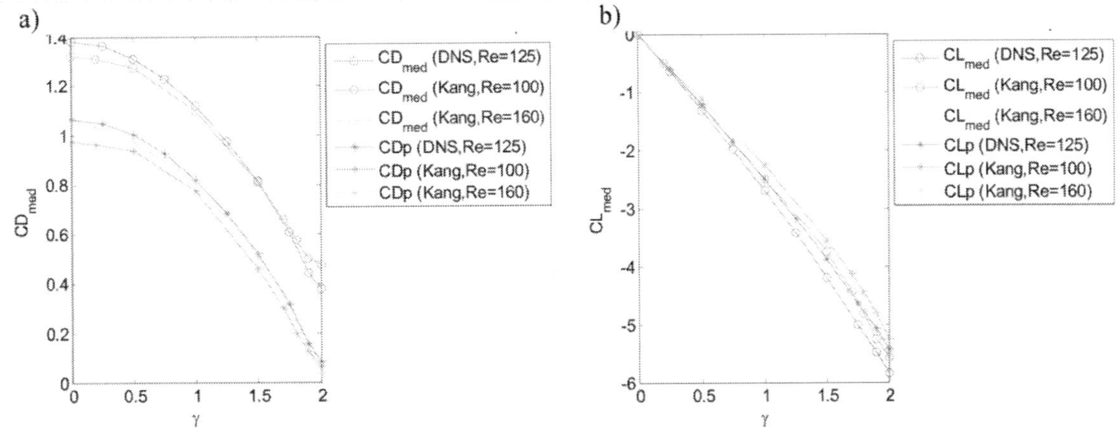

Figure 3. Effect of changes in the rotation parameter on the time averaged values of a) pressure drag (C_{Dp}) and total drag (C_{Dmed}) coefficients and b) pressure lift (C_{Lp}) and total lift (C_{Lmed}) coefficients. Kang refers to results appearing in [3]. DNS refers to present simulations.

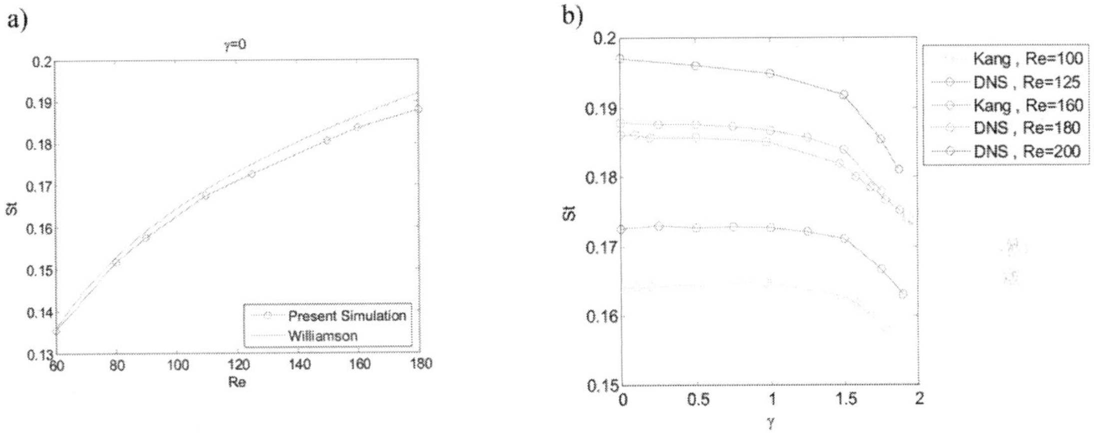

Figure 4. Strouhal number as a function of Control Parameters: a) Reynolds number; Williamson refers to results appearing in [18]. b) Rotation parameter; Kang refers to results appearing in [3]. DNS refers to present simulations.

An analysis of these figures reveals that the code satisfactorily reproduces the values of these parameters as well as their trend when the control parameters are changed.

2.2. POD ROM and Control Function

In recent articles [23, 24] the importance of the time averaged flow as the basic state that can account for stability properties of wakes under forcing conditions has been emphasized. The dynamics of instabilities in the system is proposed to be described by the free wake Landau equation and an additional term that modifies the linear growth rate and results proportional to the mean flow correction generated by the forcing. In [11-13] a scheme is proposed where the dynamics of an empirical Galerkin system to describe the flow around a cylinder is determined considering a shift (or zero) mode. This mode accounts for the difference existing between a state considered as base flow and the time averaged flow that takes place when the natural instabilities of the base flow develop.

Taking this into account, it seems rather natural, to try to determine a control function considering in the Karhunen Loeve space a correcting vector pointing from the time averaged flow of the unperturbed case ($\mathbf{u}_{\mathbf{m0}}(\mathbf{x})$) towards the mean of the perturbed state under actuation. The control mode is then associated to the mean flow modifications produced by actuation as stated in (5).

As we consider a steady rotation, the weighing (temporal) coefficient that affects spatial function φ_c may be here directly expressed as a function Γ that depends on the rotation parameter but is independent of time

$$\mathbf{u}(\mathbf{x},t) \approx \Gamma(\gamma)\varphi_c(\mathbf{x}) + \mathbf{u}_{\mathbf{m0}}(\mathbf{x}) + \sum_{i=1}^{s} a_i(t)\varphi_i(\mathbf{x}) \tag{8}$$

The new phase-space is constructed considering an ortho-normalized vector associated to the mean field correction with the following procedure of construction (Gram Schmidt) that assures orthogonality of all modes. The procedure is in some sense similar to the sequential proper orthogonal decomposition (SPOD) proposed by other authors [14, 24, 25]. In our case it may be laid out as follows:

- **Step 1**: For the unperturbed state (γ=0) we calculate the mean field $\mathbf{u}_{\mathbf{m0}}(\mathbf{x})$ and compute the relevant POD eigenmodes that describe the fluctuations of the flow. The ensemble of these modes is a root basis $\Psi_0 = (\varphi_1, \varphi_2, \varphi_3, \ldots, \varphi_n)$ that is truncated in n modes under a given criteria (energetic or spectral gap).

- **Step 2**: We compute for a perturbed flow taken as reference (i.e. γ=1) a new basis Ψ_1 (first extending basis) obtained as the relevant POD modes of the residual field ($\mathbf{u}_\mathbf{r}$) of the perturbed velocity field projection onto Ψ_0

$$\mathbf{u}_\mathbf{r}(\mathbf{x},t^k) = \mathbf{u}(\mathbf{x},t^k) - \mathbf{u}_{\mathbf{m0}}(\mathbf{x}) - \sum_{i=1}^{n} a_i(t^k)\varphi_i(\mathbf{x})$$

where the temporal coefficients are determined as

$$a_i(t^k) = \left(\mathbf{u}(\mathbf{x},t^k) - \mathbf{u}_{\mathbf{m0}}(\mathbf{x})\right)\cdot \varphi_i(\mathbf{x})$$

If the first temporal coefficient exhibits a constant value, the first eigenvector of the residual field is associated to the time averaged correction vector (control function mode) $\varphi_c(\mathbf{x})$. Thus the extending basis is composed by an ensemble of the fluctuating modes $\Psi_1 = (\varphi_{n+1}, \varphi_{n+2}, \varphi_{n+3}, \ldots, \varphi_s)$ that is truncated in (s-n) modes under a given criteria (energetic or spectral gap).

- **Step 3**: For the other perturbed states γ we calculate projection of the velocity field onto the modes of the truncated basis $\Psi_0+\Psi_1$ and $\varphi_c(\mathbf{x})$ as

$$a_i(t^k) = \mathbf{u}(\mathbf{x},t^k)\cdot \varphi_i(\mathbf{x})$$
$$\Gamma(\gamma) = \mathbf{u}_\mathbf{m}(\mathbf{x})\cdot \varphi_c(\mathbf{x})$$

The residual velocity field that remains unresolved after this projection is

$$\mathbf{u}_\mathbf{r}(\mathbf{x},t) = \mathbf{u}(\mathbf{x},t) - \Gamma(\gamma)\varphi_c(\mathbf{x}) - \mathbf{u}_{\mathbf{m0}}(\mathbf{x}) - \sum_{i=1}^{s} a_i(t)\varphi_i(\mathbf{x})$$

We evaluate the validity of the model to represent the flow considering some criteria based on the residual velocity field.

By construction, the first mode of the extending basis Ψ_l is associated to an ortho-normalized mean field correction or displacement vector, from the time averaged flow fields of the unperturbed case to the mean field of the perturbed state. The number of modes s that composes the whole basis depends on the truncation criteria adopted and the range of the control parameter analysed.

The range of the control parameter γ in which the use of this control function will be able to give accurate results can be determined considering the range in which the eigenvectors with significant constant eigenvalues do not appear. If this happens, it means that is not possible to describe the mean flow with the control function correction.

However, it is always possible to compute an additional control function and a new extending basis Ψ_i. This new basis can be obtained applying again the POD technique over the residue of the perturbed velocity field projection. Thus, an enhanced range can be achieved by increasing the number of control functions. For instance, if two control functions of the parameter control space are included we can formulate

$$\mathbf{u}(\mathbf{x},t) \approx \Gamma_1(\gamma)\varphi_{c1}(\mathbf{x}) + \Gamma_2(\gamma)\varphi_{c2}(\mathbf{x}) + \mathbf{u}_{m0}(\mathbf{x}) + \sum_{i=1}^{s} a_i(t)\varphi_i(\mathbf{x})$$

The first control function is associated to the first constant eigenvector of the first extending basis while the second one is again associated to the first constant eigenvector of a second extending basis.Both control functions are orthonormal to each other by construction. If necessary, this procedure can be extended to include a larger number of control functions.

It should be noted that as the actuation for the spinning cylinder increases we get closer to the Hopf bifurcation state. In this sense it is expected that new dynamical states for high values of the rotation parameter will be less rich in modes than those for the lower values. As a consequence, the addition of just a limited number of modes will be required to obtain a good description of the flow with the POD ROM.

The coefficients affecting the control functions correspond to the value of the time averaged velocity field projection onto the control functions, and one possible way to determine them uses the interpolation between the known states.

The Galerkin projection onto NS equation with two control function leads to the following system

$$\begin{aligned} \frac{da_i}{dt} &= A_i + \sum_{j=1}^{s} B_{ij} a_j + \sum_{j=1}^{s} \sum_{k=1}^{s} C_{ijk} a_j\, a_k \\ &+ \sum_{l=1}^{2} D_{il} \frac{\partial \Gamma_l}{\partial t} + \sum_{l=1}^{2} \left(E_{i_l} + \sum_{j=1}^{N} F_{ij_l} a_i \right) \Gamma_l \\ &+ \sum_{m=1}^{2} \sum_{l=1}^{2} G_{ilm} \Gamma_l \Gamma_m \end{aligned} \qquad (9)$$

where

$$A_i = -\int_\Delta p\varphi_i \cdot \mathbf{n}\, dx - (\varphi_i, \mathbf{u}_{\mathbf{m0}} \nabla \mathbf{u}_{\mathbf{m0}}) + \frac{1}{\mathrm{Re}}(\varphi_i, \nabla^2 \mathbf{u}_{\mathbf{m0}})$$
$$B_{ij} = -(\varphi_i, \mathbf{u}_{\mathbf{m0}} \nabla \varphi_j) - (\varphi_i, \varphi_j \nabla \mathbf{u}_{\mathbf{m0}}) + \frac{1}{\mathrm{Re}}(\varphi_i, \nabla^2 \varphi_j)$$
$$C_{ijk} = -(\varphi_i, \varphi_j \nabla \varphi_k)$$
$$D_{i_l} = -(\varphi_i, \varphi_{c_l})$$
$$E_{i_l} = -(\varphi_i, \mathbf{u}_{\mathbf{m0}} \nabla \varphi_{c_l}) - (\varphi_i, \varphi_{c_l} \nabla \mathbf{u}_{\mathbf{m0}}) + \frac{1}{\mathrm{Re}}(\varphi_i, \nabla^2 \varphi_{c_l})$$
$$F_{ij_l} = -(\varphi_i, \varphi_{c_l} \nabla \varphi_j) - (\varphi_i, \varphi_j \nabla \varphi_{c_l})$$
$$G_{ilm} = -(\varphi_i, \varphi_{c_l} \nabla \varphi_{c_m})$$

In the independent term A_i, the effect of pressure for wake flows is usually neglected. However, the loss of symmetry as a consequence of actuation may increase the error contributed by this term and careful analyses should be undertaken before deciding its elimination. Eventually, a system based on a spatio-temporal decomposition for pressure and velocities, like the one used for the velocity in equation (1), may be proposed [9].

If the effect of the pressure is not so important for a given Reynolds number, the coefficients are calculated once and for all regardless of the changes in the values of the rotation parameter.

One of the advantages of this approach is that one can have a system of equations representing the flow that easily includes the effect of the actuation.

On the other hand the system is of higher dimension than the one obtained with conventional POD ROM as it includes a larger number of modes than in the more traditional approaches.

3. Results

As an example of control by spinning the cylinder in the laminar vortex shedding regime, we analyze the case for a fixed Re=125 and different values of the non dimensional rotation parameter $0<\gamma<2$. The POD modes were obtained with snapshots equispaced in time during about 20 vortex shedding periods.

3.1. Derivation of the Control functions

The mean velocity field for the non actuated flow is represented in figure 5.

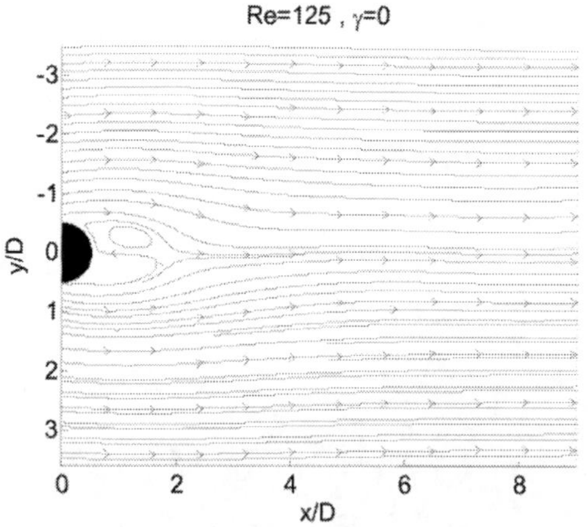

Figure 5. Time averaged flow for γ =0 and Re=125.

Journal of Physics: Conference Series **166** (2009) 012016 doi:10.1088/1742-6596/166/1/012016

We calculated two extending basis: Ψ_1 associated to values of the control parameter γ=1.0 and the other one Ψ_2 for γ=1.9, which is quite close to the value where the vortex shedding phenomena disappears. Results of the mean flow corresponding to these values of the rotation parameters are given in figures 6 and 7.

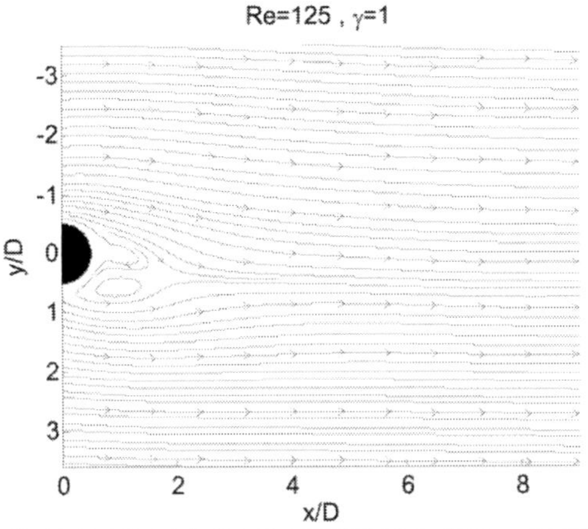

Figure 6. Time averaged flow for γ=1.0 and *Re*=125.

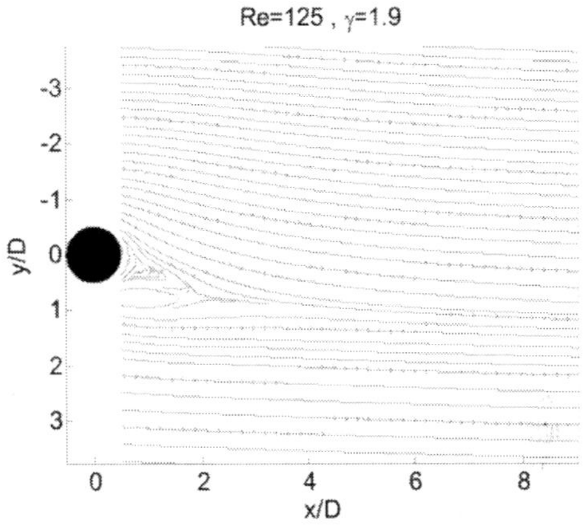

Figure 7. Time averaged flow for γ=1.9 and *Re*=125.

The set of modes of the extending basis were determined with the Gram Schmidt construction and resulted in a first mode with an almost constant temporal coefficient, followed by other modes with temporal coefficients exhibiting a periodic behavior.

The first eigenmodes are thus ascribed to the control functions. They enable to obtain the time averaged flow for different perturbed cases (figures 6 & 7) from the time averaged flow of the unperturbed case (figure 5).

The characteristics of the first and second control function modes are provided by figures 8 and 9, respectively.

For cases corresponding to other values of the control parameters the fields were projected onto the mean of the unperturbed state and onto the correction functions. In figure 10 we represent for different values of the control parameters the energy associated to the time averaged component of flow, and the residual value associated to the fraction that can not be captured.

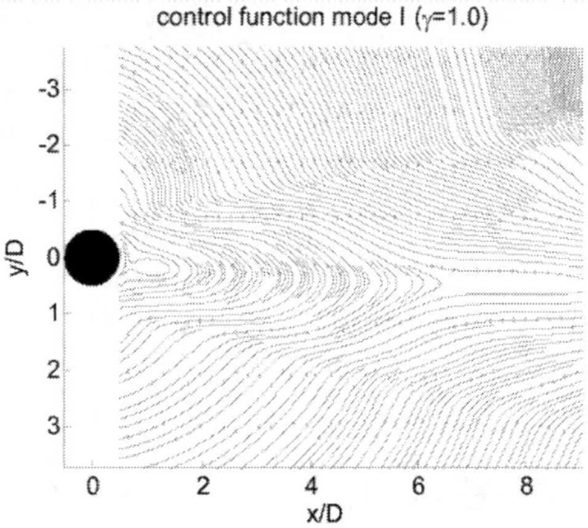

Figure 8. Control function mode $\varphi_{c1}(\mathbf{x})$, *Re*=125.

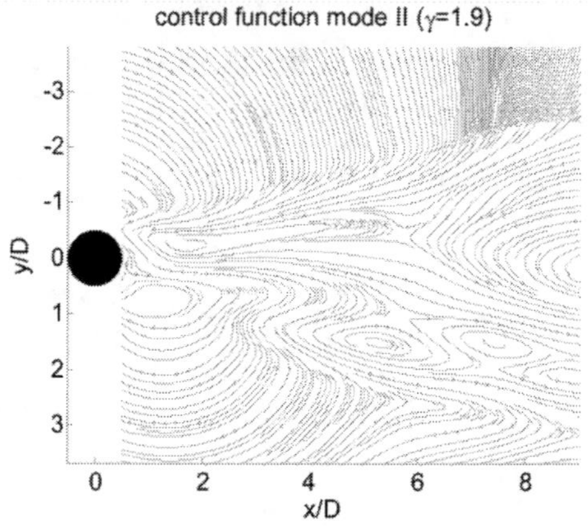

Figure 9. Control function mode $\varphi_{c2}(\mathbf{x})$, *Re*=125.

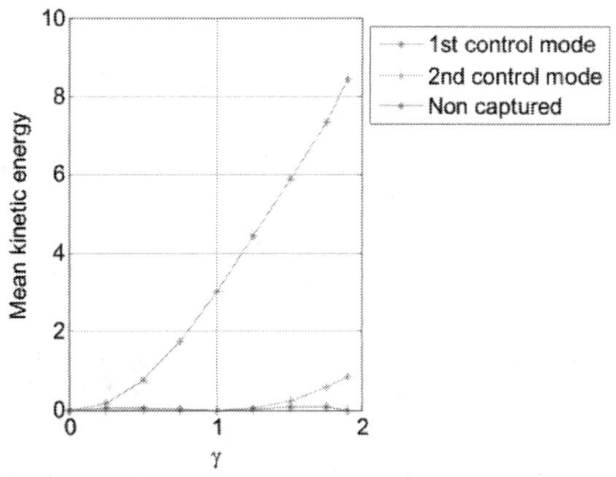

Figure 10. Energy of the flow associated to the control functions and residual mean energy not captured, *Re*=125.

In the proximity of the value of γ where vortex shedding ceases, the vector correction introduced by $\varphi_{c1}(\mathbf{x})$ is not sufficient to define the space of the perturbed time averaged flow, and supplementary corrections are required to generate the appropriate subspace.

The correction function $\varphi_{c2}(\mathbf{x})$ enables to introduce the suitable corrections onto the mean field to extend the model into the range of interest of the rotation parameter.

The energy associated to the mean velocity field of the unperturbed state (not shown in this figure) attains a constant value about one order of magnitude above those represented in the figure.

3.2. Fluctuating modes of the root Basis Ψ_0

The cartography of the unperturbed modes associated to the basis Ψ_0 (figure 11) resulted similar to the one exhibited in [10].

The modes can be grouped by pairs and the temporal coefficients of these modes present a quasi sinusoidal shape, both modes of the pair exhibiting the same frequency but with a difference in phase.

The first pair of modes has a frequency that is in agreement with the vortex shedding of the flow. For the following pair of modes, the frequency increases monotonically. These frequencies are approximately a multiple of the fundamental Strouhal frequency.

3.3. Fluctuating modes of the extending basis

The eigenvectors $\Psi_0+\Psi_1+\Psi_2$ determine a unique subspace, that is common for all actuated states in the range of the rotation parameter analysed ($0<\gamma<2.0$).

A series of charts of the fluctuating modes of the extending basis is given in figures 12 & 13.

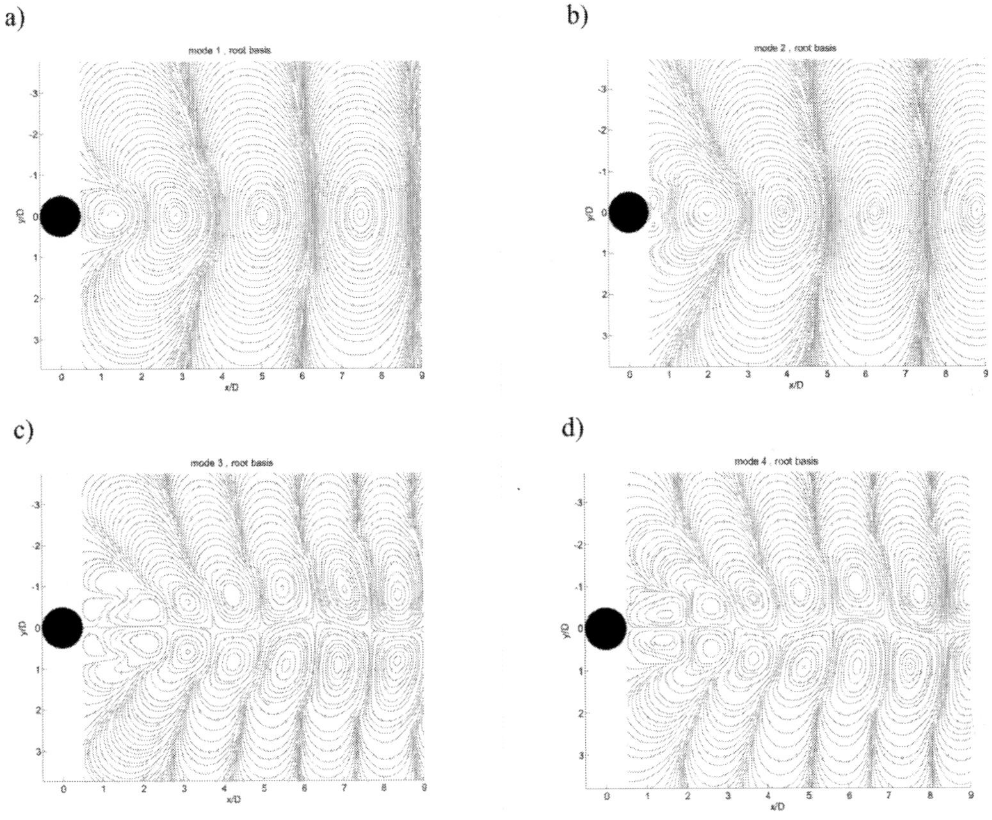

Figure 11. First fluctuating modes of the root basis, *Re*=125: a) Mode 1, b) Mode 2, c) Mode 3, d) Mode 4.

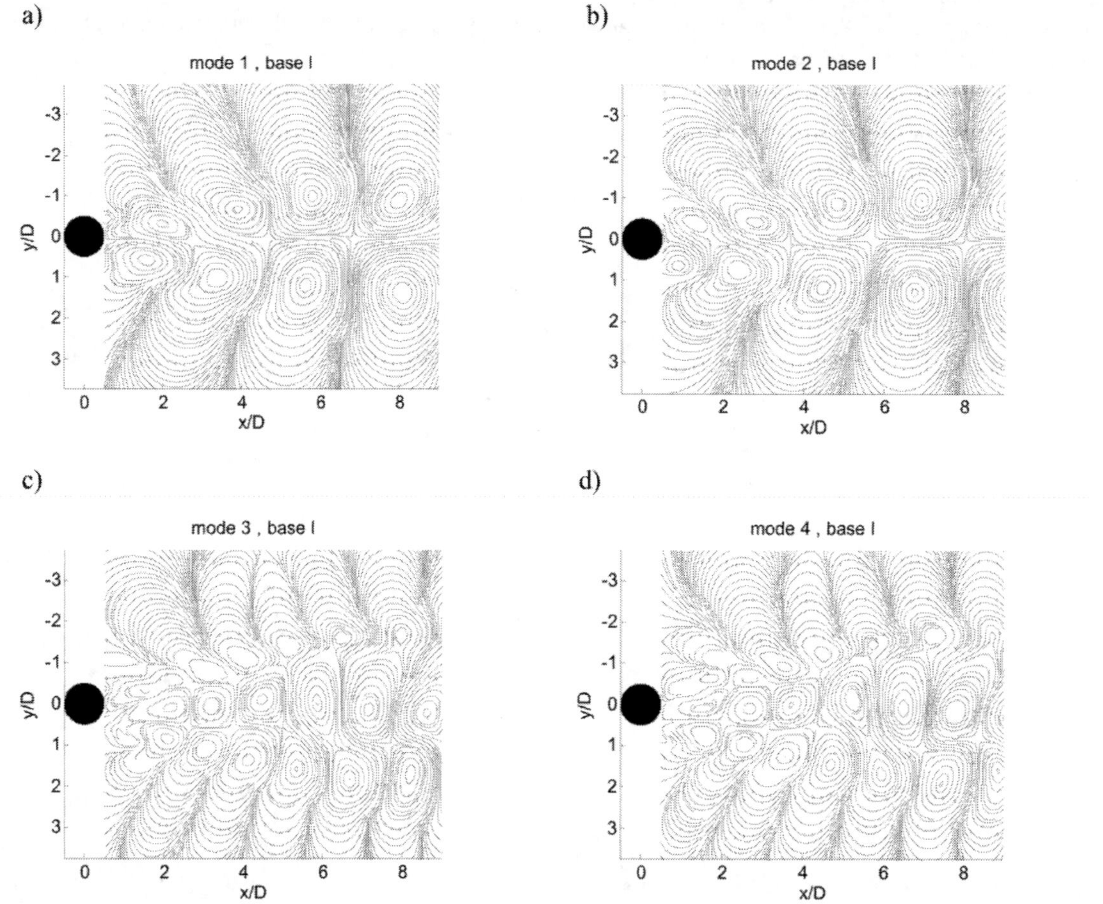

Figure 12. First fluctuating modes of the first extending basis, Re=125: a) Mode 1, b) Mode 2, c) Mode 3, d) Mode 4.

As we can see, the modes of the extending basis enable to break the symmetry of the uncontrolled basis, allowing for the representation of tilted vortex shedding, characteristic of spinning cylinder flows.

The temporal coefficients of the extending basis can also be grouped by pairs and present a quasi sinusoidal shape, both modes of the pair exhibiting the same frequency but with a difference in phase.

In figure 14 we plot, the fluctuating energy of the flow captured by the fluctuating modes of the basis, and the residual value that is associated to the fraction that can not be represented using the eigenvectors of the basis $\Psi_0+\Psi_1+\Psi_2$.

When a threshold of 95% of fluctuating kinetic energy of the flow is adopted as a criterion to truncate the set of number of modes to be retained, each extending basis adds 4 extra modes to the original basis. With this criterion, the whole basis (uncontrolled + extending basis) is composed of 12 modes.

Journal of Physics: Conference Series **166** (2009) 012016 doi:10.1088/1742-6596/166/1/012016

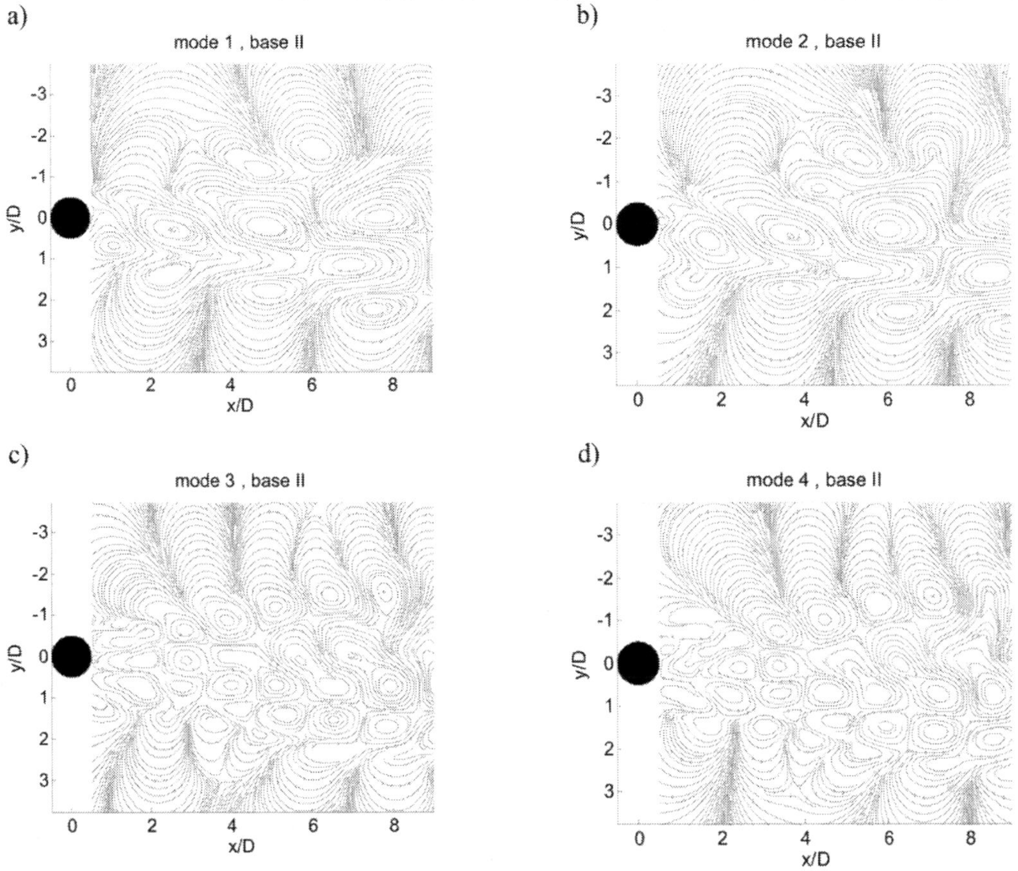

Figure 13. First fluctuating modes of the second extending basis, *Re*=125: a) Mode 1, b) Mode 2, c) Mode 3, d) Mode 4.

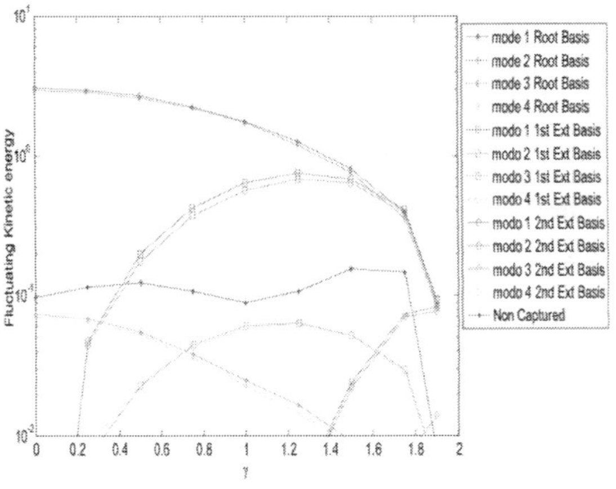

Figure 14. Fraction of the energy of the flow associated to the fluctuating modes, *Re*=125.

3.4. Weighing function for the control function

The mean flow of any state may be obtained by suitably adjusting the weighing coefficients Γ. In figure 15 we represent the functions Γ_1 and Γ_2 of equation (10) versus the parameter of rotation γ. The value is obtained considering the mean of each state and the correction introduced by the flow. The parameter Γ_1 can be approximated as a linear function of γ

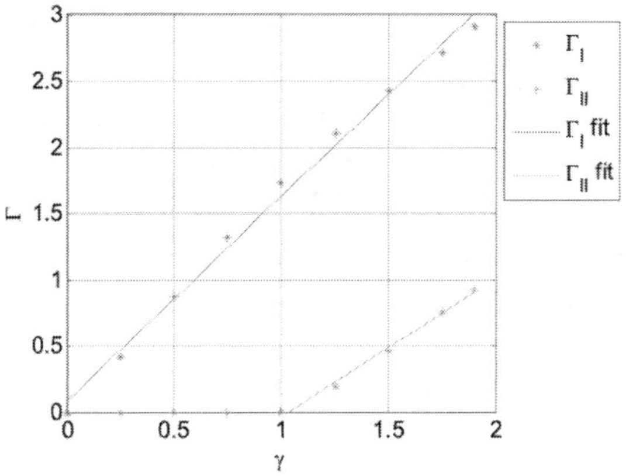

Figure 15. Coefficients Γ vs γ, *Re*=125.

The coefficient Γ_2 also exhibits an almost linear dependency with the rotation parameter, in the range where it is defined ($1.0<\gamma<2.0$)

$$\Gamma_i = m_i\ \gamma + n_i$$

Fitting coefficients for each range of validity are m_1=1.538, n_1=0.08787, m_2=1.048 and n_2=-1.084.

It is also clear that the influence of the considered reference state to define the first extending basis is not as crucial as it results in the range of rotation parameter where vortex shedding occurs; the mean velocity lies in a subspace of dimension 2 or 3.

The curve of figure 15 also enables to infer that the kinetic energy accumulated in these correcting modes (control function modes) depends quadratically on γ.

The fraction of the total energy accumulated on other modes (fluctuating modes) decreases with the increase of the control parameter until it disappears when vortex shedding ceases.

3.5. Fidelity of the model representation

An evaluation of the quality of velocity field reconstruction can in principle be estimated by the residual energy that is not captured by the truncated model. This kind of measure may hide the ability of the model to represent an arbitrary set of snapshots. It just quantifies point to point the difference between the modulus of the velocity field and the reconstructed one. Important differences of velocity field vectors associated to errors in the velocity vectors directions can not be captured by this measure.

To overcome this difficulty in [24] different alternatives have been suggested, but in our case it seems adequate to consider the Barron's angular error in the velocity field defined as [26]

$$\frac{1}{\Omega}\sum_{n}\arccos\left(\frac{\left(1+\mathbf{u}_{\mathbf{rec}}\mathbf{u}_{\mathbf{true}}\right)}{\sqrt{1+\left\|\mathbf{u}_{\mathbf{rec}}\right\|^{2}}\sqrt{1+\left\|\mathbf{u}_{\mathbf{true}}\right\|^{2}}}\right)$$

where $\mathbf{u}_{rec}$ is the reconstructed field, $\mathbf{u}_{true}$ is the field obtained by DNS simulation and Ω agrees with the number of nodes of the discretization domain.

In figure 16 we show typical pictures of instantaneous angular distance fields for two different values of the control parameter.

a) b)

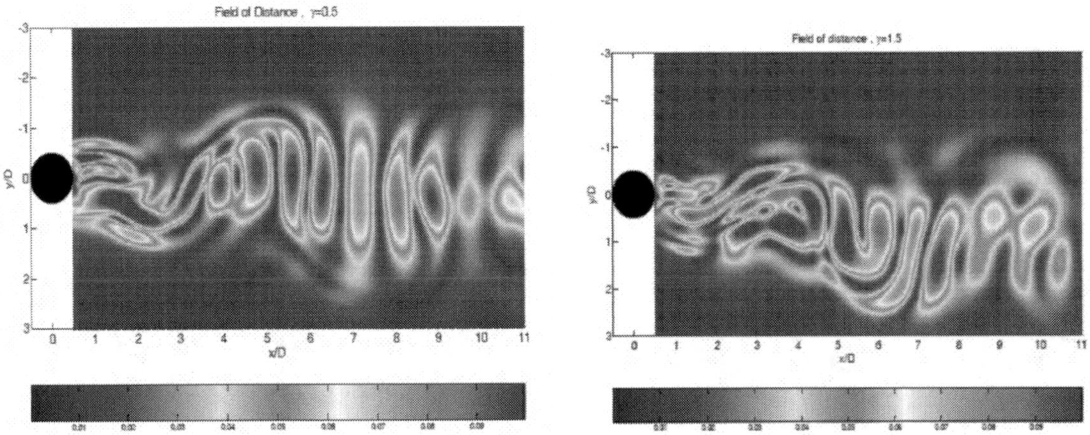

Figure 16. Instantaneous field of Barron's angular error for *Re*=125 and two control parameter values: a) γ=0.5, b) γ=1.5. Errors are in radians.

As one could expect, the largest differences are observed in the region where vortex shedding occurs. In the regions where the dynamics is not so strong, it does not appear to be any significant difference.

The Barron´s angular error, however, results quite small and the deviation angle in the worst case reaches about 5°. In figure 17 we show the evolution of this instantaneous measure with time. We can see that this distance always remains quite small and a fluctuation of the same frequency of the vortex shedding phenomena occurs for all cases.

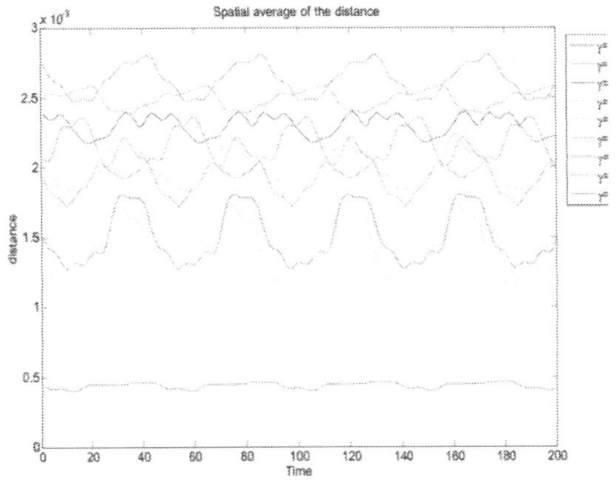

Figure 17. Temporal evolution of the distance at different values of γ for *Re*=125. Angles are in radians.

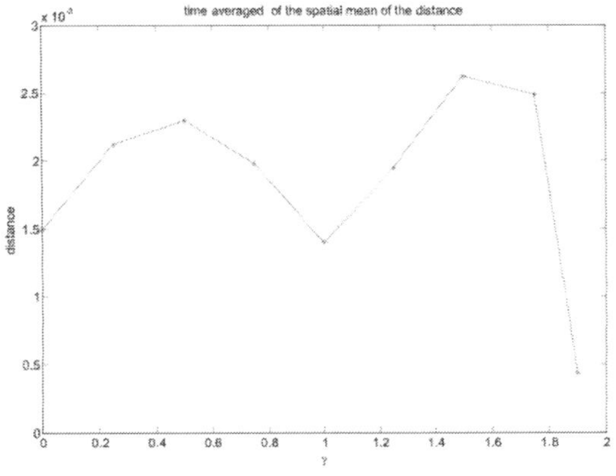

Figure 18. Time averaged distance at different values of γ for *Re*=125. Angles are in radians.

In figure 18 the time averaged value of this measure is presented. We can observe that the model slightly degrades its performance for the values of the control parameter that are far from the reference values considered to define the root and extending bases.

4. Conclusions

In this article we have analysed several tools regarding the use of empirical models that are bound to simplify the formulation of optimal control strategies for actuations produced on the boundary of external flows.

We have initially concentrated our analysis on the suitable inclusion of functions in the surrogate model that accounts for the effect on the velocity fields of a steady rotation of the cylinder around its axis. This function was educed with a Grahm Schmidt construction that gives rise to an extended basis in which the velocity field is expanded.

The control functions enable to overcome some limitations associated to the constraints of real experimental conditions data, especially in those cases where good accuracy of the velocity fields near the surface is required.

Another result of this procedure is that for a spinning cylinder at a laminar vortex shedding regime it is possible to satisfactorily reconstruct the mean velocity fields with only three vectors provided the Reynolds number is kept constant.

The obtained POD ROM is flexible enough and coefficients are calculated once and for all at the beginning of the process. This is of special interest for optimization processes where the coefficients of POD ROM have to be updated, particularly when one considers values of the control parameter that are far from the reference value.

Extension to cases where both control parameters, the Reynolds number and rotation, may vary independently or simultaneously are now under study.

5. References

[1] Moreau E 2007 Airflow control by non-thermal plasma actuators *J. Phys.* D **40** (3) 605–36
[2] Mittal S and Kumar B 2003 Flow Past a Rotating Cylinder *J. Fluid Mech.* **476** 303–34
[3] Kang S, Choi H and Lee S 1999 Laminar Flow past a rotating circular cylinder *Physics of Fluids* **11** (11) 3312–21
[4] Stojkovic D, Breuer M and Durst F 2003 On the new vortex shedding mode past a rotating circular cylinder *Physics of Fluids* **15** (5) 1257–61
[5] Protas B 2008 Vortex Dynamics Models in Flow Control Problems *Nonlinearity* **21** No 9 R203–R250
[6] Holmes P, Lumley J and Berkooz G 1996 *Turbulence, coherent structures, dynamical systems and symmetries* (Cambridge: Cambridge University Press)
[7] Graham W, Peraire J and Tang K 1999 Optimal control of vortex shedding using low order models Part I : Open loop model development *Intnl J. Numer. Meth. Engng.* **44** 945–72
[8] Fahl M Trust region methods for flow control based on reduced order modeling 2000 *PhD Thesis* University of Trier
[9] Bergmann M and Cordier L 2008 Optimal Control of the circular cylinder wake by trust region methods and POD Reduced order models *J. Comp. Phys.* **227** 7813–40
[10] Noack B R, Afanasiev K, Morzydski M, Tadmor G and Thiele F 2003 A hierarchy of low-dimensional models for the transient and post-transient cylinder wake *J. Fluid Mech.* **497** 335
[11] Tadmor G, Gonzalez J, Lehmann O, Noack B, Morzynski M and Stankiewicz W 2007 Shift modes and transient dynamics in low order, design oriented Galerkin Models *AIAA paper* **0111** (45th *AIAA Meeting*, Reno, 2007)
[12] Morzynski M, Stankiewicz W, Noack B, King R, Thiele F and Tadmor G 2007 Continuous mode interpolation for control-oriented models of fluid flows *Active Flow Control* ed. R King (Berlin: Springer Verlag) pp 260–78

[13] Christensen E, Sorensen J and Brons M 1993 Low dimensional representation of early transition in rotating fluid flow *Theoret. Comput. Fluid Dynamics* **5** 259–67
[14] Burkardt J, Gunzburger M and Lee H C 2006 Centroidal Voronoi Tessellation-Based Reduced-Order Modeling Of Complex Systems *SIAM J. Sci. Comput.* **28** No 2 459–84
[15] Akhtar I and Nayfeh A 2008 On controlling the bluff body wake using a reduced order model *AIAA paper* **4189** (AIAA 4th Flow Control Conference, 23–28, Seattle, June 2008)
[16] D'Adamo J 2007 Modelos Reducidos Para el Control de Flujos con Actuadores EHD *PhD Thesis* University of Buenos Aires
[17] Gronskis A, D'Adamo J, Artana G, Cammilleri A and Silvestrini J 2008 *J. Electrost.* **66** 1–7
[18] Williamson C H K 1996 Vortex dynamics in the cylinder wake *Annu. Rev. Fluid. Mech.* **28** 477–539
[19] Zdravkovic M 1994 *Flow Around Circular Cylinders* (Oxford: Oxford University Press vol 2)
[20] Lamballais E and Silvestrini J 2002 Direct numerical simulation of interactions between a mixing layer and a wake around a cylinder *J. Turbul.* **3** 028
[21] Silvestrini J H and Lamballais E 2002 Direct numerical simulation of wakes with virtual cylinders *Int. J. Comp. Fluid Dynam.* **16** (4) 305-14
[22] Henderson R 1995 Details of the drag curve near the onset of vortex shedding *Physics of Fluids* **7** 2102
[23] Barkley D 2006 Linear analysis of the cylinder wake mean flow *Europhys. Lett.* **75** (5) 750–6
[24] Christensen E, Brons M and Sorensen J 2000 Evaluation of proper orthogonal decomposition-based decomposition techniques applied to parameter-dependent non turbulent flows *SIAM J. Sci. Comput.* **21** (4) 1419–34
[25] Jorgensen B, Sorensen J and Brons M 2003 Low dimensional modeling of a driven cavity flow with two free parameters *Theoret. Comput. Fluid Dynamics* **16** 293–317
[26] Barron J, Fleet D and Beauchemin S 1994 Performance of optical flow Techniques *Int. J. Computer Vision* **12** (1) 43–77
[27] Deane A, Kevrekidis I, Karniadakis G and Orszaga S 1991 Low-dimensional models for complex geometry flows: Application to grooved channels and circular cylinders *Physics of Fluids* A **3** (10) 2337–54
[28] Siegel S, Seidel J, Fagley C, Luchtenburg D, Cohen K and McLaughlin T 2008 Low dimensional modeling of a transient cylinder wake using double proper orthogonal decomposition *J. Fluid Mech.* **610** 1–42
[29] Bergmann M, Cordier L, Brancher J P 2005 Optimal rotary control of the cylinder wake using proper orthogonal decomposition reduced order model *Physics of Fluids* **17**
[30] Thiria B and Wesfreid J E 2007 Stability properties of forced wakes *J. Fluid Mech.* **579** 137–61

Acknowledgements
This work has been supported by ANPCYT grant PICT 38070 and UBACYT grant number I017.

Transition to chaos of a vertical collapsible tube conveying air flow

F Castillo Flores and A Cros

Departamento de Física, Universidad de Guadalajara, 44430 Jalisco, Mexico.

E-mail: anne_cros@yahoo.com

Abstract. "Sky dancers", the large collapsible tubes used as advertising, are studied in this work through a simple experimental device. Our study is devoted to the nonlinear dynamics of this system and to its transition to chaos. Firstly, we have shown that after a collapse occurs, the air fills the tube at a different speed rate from the flow velocity. Secondly, the temporal intermittency is studied as the flow rate is increased. A statistical analysis shows that the chaotic times maintain roughly the same value by increasing air speed. On the other hand, laminar times become shorter, until the system reaches a completely chaotic state.

1. Introduction

Compliant-tube flows are the subject of various recent studies due to their applications in oil industry (see the pioneering works [1, 2]) and in different thermic systems [3] as well as in biological systems [4]. Their interest also comes from their great dynamical behaviour diversity, the nature and the complexity of their instabilities [5, 6]. For a free-ended tube, different circumferential modes are exhibited, as described in [7, 8]. These modes are characterized by the spatial deformations of the membrane. In these systems, a strong coupling exists between the flow inside the tube and the elastic pipe. Indeed, fluid pressure modifies, and is modified by, the cross-sectional area.

Different configurations and boundary conditions of the collapsible tubes conveying flow are largely described by Païdoussis [9]. A clamped-clamped system can lead to a static instability called divergence [10]. In comparison, a free-ended tube configuration shows a fluttering instability above a certain flow speed threshold, described by a Hopf bifurcation [9].

This last instability is seen daily when one uses a water-conveying hose to irrigate plants: for a high enough flow rate, the tube flutters. This "garden-hose instability", as it is called, was experimentally and theoretically studied in [11, 12, 13].

A similar system is the "sky dancer" used as advertising. It is composed of a long cylindrical membrane fluctuating vertically above an air blower. The two main differences against the studies in [11, 12, 13] are as follows: (a) the sky dancer tube wall is much thinner; this lets the pipe shape show "breaks" and folds when it is fluttering, (b) the air flow is turbulent, because of both the tube diameter size and the high flow speed.

The purpose of this work is to describe the fluctuation dynamics of a vertical tube conveying air flow, and to characterize transition to chaos when the flow rate is increased. This last analysis is made statistically. The experimental system is described in section 2, and a sequence of pipe fluctuation images is shown in section 3. A statistical study of the chaotic and laminar times is developed in section 4, followed in section 5 by a discussion. The last section is for conclusion and perspectives.

2. Experimental system

2.1. Experimental device

The tube is made of a 2.5 Mpa Young Modulus white plastic membrane, 0.1 mm in thickness. Its surface density is 0.083 kg/m^2 and its length $L = 39$ cm. The tube is heat-sealed in order to form a 4-cm diameter cylinder and is mounted on the external diameter of the air pump exit. We calculated the parameter $\beta = m/(m + M)$, used in these kind of systems, where m is the mass per unit of length of air in the pipe, and M the mass per unit of length of the pipe. In our case $M = 0.01$ kg/m, so that $\beta = 0.13$.

Figure 1 shows the experimental device. In order to capture the tridimensional instability of the pipe, the video camera is located above the system; its optical axis is centered on the pump exit. The video frequency is equal to 29.97 Hz and the frame size is 720 x 480 pixels. In order to observe sharp images, a stroboscope is placed near the video camera and pulses at the same frequency as the video camera.

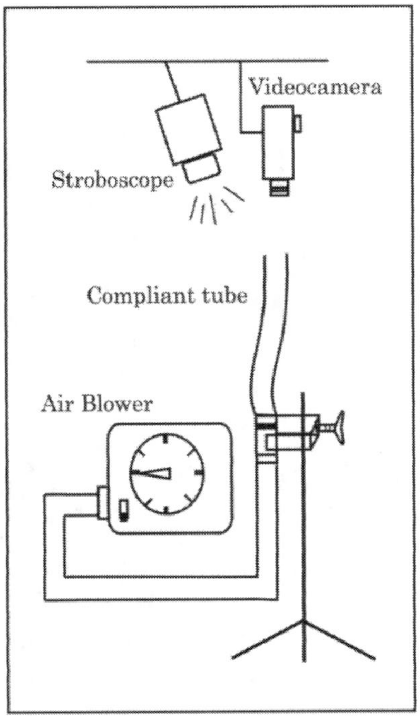

Figure 1. Experimental set-up.

The air pump exit has the same external diameter as the tube. The air speed varies from 18 m/s to more than 40 m/s, corresponding to a flow rate between 0.0226 m^3/s and more than 0.0502 m^3/s. Air speed is stable at 3%. The Reynolds number is found to be more than 51000, so that the air flow at the pump exit is turbulent.

2.2. Experimental procedure

Once the tube is mounted on the pump exit, the air blower is switched on so the air speed is increased until the tube stands up on a stable state. We wait a few seconds to avoid transient effects. After that, we capture a film during two minutes at least. We repeat the same procedure for higher flow velocities, until a completely chaotic regime is reached.

We extract all the images from the videos; this means 3603 images for each flow velocity. The left side of figure 2 shows two of these images. The picture on top shows the tube when it stands up; the pipe looks like a white circle on a gray background. The bottom picture represents the collapsed tube; it appears like a long white rectangle on a dark background.

The right column of figure 2 shows each left image binarized. All the pixels whose values are above a threshold are set to 255 (white), while the pixels whose values are below the same threshold become 0 (black). The cut-off value is constant for all the images. This value is chosen so that the tube boundaries in the binarized images match the boundaries in the original pictures. A threshold equal to 55 was set up for this experiment.

In order to analyze the transition to chaos, the system evolution is reduced to a two states time series: the stable state is when the tube stands up and the chaotic state when the tube fluctuates. In order to discriminate these two states, we count the white pixels contained inside a circle centered on the blower exit. The circle radius is chosen so that the steady state corresponds to an image where all the white pixels are included inside this circle. In our experiment, a suitable value is $r = 4.7$ cm. We calculate for each image the proportion p of the white pixels inside the circle: for instance, $p = 1$ for the figures on top (stable state), and p=0.294 ($p < 1$) for the bottom pictures (chaotic state). The evolution of p as a function of time will be shown in section 4.

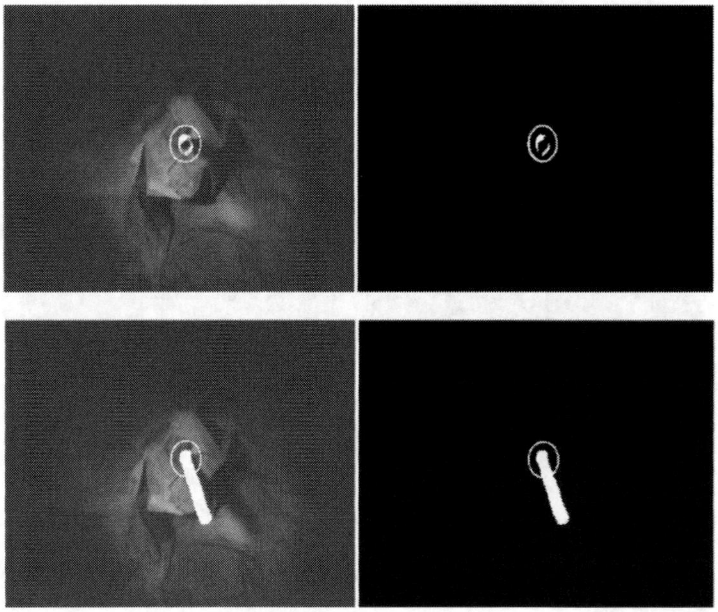

Figure 2. Raw pictures (left side) and binarized pictures (right side) for both tube states: stable state (top) and chaotic state (bottom). The proportion p of white pixels in the circle was found to be equal to $p = 1$ and $p = 0.294$ respectively.

3. Description of the fluttering behaviour

Figure 3 illustrates the tube dynamics at the threshold of the instability. In this section, the images are shown from a side view. Two consecutive pictures are separated by the time interval $\Delta t = 33$ ms.

In the first picture, a constriction appears at the base of the tube. This constriction moves downstream with time, as shown on the second picture. The third picture shows that the tube section located above the constriction point falls down. Below it, the tube is air-filled, while it is empty on top. This break point rises up, as shown in the following images, and a straight line can be drawn indicating the fold position: the constriction speed is roughly constant. Finally, when the fold reaches the top of the tube, it stands up again in a stable state.

It is important to note that the constriction velocity is different from the air speed. In this case, the air speed was 18.8 m/s, while the constriction point travelled at a speed of 1.5 m/s. In section 5 we will see that this velocity is related to the dynamical pressure needed to push the break point upwards.

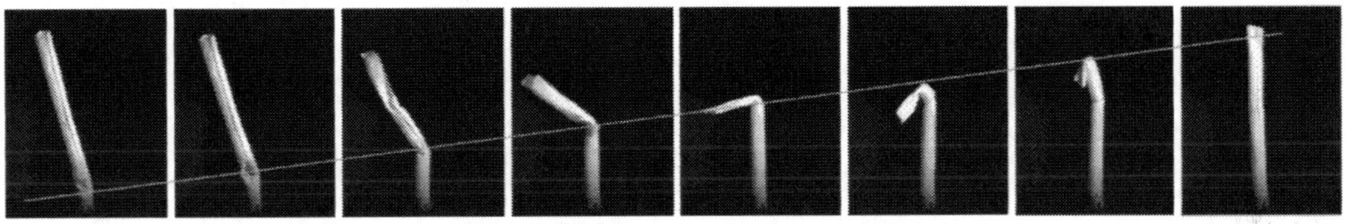

Figure 3. Time evolution of the fluctuation of 35 cm-length tube at a $v = 18.8$ m/s ($C = 0.0236$ m^3/s). Each picture is separated by the video time $\Delta t = 33$ ms, and the straight line shows how the constriction point travels downstream with a constant velocity (different from the air speed).

4. Chaos transition characterization

As explained in the section 2.2, the pictures are processed in order to extract the proportion p of white pixels inside a centered circle. Figure 4 shows the evolution of p over time for four air speed values: $v = 18.8$, 19.9, 23.3 and 25 m/s (corresponding to flow rates $C = 0.0236$, 0.0250, 0.0293 and 0.0314 m^3/s respectively). For $v = 18.8$ m/s, the tube raises up, but it is important to note that it is sensitive to perturbations. For example, approaching an object to the exit of the tube makes it bend and fluctuate. After this, the tube may or may not come back to its equilibrium position. The stable state shown in the first curve of figure 4 was sustained during more than 4 minutes without perturbation.

When the velocity is increased ($v = 19.9$ m/s, $C = 0.0250$ m^3/s), chaotic fluctuations are generated which alternate with the stable state. The binarization of the function p was made so that the corresponding points were set to zero for $p < 0.96$, while the points are set to one for $p > 0.96$. This threshold allows us to discard few white pixels which appear outside the circle because of light inhomogeneities. Moreover, if at most four isolated pixels with values one (or zero) appear in a region of zeros (or ones respectively), these isolated values are set to zeros (or ones). This is particularly convenient to delimit properly the highly fluctuating chaotic regions.

For higher flow rates, the pipe fluctuations increase. For the air speed $v = 26$ m/s ($C = 0.0327$ m^3/s), only three very short stable regions appear in the temporal evolution of p during two minutes, so that the mean value $\bar{p}$ of the binarized function is nearly zero for this value. Figure 5 shows the evolution of the function $\bar{p}$ as a function of air speed.

The mean value $\bar{p}$ can be seen as the proportion of laminar times with respect to the whole measure duration. As the system is stable for $v = 18.8$ m/s ($C = 0.0236$ m^3/s), $\bar{p} = 1$. When the air flow is increased, $\bar{p}$ decreases. For $v = 22.5$ m/s ($C = 0.0283$ m^3/s), $\bar{p} = 0.503$: it can be considered that half of time, the tube fluctuates while the other half, the tube stands up. The function $\bar{p}$ reaches the value 0.002 for $v = 26$ m/s ($C = 0.0327$ m^3/s): at this flow rate, the system appears in a permanent chaotic state.

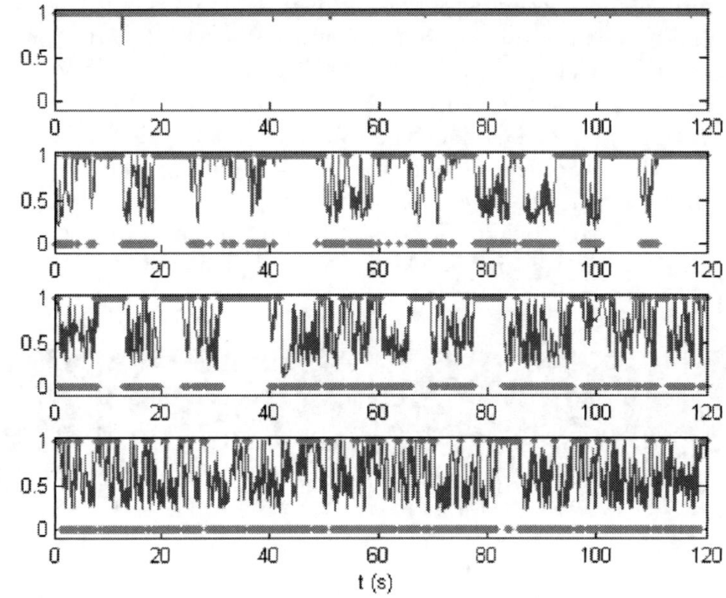

Figure 4. Time evolution of white pixel proportion p inside the circle drawn in figure 2. When $p = 1$, the tube stands up, otherwise it is fluctuating. From top to bottom, flow velocity is: $v = 18.8$, 19.9, 23.3 and 25 m/s (corresponding to flow rates $C = 0.0236$, 0.0250, 0.0293 and 0.0314 m^3/s respectively). The oscillating curves correspond to the temporal variation of p, while the rough points correspond to the binarized values of p. Time is in seconds.

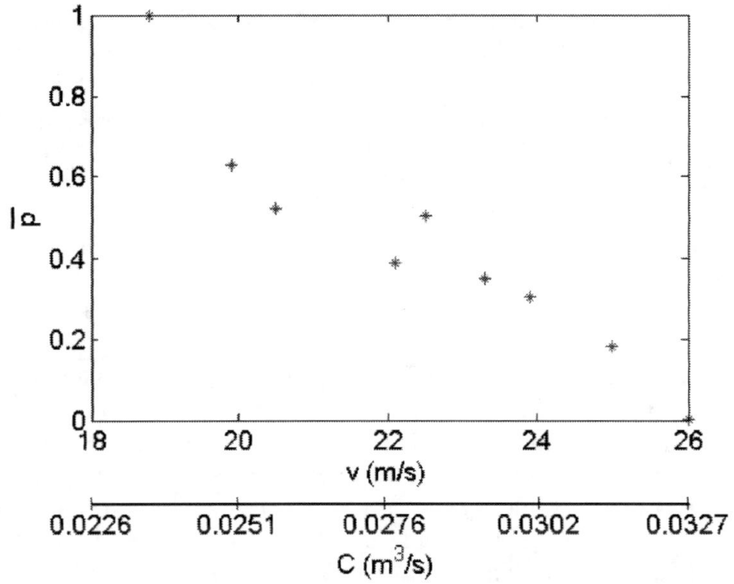

Figure 5. Evolution of the mean value $\bar{p}$ of the binarized value of the proportion p, as a function of the air speed v (m/s) and flow rate C (m^3/s). When $v = 18.8$ m/s ($C = 0.0236$ m^3/s), the tube stands up ($\bar{p} = 1$), but as the flow velocity is increased, the tube fluctuations increase and the function $\bar{p}$ decreases until it reaches $\bar{p} = 0.002$ for $v = 26$ m/s ($C = 0.0327$ m^3/s).

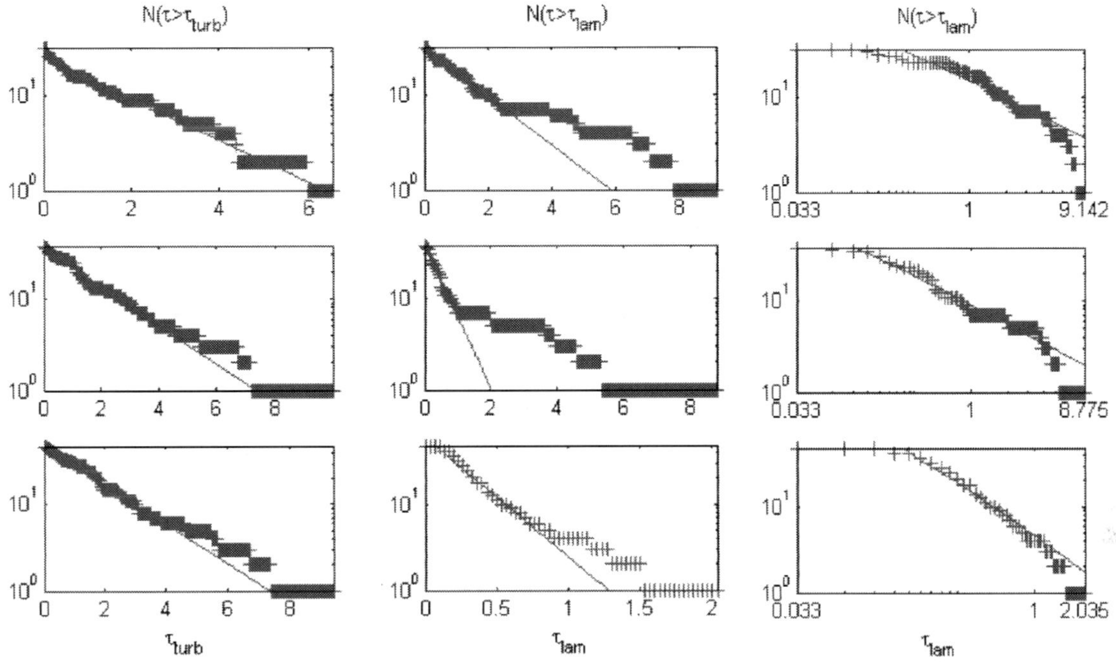

Figure 6. Cumulated histograms of: chaotic states (left column) and laminar state (centre and right column). Left and centre columns: semi-log graphics, right column: log-log graphic. For a better clarity, the y-labels are on top of each column. The different flow speed values are, from top to bottom: v = 19.9, 23.3 and 25 m/s (C = 0.0250, 0.0293 and 0.0314 m^3/s respectively). The chaotic times are best interpolated by an exponential law, while laminar time histograms appear as a product of an exponential with an algebraic law. The first points are best interpolated by an exponential law (centre column) while the intermediate points by an algebraic equation (right column).

In order to characterize this transition to chaos, we analyze statistically the laminar and chaotic times as usually done in temporal dynamical [14] and spatiotemporal [15, 16, 17] systems. We plotted in figure 6 the cumulated histograms of the chaotic times (left column) and those of the laminar durations (centre and right column). The chaotic time distributions are shown in semi logarithmic scale and can be interpolated as a straight line. So, the fit by a function of the type $N(\tau > \tau_{turb}) = N_0 \exp(\tau/T_{c-turb})$ gives a value for a characteristic chaotic time T_{c-turb} whose value can be calculated for each air speed.

The laminar time distributions are shown in the centre column in a semi logarithmic scale and in the right column in a log-log scale. The exponential law (centre column) fits better the first distribution points (short times), while the algebraic law (right column) interpolates better the intermediate values. In the case of the laminar regions, we calculated a characteristic time T_{c-lam} as for the chaotic times, as well as the exponent α of the algebraic law $N(\tau > \tau_{lam}) = N_0\tau^{\alpha}$. The evolutions of these quantities are shown in figure 7.

It can be observed on the first plot (a) that the chaotic characteristic times do not have a monotonous evolution. While T_{c-turb} = 1.96 s for v = 19.9 m/s (C = 0.0250 m^3/s), it reaches a maximum T_{c-turb} = 2.41 s for 20.5 m/s (C = 0.0258 m^3/s) and a minimum T_{c-turb} = 1.43 s for v = 22.5 m/s (C = 0.0283 m^3/s). We conclude that the fluctuation durations T_{c-turb} of the tube do not increase with time but are varying around a constant value (calculated as 1.94 s).

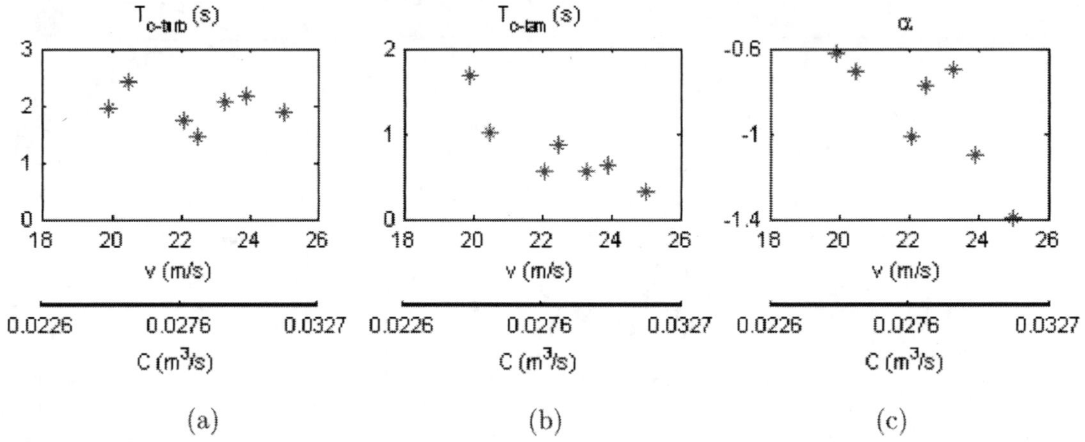

Figure 7. (a) Evolution of the characteristic time T_{c-turb} deduced from the chaotic time histogram as a function of the air speed. (b) Evolution of the characteristic time T_{c-lam} of the stable state as a function of the air speed. (c) Evolution of the laminar exponent α as a function of the air speed. (See text for comments). For a better clarity, the y-labels are on top of each graph.

The quantities deduced from the laminar time distributions present a more evident evolution with respect to v. Although fluctuations are present particularly in the interval [22, 24] m/s ([0.0276 0.302] m^3/s), the characteristic laminar time T_{c-lam} (figure 7-b) decreases from T_{c-turb} = 1.68 s for v = 19.9 m/s (C = 0.0250 m^3/s) to 0.31 s for v = 25 m/s (C = 0.0314 m^3/s). For the same air speed values the laminar exponent decreases from -0.63 to -1.40 (figure 7-c). Although we could not characterize clearly the distribution type for the laminar times, we put in evidence that these laminar regions are shorter and shorter. The system transits towards chaos via a reduction of laminar times and not by an increase of fluctuation regions.

5. Discussion

Various authors [3, 18, 19] observed that the flutter instability was sensitive to finite disturbances and had a different threshold when the flow speed was decreased. They showed that an undisturbed system present the hysteresis, which suggests that in some cases, the Hopf bifurcation is subcritical. In our experiment, the system never comes back at a rest state by decreasing the flow velocity. Once the flow rate is low enough, the gravity makes the tube fall down and the tube does not reach anymore a stable state. The existence of a subcritical bifurcation could explain our difficulty to analyze the laminar times distribution. Moreover, it was observed that some folds could form and remain near the pipe exit after a fluctuation, which influences the dynamics (generally provoking a stabilizing effect). Plastic deformations were also observed in [10].

During the last twenty years, studies about the nonlinear behaviour of collapsible tubes conveying flows have been performed. Rousselet and Herrmann [20] drew the limit cycle amplitude of the free end of a cantilevered pipe as a function of the parameter β (given in the section 2.1). For our value $\beta = 0.13$, they found a limit cycle with an amplitude $A/D \simeq 0.15$, which is very small compared to our experimental results. Modarres-Sadeghi, Païdoussis and Semler [21] found that for this value of β, the system should flutter in a plane. Nevertheless, we repeat that our experiment differs from all the works cited in the references.

As already mentioned in the introduction, the usual differential equation cannot be applied to this system because when the tube folds, its shape presents breaks so that the spatial derivatives are not continuous. The studies performed in [11, 12, 13] are realized with tubes whose wall thickness is more than 0.79 mm, which is almost an order of magnitude higher than our experimental value. Moreover, the small tube diameters (12.70 mm at most) in these works guarantees that the flow at the tube entry is laminar. It is not the case in our experiment.

We think that as the tube wall is very thin, we should not consider it as an elastic structure in the physical sense. We rather consider the tube wall as a thin weighty impermeable boundary; that makes the tube bend. The mechanism that may explain the stand-up wave velocity value is as follows.

To a first approximation, the constriction point shuts off the tube during a fluctuation. So, the air flow below the constriction may be brought largely to rest, whereas the air flow through the blower continues unabated. This causes the pressure to rise in the tube section between the blower and the constriction point, until this pressure is able to push the constriction point upwards. That is, the dynamical pressure $\frac{1}{2}\rho v^2$ should be greater than the pressure exerted by the weight of the membrane piece that shuts off the air flow. As an order of magnitude, we calculated the weight of a membrane disk-shaped piece, which has the same radius as the tube, that would block the air exit. The data in section 2.1 permit to find that the corresponding pressure exerted by the plastic piece on the bottom air cylinder is 0.8 Pa. In order to push this plastic disk, the velocity corresponding to the dynamical pressure must reach the value of 1.2 m/s, which is slightly lower than our experimental value.

This mechanism permits to find a stand-up wave velocity which is of the same order of magnitude than in our experiment. Nevertheless, this first approximation makes us underestimate the stand-up wave velocity. By considering the entire folded section weight over the constriction point, we should find indeed a greater dynamical pressure able to push the constriction point upwards. This comment is left for further investigations.

6. Conclusion and perspectives

This work belongs to an effervescent area which is the nonlinear dynamics of collapsible tubes conveying fluid. It differs in several points from previous studies; but it permits to study chaos transition through a simple experimental system; this is a reduced model of a "sky dancer". We showed that the wave that makes the tube stand up after a collapse has a velocity which is related to the growing pressure in the closed, inflating tube section, located between the blower end and the constriction point. It was also shown that the fluctuation times were roughly constant as the air flow speed is increased. On the contrary, the decrease of laminar times constitutes the essential ingredient in the transition to chaos. In a future work we may study the same system in a different configuration, in order to analyze the dynamical pressure role in the stand-up wave. Moreover, we will describe the tube trajectory in a phase space as done in recent studies [22, 23, 24, 25], in order to perform the analysis of transition to chaos from a different point of view.

Acknowledgments

We are grateful to P. Le Gal for suggesting the idea of the experiment and for helpful comments. We especially thank C. Bertram for his fruitful analysis concerning the stand-up wave mechanism.

References

[1] Bourrières F J 1939 Sur un phénomène d´oscillation auto-entretenue en mécanique des fluides réels *Publications Scientifiques et Techniques du Ministères de l´Air* 147.

[2] Ashley H and Haviland G 1950 Bending vibrations of a pipe containing flowing fluid *J. Appl. Mech.* **17** 229-232

[3] Karagiozis K N, Païdoussis M P, Misra A K and Grinevich E 2005 An experimental study of the nonlinear dynamics of cylindrical shells with clamped ends subjected to axial flow *J. Fluids Struct.* **20** 801-816

[4] Marzo A, Luo X Y and Bertram C D Three-dimensional collapse and steady flow in thick-walled flexible tubes *J. Fluids Struct.* **20** 817-835

[5] Bertram C D 1986 Unstable equilibrium behaviour in collapsible tubes *J. Biomech.* **19** 61-69

[6] Jensen O E 1990 Instabilities of flow in a collapsed tube *J. Fluid Mech.* **220** 623-659

[7] Païdoussis M P 2004 *Fluid-Structure Interactions, Slender Structures and Axial Flow* vol. 2. (London: Academic Press)

[8] Païdoussis M P and Denise J P 1972 Flutter of thin cylindrical shells conveying fluid *J. Sound & Vibration* **20** 9-26

[9] Païdoussis M P 1998 *Fluid-Structure Interactions, Slender Structures and Axial Flow* vol. 1. (London: Academic Press)

[10] Dodds H L Jr and Runyan H L 1965 Effect of high velocity fluid flow on the bending vibrations and static divergence of a simply supported pipe *NASA Technical Note* D-2870

[11] Gregory R W and Païdoussis M P 1966 Unstable oscillation of tubular cantilevers conveying fluids. II. *Experiments Proc. Royal Soc. Series A* **293** 528-542

[12] Greenwald A S and Dugundji J 1967 Static and dynamic instabilities of a propellant line *MIT Aeroelastic and Structures Research Lab, AFOSR Sci. Report: AFOSR* 67-1395

[13] Doaré O and de Langre E 2002 The flow-induced instability of long hanging pipes *Eur. J. Mech. Solid* **21** 857-867

[14] Bergé P, Pomeau Y and Vidal C 1984 *L´ordre dans le chaos* (Hermann)

[15] Chaté H and P Manneville 1988 Spatiotemporal intermittency in coupled map lattices *Physica D* **32** 409

[16] Ciliberto S and P Bigazzi 1988 Spatiotemporal intermittency in Rayleigh-Bénard convection *Phys. Rev. Lett.* **60** 286

[17] Cros A and Le Gal P 2002 Spatiotemporal intermittency in the torsional Couette flow between a rotating and a stationary disk *Phys. Fluids* **14** 3755-3765

[18] Chen S S and Jendrzejczyk J A 1985 General characteristics, transition, and control of instability of tubes conveying fluid. *J. Appl. Mech.* **43** 325-329

[19] Bajaj A K, Sethna P R and Lundgren T S 1980 Hopf bifurcation phenomena in tubes carrying fluid *SIAM J. Appl. Math.* **39** 213-230

[20] Rousselet J and Herrmann G 1981 Dynamics behaviour of continuous cantilevered pipes conveying fluid near critical velocities. *J. Appl. Mech.* **48** 943-947

[21] Modarres-Sadeghi Y, Païdoussis M P and Semler C 2008 Nonlinear dynamics of extensible fluid-conveying pipes, supported at both ends *Int. J. Non Lin. Mech.* **43** 18-25

[22] Bertram C D, Raymond C J and Pedley T J 1991 Applications of nonlinear dynamics concepts to the analysis of self-excited oscillations of collapsible tube conveying a fluid *J. Fluid Struct.* **5** 391-426

[23] Païdoussis M P and Moon F C 1988 Nonlinear and chaotic fluidelastic vibrations of a flexible pipe conveying fluid *J. Fluids Struct.* **2** 567-591

[24] Païdoussis M P, Li G X and Moon F C 1989 Chaotic oscillation of the autonomous system of a constrained pipe conveying fluid *J. Sound Vib.* **135** 1-19

[25] Païdoussis and Semler C 1993 Nonlinear and chaotic oscillations of a constrained cantilevered pipe conveying fluid: a full nonlinear analysis *Nonlinear Dyn.* **4** 655-670

Journal of Physics: Conference Series **166** (2009) 012018 doi:10.1088/1742-6596/166/1/012018

Further applications of scheme for reducing numerical viscosity: 3D hypersonic flow

S A Elaskar[1-2], O A Falcinelli[1], J P Tamagno[1] and J P Saldía[1]
[1]Dpto. de Aeronáutica. Universidad Nacional de Córdoba. [2]CONICET
Av. Velez Sarfield 1611. Córdoba (5000), Argentina.

E-mail: selaskar@efn.uncor.edu

Abstract. A Total Variation Diminishing (TVD) scheme implemented in a non structured finite volume formulation for solving the 3D Euler equations is presented. To simultaneously, achieve adequate accuracy in smooth flows, high resolution at flow discontinuities and to avoid spurious oscillations, different flux limiter functions are applied in a wave-to-wave basis. By appropriately using compressive limiter functions with waves of the family two to four (linear waves), and diffusive limiter functions with waves of the family one and five (non linear waves), important reductions of the numerical viscosity can be achieved. This sort of adaptive scheme has satisfactorily been applied to the shock tube problem and to the slip interface between two parallel flows. It has also been used when computing the flow over a spherically blunted bi-conic body at free stream Mach numbers varying from five to fifteen. The results obtained with the blunted body are presented here, and they show that the new adaptive scheme besides being less diffusive, it does not lose robustness regarding other TVD methods.

1. Introduction

When solving the fluid mechanic equations numerically using finite volume techniques, the necessity of computing convective fluxes arises. Traditionally, polynomial functions were used to approach the variations of variables. This approach gives excellent results when the variables undergo smooth variations but it has serious difficulties if the solution contains discontinuities. In these cases, the schemes that use second or higher order approximations present inconveniences during the convergence process and the solutions have oscillations next to the discontinuity. On the other hand, the schemes that use first order approximations generate solutions without oscillations but, the discontinuities may poorly be resolved (scheme highly diffusive).

To deal with this problem, flux limiter functions were built as linear combinations of first and second order approximations [1,2]. These schemes are known as TVD although, strictly speaking, the TVD condition has formally only been demonstrated for scalar convection equations. If in the linear combination, the first order approach has more weight than that of second order, the resultant scheme is diffusive and reciprocally, if the second order approach has more weight the scheme is compressive. When extending these concepts from systems of linear equations to non linear, the certainty that they continue being TVD is lost [3], and it becomes necessary to take decisions about as how to carry out such extension. By instance:

- To apply only one limiter function or so many as it is the number of equations.
- To apply the limiter functions according to the spectral decomposition of related variable jumps [3].
- What variable must be kept in mind for the calculation of the scalar limiter function.

For the more robust schemes the number of limiter functions is equal to the number of equations; and in addition, a spectral decomposition is used [3].

For the three-dimensional Euler equations system, the spectral decomposition makes that three lineally degenerate families of waves appear [4]. Discontinuities associated with these eigenvectors are very difficult to solve exactly except for schemes that use higher compressive limiter functions, however, these schemes are not very robust in solving discontinuities associated with the non linear wave families [2].

In some papers, looking for the best way to capture contact discontinuities without the characteristic smearing linked to degenerate linear waves, different schemes were proposed. One of the most popular was introduced by Harten [5] using an Artificial Compression Method (ACM) that renders the local slope of the minmod limiter steeper; however, if the field becomes genuinely nonlinear the limiter needed should not be so compressive [6]. The ACM implemented on superbee limiter functions [7,8] was not utilized in solving the gasdynamics equations [6], probably because it was recognized that it was not a high-resolution gasdynamics shock-capturing technique [3]. The ACM was also implemented on minmod limiter functions to solve the scalar wave and Burgers equations [6], and in second order central difference schemes for Euler equations [9]. More recent ACM applications are given in Ref. [10].

In this work an alternative scheme is presented, which has the capacity to solve satisfactorily discontinuities associated with lineally degenerate wave families and without losing robustness in dealing with gasdynamics shock waves. This new scheme was implemented in a three-dimensional computational code to solve the Euler equations using a non-structured finite volume technique.

2. Description of the scheme

The three-dimensional Euler equations can be written as:

$$\frac{\partial U}{\partial t} + \nabla \cdot F = 0 \tag{1}$$

U is the vector of conservative variables, and F is the 3D vectorial flow.

The temporal change of the conservative variables can be expressed as:

$$U^{n+1} = U^n - \frac{\Delta t}{Vol} \cdot \sum_{l=1}^{lfaces} F_l^* \bullet n_l \cdot A_l \tag{2}$$

where the flux of the conservative variables "F" has been replaced by the numerical flux tensor "F^*".

Equation 2 allows the use of a locally aligned system of coordinates whose unit vector i coincides with the normal to the face "l" of the cell, and the unit vectors j and k are tangential directions. To achieve second order accuracy the numerical flux at the interface between cells "l" and "l+1" in the direction normal to the face "l" [11], is calculated by:

$$f^*_{i+1/2} = \frac{f_i + f_{i+1}}{2} + \frac{1}{2} \cdot \sum_m \Phi^m_{i+1/2} \cdot \overline{r^m_{i+1/2}} \tag{3}$$

where f_i and f_{i+1} are the physical fluxes normal to the face in each cells, $\overline{r^m_{i+1/2}}$ is the *m-th* right eigenvector, and $\Phi^m_{i+1/2}$ is, in the original Harten-Yee scheme [12-14], defined as:

$$\Phi^m_{i+1/2} = g^m_i + g^m_{i+1} - \left|\lambda^m_{i+1/2} + \gamma^m_{i+1/2}\right| \cdot \alpha^m_{i+1/2} \tag{4}$$

being:

$$g_i^m = \frac{S}{2} \cdot \max\left[0, \min\left(\left|\lambda_{i+1/2}^m \cdot \alpha_{i+1/2}^m\right|, S \cdot \left|\lambda_{i-1/2}^m\right| \cdot \alpha_{i-1/2}^m\right)\right] \tag{5}$$

$$S = sign\left(\lambda_{i+1/2}^m\right) \tag{6}$$

$$\gamma_{i+1/2}^m = \begin{cases} \dfrac{g_{i+1}^m - g_i^m}{\alpha_{i+1/2}^m} & si \quad \alpha_{i+1/2}^m \neq 0 \\ 0 & si \quad \alpha_{i+1/2}^m = 0 \end{cases} \tag{7}$$

where $\alpha_{i+1/2}^m$ is the jump of the conserved variables across the interfaces between cells "i" and "i+1", and $\lambda_{i-1/2}^m$ is the m-th eigenvalue of the Jacobian matrix. Since the local Riemann problem is solved with rotated data, the eigensystem is calculated in the locally aligned coordinate frame.

The limiter function given in Eq. (5) is known as minmod [1-4]. The minmod selects the minimum possible value, so that the scheme is TVD. The other end is the limiter function "superbee" that ponders the contribution of the high order flux [2]. The implementation of the "superbee" function leads to an excessively compressive scheme which it is not very robust for general practical aerospace applications.

In the numerical solution of the three-dimensional Euler equations five wave families appear. If the five wave families are enumerated in correspondence with their speed, being one the slowest and five the faster, it can be demonstrated that for waves of the families two to four, the characteristic velocities at both sides of the discontinuity are the same and equal to the velocity discontinuity [2,3]. This property makes very difficult to solve theses waves accurately, unless they are solved diffusely.

In this work, it is explored the possibility of implementing different limiter functions for different wave families. The objective is to improve the numerical resolution of the discontinuities associated with the families two to four using compressive limiter functions (superbee), and without losing robustness due mainly to the use of diffusive limiter functions (minmod) for the wave families one and five.

To introduce in the numerical fluxes calculations the limiter function superbee the Eq. (5) is replaced by the following expression:

$$g_i^m = \begin{cases} 0 & \text{if} \quad \alpha_{i+1/2}^m \cdot \alpha_{i-1/2}^m < 0 \\ \max\left[0, \min(2 \cdot r, 1), \min(r, 2)\right] \cdot \dfrac{1}{2} \cdot \left|\lambda_{i-1/2}^m\right| \cdot \alpha_{i-1/2}^m & \text{if} \quad \alpha_{i+1/2}^m \cdot \alpha_{i-1/2}^m \geq 0 \end{cases} \tag{8}$$

being:

$$r = \frac{\left|\lambda_{i+1/2}^m\right| \cdot \alpha_{i+1/2}^m}{\left|\lambda_{i-1/2}^m\right| \cdot \alpha_{i-1/2}^m} \tag{9}$$

To improve the overall scheme robustness, the implementation of different limiter functions is carried out only in those cells interfaces where the greater relative intensities of the discontinuities in central waves are registered, and using the conventional Harten-Yee TVD scheme in all other cases. Notice that the comparison among the intensity of the waves cannot be made using directly the coefficients of the spectral decomposition $\left(\alpha^m_{i+1/2}\right)$ since these coefficients depend on the module assigned to each eigenvector.

In the local coordinate system adopted for computing the numerical fluxes across each face, the corresponding eigenvectors are given by:

$$r_1 = \begin{bmatrix} 1 \\ u-a \\ v \\ w \\ H-u\cdot a \end{bmatrix} \quad r_2 = \begin{bmatrix} 1 \\ u \\ v \\ w \\ \dfrac{u^2+v^2+w^2}{2} \end{bmatrix} \quad r_3 = \begin{bmatrix} 0 \\ 0 \\ 1 \\ 0 \\ v \end{bmatrix} \quad r_4 = \begin{bmatrix} 0 \\ 0 \\ 0 \\ 1 \\ w \end{bmatrix} \quad r_5 = \begin{bmatrix} 1 \\ u+a \\ v \\ w \\ H+u\cdot a \end{bmatrix} \tag{10}$$

where H is the stagnation enthalpy. It can be deduced from Eq. (10) that $\alpha^1_{i+1/2}$, $\alpha^2_{i+1/2}$ and $\alpha^5_{i+1/2}$ measure the density jump in the waves 1, 2 and 5 respectively and that $\alpha^3_{i+1/2}$ y $\alpha^4_{i+1/2}$ measure the momentum jump in waves three and four. To compare these jumps it becomes necessary the select reference values for the density and velocity. Thus:

$$I_1 = \frac{\left|\alpha^1_{i+1/2}\right|}{\rho_{ref}} \;,\; I_2 = \frac{\left|\alpha^2_{i+1/2}\right|}{\rho_{ref}} \;,\; I_3 = \frac{\left|\alpha^3_{i+1/2}\right|}{\rho_{ref} u_{ref}} \;,\; I_4 = \frac{\left|\alpha^4_{i+1/2}\right|}{\rho_{ref} u_{ref}} \;,\; I_5 = \frac{\left|\alpha^5_{i+1/2}\right|}{\rho_{ref}} \tag{11}$$

In this investigation is taken as density reference $\rho_{ref} = 0.5\ (\rho_i + \rho_{i+1})$ and as the velocity reference the average of the sound velocities at the cells $u_{ref} = 0.5\ (c_i + c_{i+1})$.

Finally, if the maximum of I_1, I_5 is higher than the maximum of I_2, I_3, I_4 the conventional Harten-Yee TVD scheme is used; otherwise, the values of g_i^2, g_i^3, g_i^4 are calculated with the limiter function superbee and g_i^1, g_i^5 with the limiter function minmod.

3. Implementation

3.1 The limiter functions evaluation

For the evaluation of g_i^m and g_{i+1}^m in Eq. (4), it is necessary to calculate the spectral decompositions of the conservative variables increments at the interfaces "i-1/2", "i+1/2" and "i+3/2". In the context of three-dimensional not structured meshes of tetrahedrons, the identification of the cells "i" and "i+1" is intuitive (they are two cells that share a face) but the determination of the points "i-1" and "i+2" is not direct. If two tetrahedrons that share a face are analyzed, the nodes not belonging to the common face can be used as representative points for "i-1" and "i+2". Then, these points can be used as imaginary cells. In this work these ideas have been implemented, being the nodal values calculated as a pondered average of the conservative variables between all cells that are in contact with the nodes "i-1" and "i+2". Such pondered average is given by [15]:

$$U_{\text{node k}} = \frac{\displaystyle\sum_{i=1}^{n} \frac{U_{\text{cell i}}}{d_{\text{GC}_{\text{cell i}}\text{-node k}}}}{\displaystyle\sum_{i=1}^{n} \frac{1}{d_{\text{GC}_{\text{cell i}}\text{-node k}}}} \tag{12}$$

where $d_{GC_{cell\ i}-node\ k}$ is the distance that separates the gravity center of the cell i from the node k, and n is the cell number in contact with the node k.

3.2 Boundary conditions
The treatment of the boundary conditions is carried out through the imaginary cells technique [2,3]. Five different types of boundaries are considered: 1 – Subsonic inlet, 2 - Supersonic inlet, 3 – Subsonic exit, 4 - Supersonic exit, 5 - Non penetration (solid boundary and symmetry).

4. Test cases
In order to verify the accuracy and robustness of the adaptive scheme three test cases were simulated. The first one is the flow inside of a shock tube. This example was used to explore the capacity of the adaptive scheme which has been described, to model contact discontinuities; the flow does not have velocity regarding the contact discontinuity (discontinuity in waves of the family 2). The second test is the simulation of a slip interface layer between two flows with different velocities and densities, but equal pressures. This test was chosen to study the capacity of the scheme to solve flows in which the discontinuity is in the velocity tangent to a given interface (discontinuity in waves of the families 3 and 4). The third case is the flow around a spherically blunted bi-conic body at several supersonic Mach numbers. The object of this third case was to evaluate the robustness of the scheme to capture strong shock waves (discontinuity in waves of the families 1 and 5).

Although the main objective of this work is to present computer results pertaining to the third case, some brief comments on the other two cases are appropriate. A more detailed description of test cases one and two may be found in Ref [15].

4.1 Flow inside a shock tube (ST)
It is well known that an analytical solution for the flow inside a ST can be built considering three types of waves: shocks, expansion fans and contact discontinuities [2-4]. Figures 1 and 2 show analytical and computed solutions in terms of the density plotted as a function of x/L (being x the distance along the tube and L its total length). In Figure 1, the analytical solution is compared with the results obtained applying both, the adaptive TVD scheme here proposed and the conventional Harten-Yee TVD scheme. It can be appreciated that the simulation of the non linear waves is equally good for both schemes TVD, however, the capture of the contact discontinuity has been improved notably. In Figure 2, the results obtained applying the adaptive scheme is compared with those obtained using superbee limiter functions in all waves. It shall be remarked that the results obtained using the flux limiters superbee have a tendency to show oscillations, which are not present if the new scheme is used.

4.2 Slip surface
The second test case is depicted by an air layer flowing over another one. The velocities and densities in both flows are different, but the pressures are equal (no forces can be supported by the slip surface). The straightforward solution predicts the slip of a flow on another without interferences. However, due to the numerical viscosity, the computed solution produces an apparent mixture zone that gets wider downstream the end of an assumed splitter plane. The spreading of such unphysical mixing region can be interpreted as a measure of the accuracy of the numerical method. In Figure 3 is presented the absolute value of the percentage of error incurred in the velocity calculated at a lateral plane of the computational domain [15]. The results using the traditional Harten-Yee TVD scheme, are shown in the upper part, those obtained with the new scheme in the middle, and in the lower part, are shown the results calculated using superbee limiter functions in all waves.

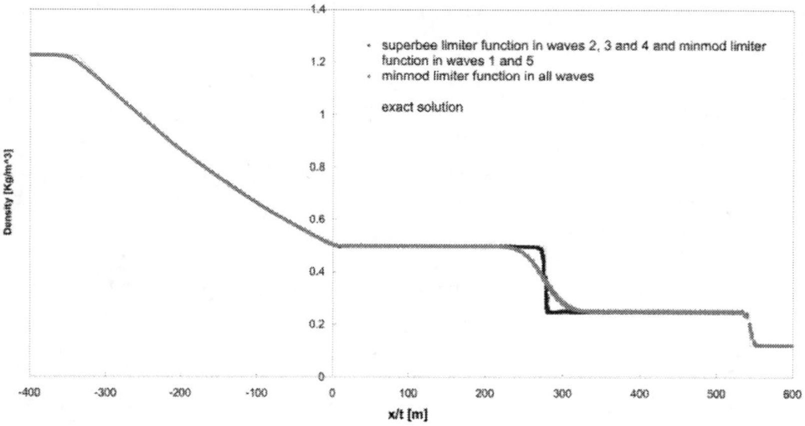

Figure 1. Shock tube results. Comparison between analytical solution, Harten-Yee TVD and proposed scheme.

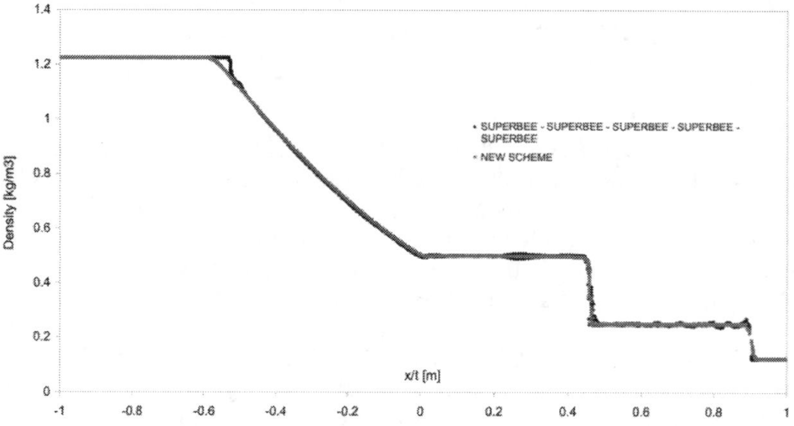

Figure 2. Shock tube results. Comparison between scheme using superbee limiter function for all waves and the new scheme.

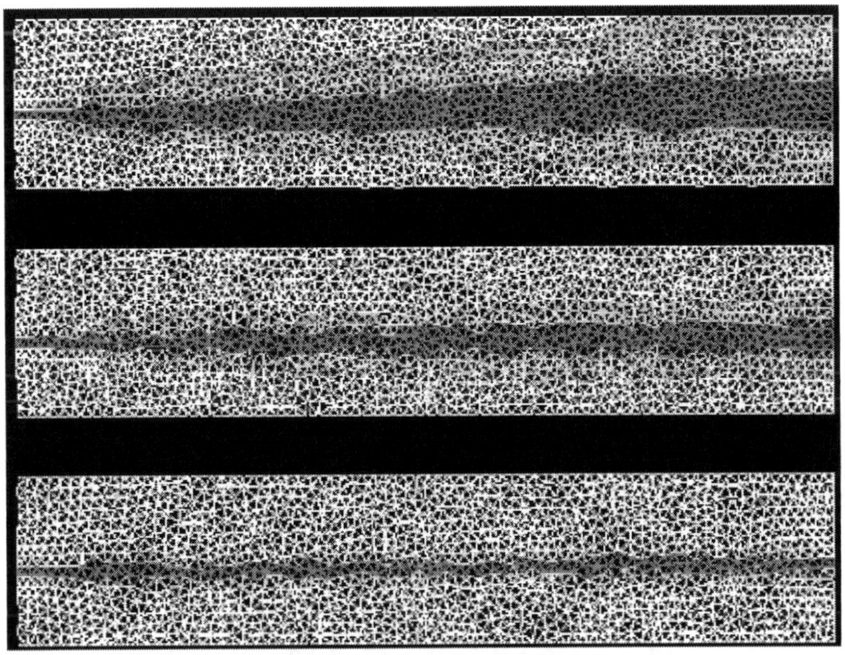

Figure 3. Slip surface results. Error: white $0\% \leq e \leq 2\%$, green $2\% \leq e \leq 5\%$, yellow $5\% \leq e \leq 10\%$, light blue $10\% \leq e \leq 20\%$, red $20\% \leq e$.

As it can be deduced from Figure 3, the proposed adaptive scheme is notably less diffusive that the conventional Harten-Yee method, although more diffusive than the scheme using superbee in all five waves of the Euler equations.

4.3 Spherically blunted bi-conic body at supersonic speeds

It was analyzed the flow around the spherically blunted bi-conic configuration shown in Figure 4. The nose radius is of 0.15m, the aft part of the nose is conical with an angle of 10° and a length of 0.3m, and there after, there is another conical part with an angle of 20° and 0.2m of length. Two Mach numbers, 5 and 15 were considered for numerical simulations.

By means of these numerical studies, the following properties of the computer program were intended to be analyzed:

- The quality of the results describing the flow on the spherical nose.
- The ability for capturing accurately the shock wave between the first and second cone.
- The preservation of the axial symmetry of the flow despite using a 3D code.
- The potential of the new scheme for capturing strong shock waves. It is realized nevertheless, that the results at M = 15 can be questioned because real gas effects are not accounted for.

The mesh, shown in Figure 5, has 110000 cells and 23000 nodes. The boundary conditions are: (1) supersonic flow enters through the control volume front; (2) supersonic flow exits through the back plane and (3) non penetration on the body and in the planes y-x and z-x. The flow that enters is air and has a density of 1.225Kg/m3, and a pressure of 100000Pa. The free stream components in the directions of the z- and y-axis are null, and in the direction of the x-axis are 1690.3 m/s for M = 5, and 5070.9 m/s for M = 15.

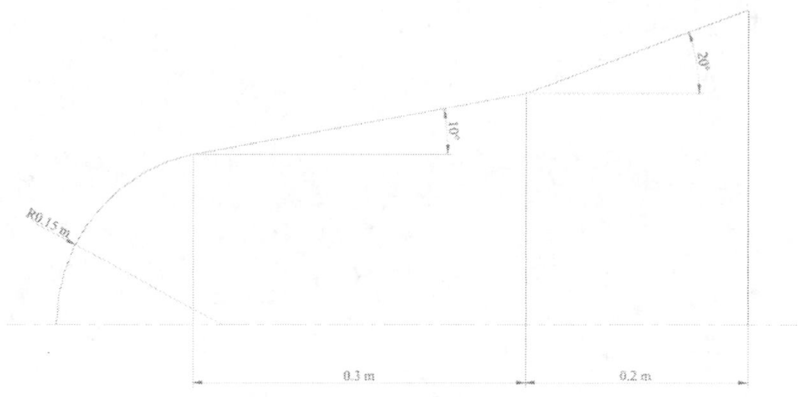

Figure 4. Spherically blunted bi-conic body

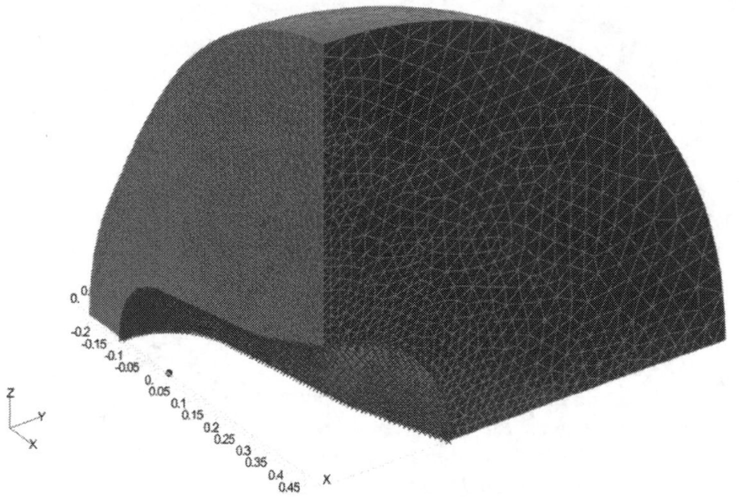

Figure 5. Mesh used in the blunted bi-conic body case.

In Figures 6 and 7 pressure distributions on the body for M = 5, are presented. Figure 6 shows the pressure distribution on the body, however due to the picture scale only the nose shock appears clearly defined. In Figure 7 the pressures plotted are of all the cells in contact with the body (from 0° to 90°). It can clearly be perceived the axial symmetry of the results and the accurate capture of shock waves. The distributions of pressure agree satisfactorily if they are compared with results presented by other authors [16].

In Figures 8 and 9, the pressure distributions are plotted for M = 15. In Figure 8, again is shown the pressure distribution produced by the primary or nose shock. The resolution of the discontinuities is satisfactory, oscillations are not present and the axial symmetry of the flow is maintained appropriately.

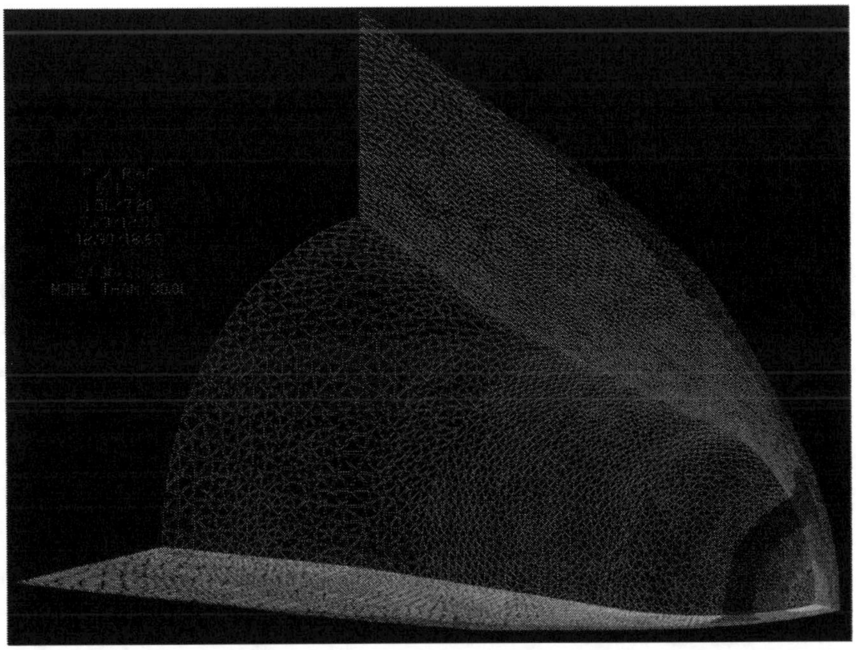

Figure 6. Pressure distribution over axial-symmetric body. M = 5. Primary shock.

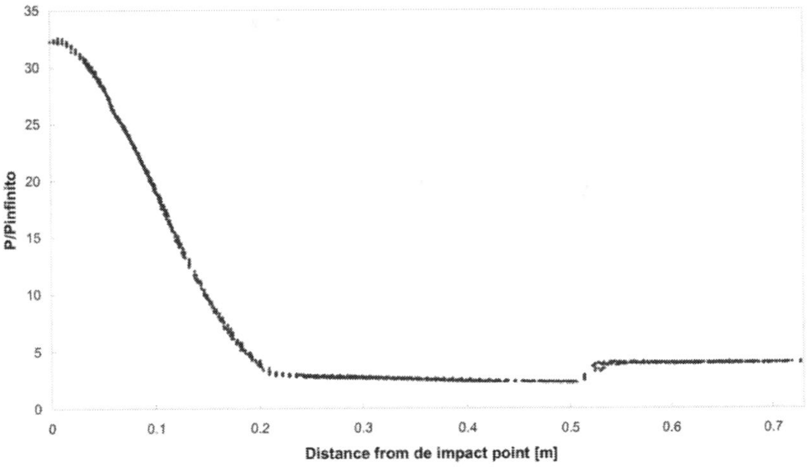

Figure 7. Pressure distribution over axial-symmetric body. M = 5 external flow.

It can be appreciated from Figures 7 and 9 that the spread of the pressure results, due the cells arrangements on the body at each x-station, is very small compared with related local values.

Figure 10 shows the shock in front of the blunt nose of the body at M = 15 as it is determined by the new adaptive scheme and by the traditional Harten-Yee TVD technique. It can be observed that the two figures look very much alike. Consequently, the new scheme captures strong shock waves as well as it does the traditional Harten-Yee. Furthermore, Figure 11 which is a close up of the blunt body nose region, shows that the shock is captured between two grid cells only.

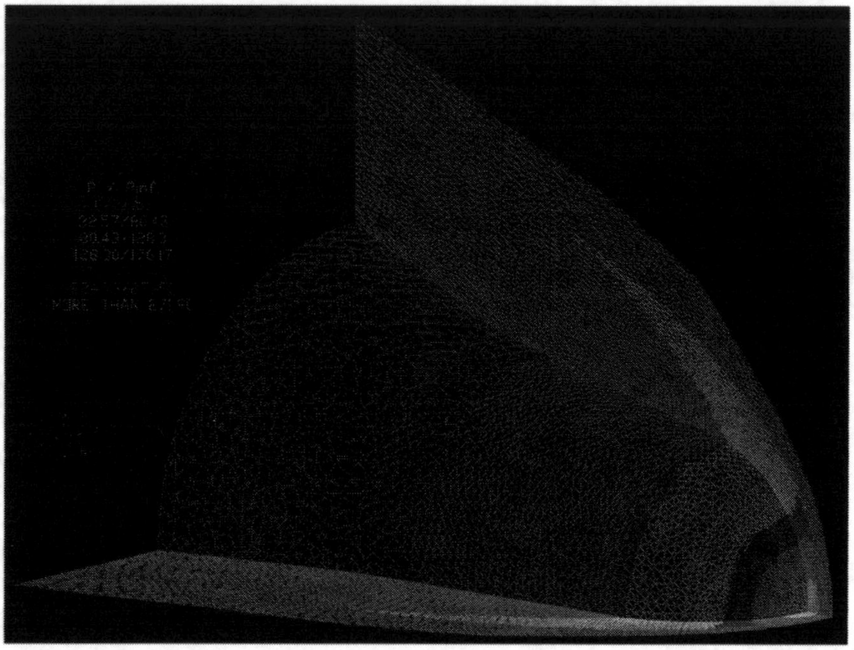

Figure 8. Pressure distribution over axial-symmetric body. M = 15. Primary shock.

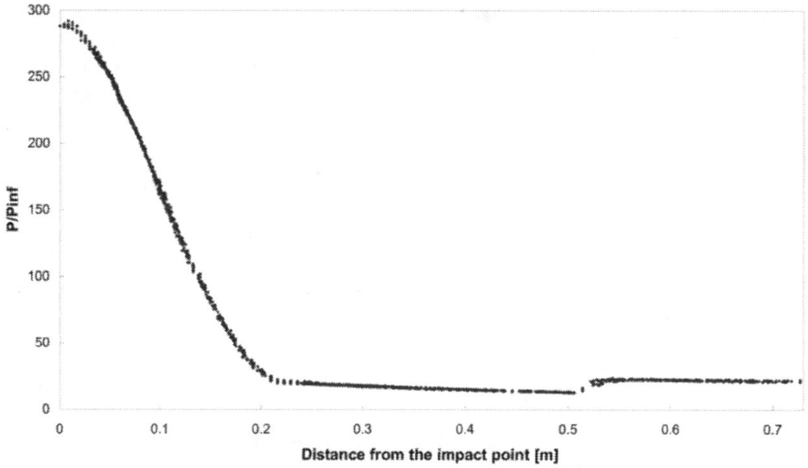

Figure 9. Pressure distribution over axial-symmetric body. M = 15 external flow.

It is important to notice that a numerical scheme which uses only superbee flux limiters to simulate blunted bodies problem, introduces oscillations, produces negative pressures near discontinuities and it ends up being unstable. To emphasize this last point, two cases with different ways of introducing initial conditions were considered. In the first one, starting constant values typical of a free stream flow were used, and in the second one, the simulation was started from a "steady-state results" obtained after using the new adaptive scheme. In both cases, the scheme using superbee functions became unstable.

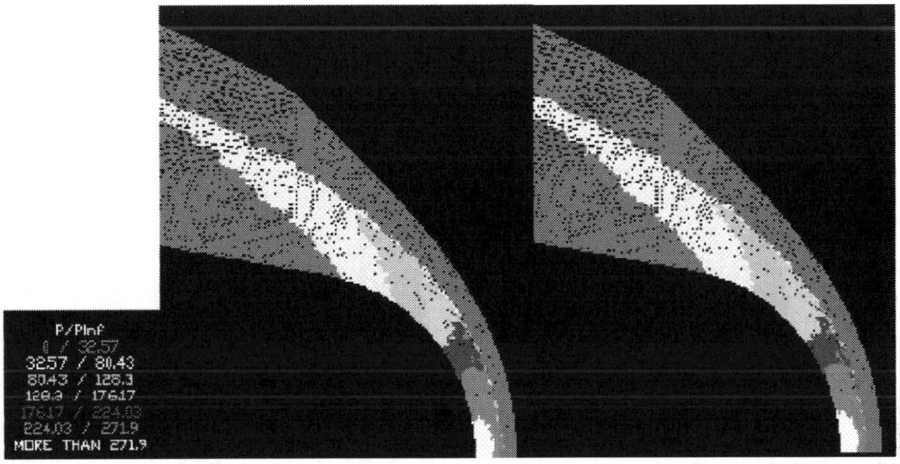

Figure 10. Shock wave over axial-symmetric body at M = 15. Right figure: Traditional Harten-Yee. Left figure: adaptive scheme

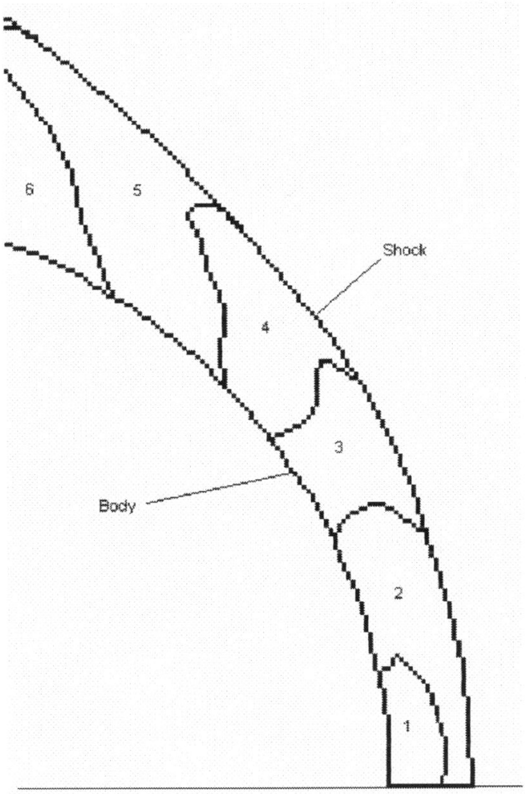

Figure 11. Close up of the blunt body nose region. (1) $p/p_\infty > 272$; (2) $224 < p/p_\infty < 272$; (3) $176 < p/p_\infty < 224$; (4) $128 < p/p_\infty < 176$; (5) $80 < p/p_\infty < 128$; (6) $33 < p/p_\infty < 80$

5. Conclusions

In previous test cases, it has been proven the capacity that the adaptive TVD scheme here described has to model contact discontinuities (ST test case), and to simulate interfaces between two flows with different velocities and densities (slip surface test case). Now, its potential to capture strong pressure discontinuities (shock waves) in supersonic flows has been tested.

After comparing results obtained with the proposed TVD scheme, with those obtained using the conventional TVD of Harten-Yee and with other solver that uses superbee functions as limiters for all numerical fluxes, the following conclusions can be written.

- From the **Shock Tube** test case: The adaptive TVD scheme diminishes the numerical viscosity with regards to the conventional Harten-Yee TVD and it does not introduce oscillations.
- From the **Slippery Surface** test case: The adaptive TVD scheme works more efficiently than the conventional Harte-Yee TVD. However, the scheme that only uses superbee limiters functions produces less numeric diffusion than the other two.
- From the **Blunted Body** test case: The adaptive TVD scheme proposed capture strong shocks as well as it does the conventional TVD of Harten-Yee. The much higher resolution of discontinuities (the shock in the nose region is captured between two grid cells, only), does not affect its robustness. No difficulties have ever been experimented in computing the flow around a blunted bi-conic body in the free stream Mach number range of 5 to 15. Attempts to apply the numerical scheme that only uses superbee limiters functions failed (generate oscillations, produces negative pressure and it finishes being unstable).

Acknowledgement

This work has been supported by means of grants PIP No 5692 of the "Consejo Nacional de Ciencia y Tecnología (CONICET)" of Argentina, PICTO2005-UNRC: 30339-162 of the "Secretaría de Ciencia, Tecnología e Innovación Productiva (SECYT)" of Argentina, and of the "Universidad Nacional de Córdoba".

References

[1] Sweeby P 1984 High resolution schemes using flux limiters for hyperbolic conservation laws *SIAM Journal on Numerical Analysis* **21** 995-1011.

[2] Hirsch C 1992 *Numerical Computation of Internal and External Flows, Vol.2 Computational Methods for Inviscid and Viscous Flows* (London: John Wiley & Sons Ltd.).

[3] Toro E 1999 *Riemann Solvers and Numerical Methods for Fluid Dynamics* (Berlin: Springer-Verlag).

[4] Leveque R 1992 *Numerical Methods for Conservation Law* (Basel: Birkhäuser Verlag).

[5] Harten A 1978 The artificial compression method for computation of shocks and contact discontinuities: III self adjusting hybrid schemes *Mathematics of Computation* **32** 363-99.

[6] Rider W 1992 The design of high-resolution upwind shock-capturing methods *Los Alamos National Laboratory*.

[7] Rider W and Woodruff S 1991 High-order solute tracking in two-phase thermal hydraulics LA-UR-91-2263 *Los Alamos National Laboratory*.

[8] RELAP5-3D© Code Development Team 2005 RELAP5-3D© code manual. Volume 1: code structure, system models, and solution methods *Idaho National Engineering and Environmental Laboratory*.

[9] Lie K and Noelle S 2003 On the artificial compression method for second-order nonoscillatory central difference schemes for systems of conservation laws *SIAM Journal of Scientific Computation* **24** 1157-74.

[10] Lo S Blaisdell G and Lyrintzis A 2007 High-order shock capturing schemes for turbulence calculations *AIAA Paper* N°2007-827.

[11] Udrea B 1999 An advanced implicit solver for MHD *PhD Thesis, University of Washington.*

[12] Harten A 1982 On to class of high resolution total-variation-stable finite difference schemes *Technical Report NYU.*
[13] Harten A 1983 High resolution schemes for hyperbolic conservation laws *Journal of Computational Physics* **49** 250-7.
[14] Yee H Warming R and Harten A 1985 Implicit total variation diminishing (TVD) schemes for steady-state calculations *Journal of Computational Physics* **57** 327-60.
[15] Falcinelli O Elaskar S and Tamagno J 2008 Reducing the numerical viscosity in non structured three-dimensional finite volumes computations *Journal of Spacecraft and Rocket, AIAA* **45** 506-9.
[16] Pirolo J and Wardlaw Jr. A 1991 High temperature effects for missile type bodies using the Euler solver ZEUS *AIAA Paper* N°91-3259.

Heat transfer on a naturally cross-driven ventilated triangular cavity with openings.

M E Berli[1, 2], J Di Paolo[2] and F A Saita[1]
[1] INTEC, Universidad Nacional del Litoral, Güemes 3450, 3000, Santa Fe, Argentina.
[2] GIMEF, Departamento Ingeniería Industrial, FRSF - UTN, Lavaise 610, 3000, Santa Fe, Argentina

E-mail: mberli@santafe-conicet.gov.ar

Abstract. This work addresses the problem of heat transfer through the roof of a family home with the purpose of improving air-conditioning energy savings during hot summer days. To this end, the air natural convection in a right-angle triangular cavity, resembling an attic, is numerically analyzed. The air cavity is assumed as naturally ventilated through two openings that induce an external air-stream flowing into and out of the cavity. The governing equations for both flow and heat transfer are simultaneously solved with appropriate boundary conditions. Finite element technique is employed to transform the original set of differential equations into a non-linear discrete one, which is finally solved by Newton iteration. Two major assumptions are made, namely: radiation heat transfer is negligible and the air flow is in laminar regime; in addition, the usual Boussinesq approximation is employed. Under these assumptions the numerical predictions show that the amount of heat transferred through the ceiling rapidly diminishes as the flow through the cavity increases. Therefore; whether or not the ceiling is insulated, the energy transfer is reduced by more than 50% and energy savings are considerably improved.

1. Introduction

The roof of a family home is usually the component that transfers more heat than any other wall of the house. The large temperature gradient that exists between the roof and the ceiling on days of extremely hot temperatures, introduces important amounts of energy that needs to be eliminated to keep the interior comfortable. In this sense, an air cavity located between the roof and the ceiling might produce important energy savings if it is properly designed. To this end is crucial to know the mechanisms and details by which the airflow in the cavity moves and transports heat from the roof toward the ceiling and vice-versa.

Several authors have studied buoyancy driven flows in cavities motivated by a number of engineering applications. Rectangular cavities ([1], [2], [3], [4], [5], [6]), triangular cavities ([7], [8]) and also cylindrical cavities [9] have received special attentions. The latter geometry has also been applied to study sterilization processes of canned foods [10].

Recently, Basak et al. [8] numerically solved a natural convection model in a triangular cavity employing the penalty finite element technique. The goal was to study the temperature distribution and the heat transfer through the base of a closed triangle that very much resembles a roof with an attic space. Their predictions show that the amount of heat transferred through the base increases with the Rayleigh number, i.e. with the flow speed; they also conclude that the Prandtl number has much less influence on heat transfer than the Rayleigh number.

Hirunlabh et al. [11] analyzed the performance of Roof Solar Collectors (RSC) in several different design configurations; they studied the airflow rate through a channel of constant section located just beneath the RSC and found that the induced flow rate depends on both roof inclination and solar intensity radiation. Nonetheless, they also found that not even those configurations that maximize the air convection provide sufficient natural ventilation as to satisfy resident's comfort. Ciampi et al. [12] studied a roof configuration similar to the one analyzed by Hirunlabh; they paid particular attention to microventilation, i.e. to roof with small-sized-thickness ducts in which the airflow is laminar. The airflow rate in the channel depends not only on the heat field but also on several other factors like duct geometry, fluid dynamic head losses, etc, and it is estimated through an equation obtained from a simplified one-dimensional flow model. These authors found that microventilation might produce energy savings as large as 30%, this percentage can be improved with larger flow passage sections and larger roof inclination angles.

Asan and Namli [13] numerically analyzed the two-dimensional laminar convection in a closed triangular cavity shaped by an inclined roof and a suspended ceiling. Assuming summer day boundary conditions, they used the stream function-vorticity formulation to solve the equation for different values of the Rayleigh number. These authors concluded that the height-base ratio of the triangle is the parameter that most affects the heat transfer through the base; in fact, as this ratio increases they found that heat transfer decreases.

Oztop et al. [14] solved a very similar problem, i.e. they used the same geometry, boundary conditions and numerical formulation as Asan and Namli, but they also considered a roof with and without eaves. The predictions obtained by Oztop et al. show that heat transfer strongly depends on both the aspect ratio and the length of the overhanging portion of the roof.

The analyses performed in all the works previously mentioned were carried out on closed cavities; the goal of the present work is to assess the influence that natural ventilation of the cavity might have on the amount of heat conducted through the ceiling toward the interior of the house. For that purpose we will employ geometries similar to those used by Asan and Namli —which are also similar to the roof geometries analyzed by Oztop et al.— but we will also consider openings located on the lateral walls supporting the roof. Thus, natural ventilation will occur driven by buoyancy forces originated in the temperature difference that exist between the roof and the ceiling. The temperature gap will be extreme on summer days when the roof is exposed to solar radiation and more energy will be consumed by the air conditioning system to keep the interior comfortable.

In the next section we present the main features of the model under analysis, i.e. the geometry of the cavity, the governing equation, and the appropriate boundary conditions to be used when the ceiling is insulated and when it is not. In a short third section we summarize the numerical technique employed and we validate the code by comparing our numerical predictions with results already published. In section fourth we present the results obtained while the last section is devoted to conclusions.

2. The model

Figure 1 sketches the shape of the domain where the governing equations have to be solved; the roof, the ceiling and the lateral walls with openings delimit the domain. The adopted length of the openings was 30 cm, which is the standard thickness of an exterior wall made out of bricks, and the adopted height was 15 cm; the parameter Am is the average height of the cavity, i.e. the distance between the roof and the ceiling at the center of the domain length. The air in the cavity is set into motion driven by buoyancy forces; i.e. the density gradients originated in the temperature difference that exists between the roof and the ceiling induces the air motion. The phenomenon just described is known as natural convection; this type of flow is usually solved with the aid of the so-called Boussinesq approximation that considers the density differences small enough to be neglected everywhere, except for the term where the density affects the gravity force; in that case the density is supposed to linearly change with temperature. In addition to incompressibility, which is already implied in the Boussinesq approximation, we will assume steady-state laminar flow regime and negligible thermal radiation effects as well as viscous dissipation [15].

2.1. Governing equations

The set of dimensionless governing equations for natural convection flow in a 2-D cavity is given by the equation of continuity (1), the x- and y- components of the momentum equation (2), (3), and the thermal energy equation (4)

$$\frac{\partial u}{\partial x}+\frac{\partial v}{\partial y}=0 \tag{1}$$

$$u\frac{\partial u}{\partial x}+v\frac{\partial u}{\partial y}=-\frac{\partial p}{\partial x}+\left(\frac{1}{\mathrm{Re}}\right)\left(\frac{\partial^2 u}{\partial x^2}+\frac{\partial^2 u}{\partial y^2}\right) \tag{2}$$

$$u\frac{\partial v}{\partial x}+v\frac{\partial v}{\partial y}=-\frac{\partial p}{\partial y}+\theta+\left(\frac{1}{\mathrm{Re}}\right)\left(\frac{\partial^2 v}{\partial x^2}+\frac{\partial^2 v}{\partial y^2}\right) \tag{3}$$

$$u\frac{\partial \theta}{\partial x}+v\frac{\partial \theta}{\partial y}=\left(\frac{1}{\mathrm{Re}\,\mathrm{Pr}}\right)\left(\frac{\partial^2 \theta}{\partial x^2}+\frac{\partial^2 \theta}{\partial y^2}\right) \tag{4}$$

In equations (1)-(4) $Re=\rho U_C\, L/\mu$ and $Pr=\nu/\alpha$ are the Reynolds and Prandtl numbers respectively; the employed values of the physical properties of the air (density, viscosities, thermal diffusivity and thermal expansion coefficient) are obtained at the mean temperature between the roof and the ceiling ($(Tr+Tc)/2$). Lengths are measured in terms of the ceiling length (L), temperatures are measured in terms of the temperature difference between the roof and the ceiling (Tr-Tc) and velocities are measured in terms of the characteristic velocity (U_C) which is here defined as $(\beta\, L(Tr-Tc)\, g)^{1/2}$; being β and g the thermal expansion coefficient and the gravity constant, respectively.

When a thermal insulating layer is located between the ceiling and the cavity, as is shown on the left of figure 1, the thermal energy equation for the 2-D heat transfer in the solid (5) must be added to the equation set shown above.

$$0=\left(\frac{\partial^2 \theta}{\partial x^2}+\frac{\partial^2 \theta}{\partial y^2}\right) \tag{5}$$

2.2. Boundary conditions

The boundary conditions employed for both momentum and energy equations are summarized in figure 1. When an insulated ceiling is considered, the appropriate additional boundary condition to be imposed is shown in a shaded rectangle appearing at the bottom left side of figure 1. As this figure shows, on solid walls the non-slip condition is imposed. At the openings we assume zero traction vector (***T.n***$=0$), the atmospheric pressure is arbitrarily set equal to zero and we neglect heat conduction ($\partial\theta/\partial n=0$). At the air entrance, in addition to the air temperature, the y- and x-components of the velocity are imposed; they are set as zero and constant (($\partial u/\partial y=0$), respectively.

The temperature of both the roof and the ceiling is constant (Tr and Tc) and their dimensionless values are one and zero, respectively. The lateral walls are supposed adiabatic, thus ($\partial\theta/\partial n=0$). Finally, if an insulating layer is set on the ceiling, the flow of thermal energy between the air and the

Journal of Physics: Conference Series **166** (2009) 012019 doi:10.1088/1742-6596/166/1/012019

material must be continuous; thus, if b is the dimensionless thickness of the insulating layer, at y = b the following relation applies

$$K_I \frac{\partial \theta}{\partial x} = K_A \frac{\partial \theta}{\partial y} \tag{6}$$

where K_I and K_A are the coefficients of thermal conductivity of the insulating material and the air, respectively.

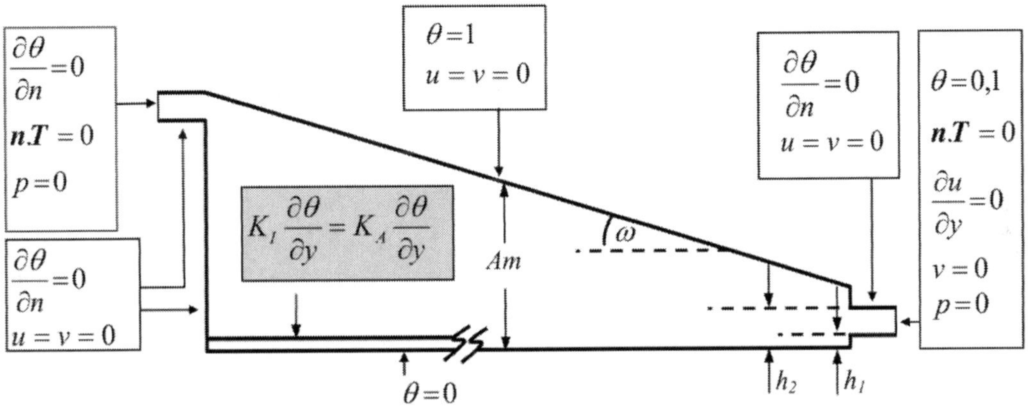

Figure 1. Schematic view of the cavity analyzed and boundary conditions imposed. The left half of the figure shows an insulated ceiling while the right half shows a ceiling without insulation.

3. Numerical technique

The highly nonlinear and strongly coupled system of governing equations can only be solved by numerical means; for that purpose we employed the Galerkin/finite element approach. This is a combination of the Galerkin's weighted residual method with trial functions having support on a few elements into which the flow domain is subdivided. In this work we used the so-called mixed interpolation technique on quadrilaterals; i.e. we adopted biquadratic trial functions for the expansion of velocities and temperature, while bilinear basis functions were employed for pressure. The resulting set of nonlinear algebraic equations was solved with the aid of isoparametric mapping and Newton iteration. The solution path along Reynolds numbers was traced using zero order continuation; i.e. the results obtained for a previous Reynolds value were used as a first guess to start the new Newton iteration for a Reynolds value not too much different from the preceding one.

The numerical algorithm was implemented in Matlab 6.0 on a Window XP platform and the computing time needed to solve 13000 degrees of freedom in a PC Pentium IV of 3.0 Ghz and 1.0 Gb of Ram was about 20 minutes.

4. Results

4.1. Validation of the model.

In order to validate our model we made comparisons with the results presented by Barak et al. [8] and by Haese and Teubner [16]; these comparisons are summarized in figure 2.

Figure 2(a) portrays the isotherms that result in a closed triangular cavity due to air natural convection; the shape of the triangular cavity is just the lower half of a square divided along its diagonal. The horizontal wall, which coincides with the abscissa, is at $\theta = 1$ while the inclined wall is at $\theta = 0$ and the vertical wall is adiabatic. Figure 2(b) shows the temperature variation along the vertical wall when $Pr = 0.71$ and $Re = 31.6$; clearly the predictions of our model completely agree with those obtained by the authors just mentioned.

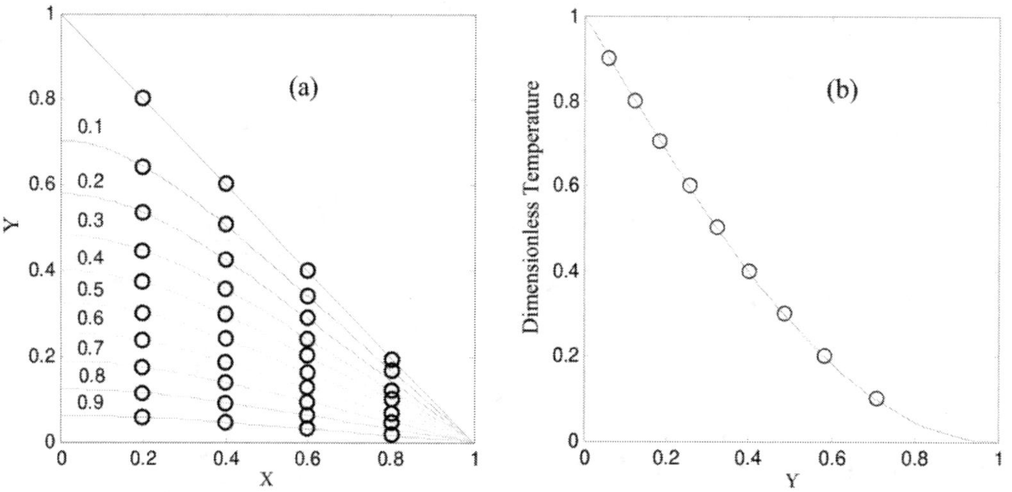

Figure 2. Comparison of the predictions of the present work (full lines) with those obtained by Basak et al. [8] and Haese and Teubner [16] indicated with circles. (a): isotherms in a cavity with a hot horizontal wall, a cold inclined wall and an adiabatic vertical wall. (b): Temperature along the adiabatic vertical wall.

The results to be presented in the following sections were obtained assuming flow motions in laminar regime. Though the natural convection of air in cavities can reach turbulent regimes, particularly for large cavities and strong temperature gradients, analyses assuming laminar conditions are commonly employed to infer the behavior at larger *Re* ([8], [12], [13]).

4.2. A cavity without insulation.

Figure 3(a) compares the amount of heat transferred through the ceiling when air flows across the cavity, and when the openings are closed producing just an inner circulation. The amount of heat transferred is evaluated by computing the average Nusselt number as follows (see [8])

$$Nu = \left(\frac{1}{l}\right)\int_0^l \left.\frac{\partial \theta}{\partial y}\right|_{y=0} dx\,, \tag{7}$$

where l is the dimensionless ceiling length, which in our case is just one.

The results shown in figure 3 pertain to a cavity with $Am/L = 0.22$ and a roof with an inclination angle of 15º; they indicate that both open a closed cavities transfer almost the same amount of heat at low values of *Re*, that is so because the air motion is slow and the heat transfer process is dominated by conduction. However, as the *Re* increases both systems behave completely different; while the Nusselt number increases slowly for a closed cavity, it diminishes rather fast for an open cavity. Figure 3(a) shows that the flow rate through the openings increases noticeably with *Re* and most of the heat introduced into the cavity is removed by convection. The consequence is that the two curves depart from each other indicating that a ventilated cavity becomes more convenient as *Re* increases.

Figure 3(b) shows the reduction in the amount of energy that gets across the ceiling when a closed cavity is substituted by a ventilated one. That reduction increases with *Re* and exceeds 50% when *Re* is about 2600.

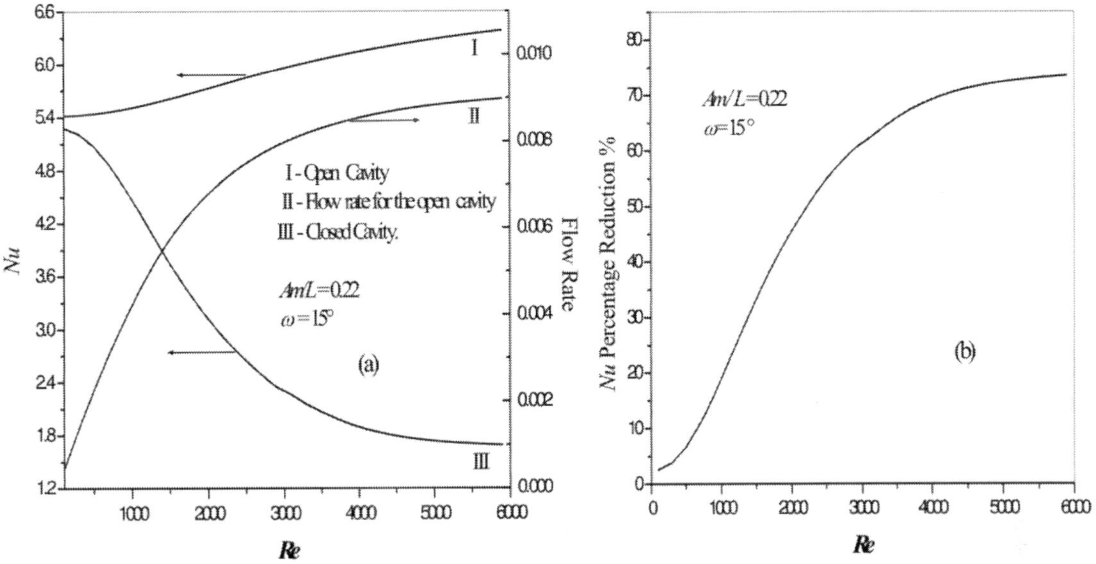

Figure 3. Comparison of the amount of heat transferred through the ceiling for open and closed cavities. (a): Average Nusselt and dimensionless flow rate versus *Re*. (b): Nusselt percentage reduction versus *Re* when an open cavity substitutes a closed one.

In Figure 3(a) the dimensionless flow rate per unit length of the opening is given by

$$q = \int_{h1}^{h2} \mathbf{v}.\mathbf{n}\, dy \qquad (8)$$

being $\mathbf{v}$ the dimensionless velocity vector, $\mathbf{n}$ the unit vector normal to the flow passage and h_2-h_1=h (see figure 1) the dimensionless height of the opening.

4.2.1. Influence of the cavity volume (Am/L). Figure 4 shows how the volume of the cavity affects the amount of heat transferred through the ceiling when the inclination angle of the roof is 15°. No matter if the cavity is open or closed, the average Nu decreases as the volume of the cavity increases —i.e. when the ratio *Am /L* increases. Thus, in figure 4 we observe that the curves of constant *Re* have negative slope. For closed cavities, the curves of constant *Re* move upwards as the Reynolds number increases; this result is expected since the heat transported from the roof toward the ceiling should increase with convection. On the other hand, when the cavities are open, more heat should be taken out of the cavities by convection as the Reynolds number increases; therefore, in this case the curves of constant *Re* move downwards. In addition, the results shown in figure 4 indicate that heat transfer is much more sensitive to the Reynolds number in open cavities than in closed ones.

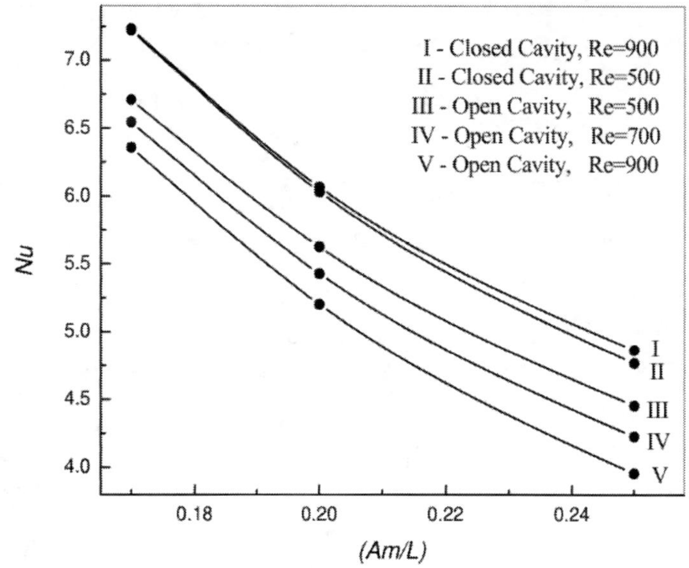

Figure 4: Average Nu versus cavity size (Am/L) for several Reynolds values.

4.2.2. Influence of the roof inclination. Roofs are commonly designed with angles between 10° and 15°, unless the local climate conditions favors the use of larger inclination angles or the space just below the roof will be used for storage purposes. Figure 5 portrays predictions of *Nu* versus *Re* for three different inclination angles; they are 18°, 15° and 10° respectively. These predictions indicate better performances for less inclined roofs; thus, in the range of *Re* explored, the average Nusselt number obtained with an inclination of 10° is approximately 12% smaller than the one obtained with an inclination of 15°. However, in order to favor both drainage and self-cleaning, the inclination angles usually adopted are not too small; for that reason in what follows we use $\omega = 15°$.

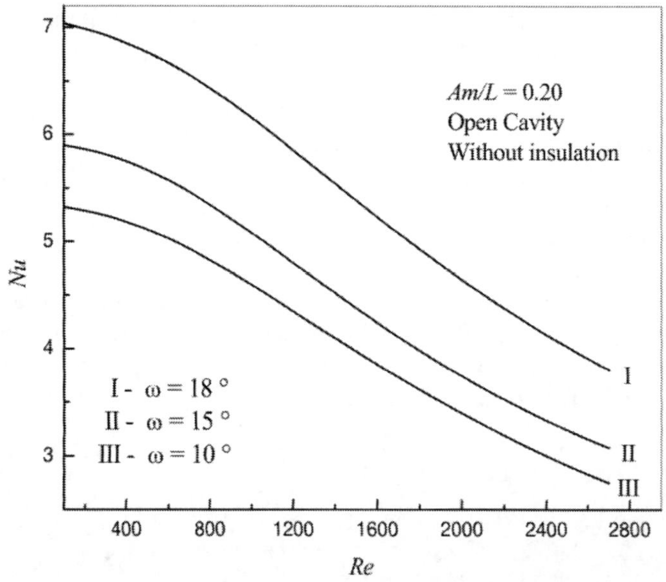

Figure 5. Influence of the inclination on the average Nu for an open cavity without insulation.

4.3. A cavity with an insulated ceiling.

We already pointed out that *Am* is the average distance between the roof and the ceiling (see figure 1); therefore the value of the Am / *L* does not change when a layer of certain insulating material is set on the ceiling as is shown on the left side of figure 1. Also, the adopted dimensionless thickness of the layer (b) is 0.02, which amounts to 8cm in thickness when *L* is 4 m.

In figure 6 we summarize the average Nusselt values obtained for a cavity with and without insulation when $Re \leq 6.1\ 10^3$. On the top of figure 6 the curves denoted as (I) and (II) pertain to a closed cavity and depict the results for the non-insulated cavity and the insulated one, respectively. In this case we have considered an insulation material having exactly the same thermal conductivity as the air; therefore, the ratio between K_I and K_A appearing in (8) is just one ($K = K_I / K_A = 1$). Clearly, at very small values of *Re* the curves coincide since the mechanism of heat transfer is exclusively by conduction. As the values of *Re* increase the effect of convection becomes evident and the insulated cavity performs slightly better allowing less heat to be transferred through the ceiling.

The curves denoted as (III), and (IV) correspond to an insulated and a non-insulated open cavity, respectively. The remarkable outcome is that the non-insulated case performs slightly better than the insulated one when the Reynolds number is smaller than 4.0 103. Both curves are coincident at very small values of the Reynolds number; when Re exceeds 5.0 102 they start to separate from each other until they meet again when the Reynolds value is about 4.1 103. Finally, when Re becomes larger than 4.1 103 the two curves change their relative position indicating that at higher Reynolds the use of an insulated ceiling becomes advantageous. This result was expected since the heat transported toward the ceiling by convection must increase with Re; in addition, this behavior should be enhanced significantly when air flow in the cavity is turbulent.

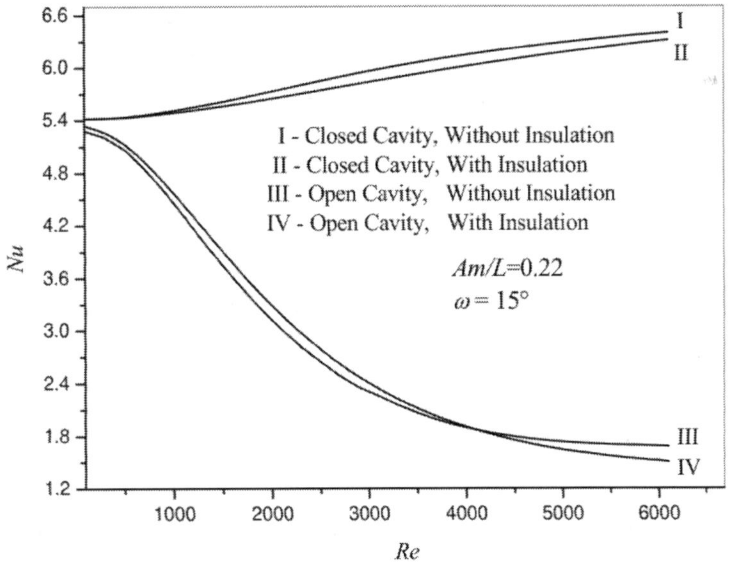

Figure 6. Effect on the average Nu of an insulating layer located on the ceiling of both closed and open cavities.

Figure 6 portrays some interesting results when an open cavity is considered; one of them is that at relatively low values of the Reynolds number, the heat transfer through the ceiling strongly decreases as the air flow rate across the cavity increases, i.e. as the *Re* increases. Actually, when *Re* is about $6.0\ 10^3$ the heat transfer is just 28% of the amount of heat transferred at *Re* smaller than 200. However, if *Re* is further increased the curve of *Nu* becomes almost horizontal indicating that no additional benefits are gained as long as the flow keeps its laminar regime. It is not known the exact value of the Rayleigh number at which turbulent regime appears in this type of cavity, but it is believed that it might happens when the values are between 10^7 and 10^8 [17]. These values occur when the Reynolds number we use in this work is in the range: 3.7 10^3 - 1.2 10^4.

Another result that must be noticed is the small effect produced by the insulating layer located on the ceiling; the reason for this behavior is a consequence of both temperature and flow fields in the cavity. The temperature field shows that the air inside the cavity is rather stratified, being the warmer layers located near the roof and the cooler ones near the ceiling. Though the warmer layers move at considerable speeds toward the exit openings, the cooler layers are almost stagnant; therefore the heat in this region is mostly transported by conduction and the insulating layer produces just a small reduction in the heat transported.

5. Conclusion

In this work we have studied the influence of a ventilated attic on the heat transferred from the roof and through the ceiling, toward the interior of a building of a family house. We numerically solved the complete set of governing equations that couples flow and heat transfer in natural convection. For that purpose, and in addition to the usual Boussinesq approximation, we made two important assumptions: namely, the air flow is in laminar regime and radiation has negligible effect on heat transfer. Under these assumptions we found that on hot summer days, an attic with suitable openings can significantly reduce heat transfer and produce considerable energy savings in air conditioning.

If one assumes that an appropriate radiant barrier is employed, the results obtained should also be valid regardless of how important the radiation transfer might be. Nonetheless, in future analyses we will consider the influence of radiation and we will also explore the turbulent regime —which might occur if the roof temperature is high enough— where the heat transfer by convection should be more important and the employment of insulating layers must be necessary to preserve significant energy savings.

Acknowledgements

This work was supported by grants in aid from Universidad Nacional del Litoral, Concejo Nacional de Investigaciones Científicas y Técnicas (CONICET) and Agencia Nacional de Promoción Científica y Tecnológica of Argentina.

References

[1] Nawaf, S H 2007. Conjugate natural convection in a vertical porous layer sandwiched by finite thickness walls. *International Journal of Heat and Mass Transfer.* **34** 210–16.

[2] Das M K, Reddy K S 2006. Conjugate natural convection heat transfer in an inclined square cavity containing a conducting block. *International Journal of Heat and Mass Transfer.* **49** 4987–5000.

[3] Muftuoglu A, Bilgen E 2008. Heat transfer in inclined rectangular receivers for concentrated solar radiation. *International Journal of Heat and Mass Transfer.* **35** 551-56.

[4] Bilgen E, Yedder R B. 2007. Natural convection in enclosure with heating and cooling by sinusoidal temperature profiles on one side. *International Journal of Heat and Mass Transfer.* **50** 139–50.

[5] Altac Z, Kurtul Ö 2007. Natural convection in tilted rectangular enclosures with a vertically situated hot plate inside. *Appl. Ther. Eng.* **27** 1832– 40.

[6] Lamsaadi M, Naïmi M, Hasnaoui M 2006. Natural convection heat transfer in shallow horizontal rectangular enclosures uniformly heated from the side and filled with non-Newtonian power law fluids. *Energy Conversion and Management.* **47** 2535– 51.

[7] Varol Y, Oztop H. F, Tuncay Yilmaz. 2007. Natural convection in triangular enclosures with protruding isothermal heater. *Int. J. of Heat and Mass Transf.* **50** 2451–62.

[8] Basak, T, Roy, S, Thirumalesha, Ch 2007. Finite element analysis of natural convection in triangular enclosure: Effects of various thermal boundary conditions. *Chemical Engineering Science.* **62** 2623– 40.

[9] Cesinia G, Paroncinia M, Cortellab G, Manzanb M 1999. Natural convection from a horizontal cylinder in a rectangular cavity. *Int. J. of Heat and Mass Transf.* **42** 1801-11.
[10] Kannan A, Gourisankar Sandaka PCh 2008. Heat transfer analysis of canned food sterilization in a still retort. *Journal of Food Engineering*. **88** 213-28.
[11] Hirunlabh, J, Wachirapuwadon, S, Pratinthong, N, Khedari, J 2001. New configurations of a roof solar collector maximizing natural ventilation. *Building and Environment.* **36** 383– 91.
[12] Ciampi, M, Leccese, F, Tuoni, G 2005. Energy analysis of ventilated and microventilated roofs. *Solar Energy.* **79** 183–92.
[13] Asan, H, Namli, L 2000. Laminar natural convection in a pitched roof of triangular cross-section: summer day boundary conditions. *Energy and Buildings.* **33** 69-73.
[14] Oztop H F, Varol Y, Koca A 2007. Laminar natural convection heat transfer in a shed roof with or without eave for summer season. *Applied Thermal Engineering.* **27** 2252–2265.
[15] Jaluria, Y 1980. *Natural Convection. Heat and Mass Transfer.* Vol V, ed D B Spalding (Imperial College of Science and Technology, London, England: Pergamon Press). P 326.
[16] Haese, P M, Teubner, M D 2002. Heat exchange in an attic space. *International Journal of Heat and Mass Transfer.* **45**(5) 4925– 36.
[17] Papanicolaou, E and Belessiotis, V 2005. Doubble-diffusive natural convection in an asymmetric trapezoidal enclosure: unsteady behaviour in the laminar and the turbulent-flow regime. *International Journal of Heat and Mass Transfer.* **48** 191-209.

Computational Plasmadynamics applied to parametric studies

L Bilbao
INFIP-CONICET, and Physics Department (FCEN-UBA),
Ciudad Universitaria, Pab. I, 1428 Buenos Aires, Argentina

E-mail: bilbao@df.uba.ar

Abstract. A Three-Dimensional Finite Volume Arbitrary Lagrangian-Eulerian simulation code was developed to study different plasma physics problems in 3D+t. The code is based on a complex multi-component species program with transport and radiation terms written and applied to plasma and fusion physics problems. The integration domain is represented with a structured irregular mesh, with fixed connectivity made of hexahedral cells. Coordinates and velocities are assigned to cell vertices. After each calculation cycle, mesh vertices are moved arbitrary over the fluid. The adaptive method consists of shifting mesh vertices over the fluid in order to keep a reasonable mesh structure and increase the spatial resolution where the physical solution demands. The code was a valuable tool for parametric study of different physical problems, mainly optimization of Plasma Focus Machine, detonation and propagation of thermonuclear reactions, and Kelvin-Helmholtz instabilities in the boundary layer of the terrestrial magnetopause.

1. Introduction

A new Arbitrary Lagrangian-Eulerian (ALE) Finite Volume algorithm was developed based on a previously reported code [1]. As in the previous version, Three-Dimensional (3D), time depend, multi-component, two-temperature code was used. A new algorithm for the hydrodynamics was implemented in order to improve the computational efficiency plus the improved capability of adapting the mesh to the solution.

The code includes ion viscosity, thermal conduction (electrons and ions), magnetic diffusion, thermonuclear or chemical reaction, Bremsstrahlung radiation, EOS (from the ideal gas to the degenerate electron gas).

The scope for developing the present numerical method was to perform parametric studies for optimization of several problems in plasma physics. Nowadays there exist several efficient numerical codes in the subject. However the construction of own computational codes brings the following important advantages: (a) To get a deeper knowledge of the physical processes involved and the numerical methods used to simulate them; and, (b) More flexibility to adapt the code to particular situations in a more efficient way than a closed general code would.

In the next paragraph the main improvements will be highlighted.

2. Numerical resolution

We have used an ALE Finite Volume method with hexahedral cells that move at arbitrary velocity. The notation and definition of the geometry is described elsewhere [1]. The main advantage of this method is that the geometry is imposed to the cells and not to the operators.

This method consists of integrating conserved quantities over a cell. For a general fluid property Ψ, the following identity holds

$$\frac{d}{dt}\int_{V(t)} \Psi dV = \frac{\partial}{\partial t}\int_{V(t)} \Psi dV + \int_{S(t)} \Psi \mathbf{w}\cdot d\mathbf{S} \tag{1}$$

where integration is performed over a cell of arbitrary volume *V(t)* with boundary *S(t)* moving at an arbitrary speed **w**. By choosing **w** = 0, that is both *V* and *S* remain fixed, a full Eulerian formulation is retrieved. On the other hand, setting **w** = **v**(**x**,*t*) –fluid velocity- leads to the standard Lagrangian formulation. The adaptive method consists of choosing an appropriated **w**(**x**,*t*) over the fluid in order to keep a reasonable mesh structure and increase the spatial resolution where the physical solution demands.

For the ideal MHD, the equations for mass (*M*), momentum (**P**), internal energy (*E*), and, magnetic flux (*Φ*), are

$$\frac{dM}{dt} = \frac{d}{dt}\int_{V(t)} \rho dV = -\int_{S(t)} \rho(\mathbf{v}-\mathbf{w})\cdot d\mathbf{S} \tag{2}$$

$$\frac{d\mathbf{P}}{dt} = \frac{d}{dt}\int_{V(t)} \mathbf{v}\rho dV = -\int_{S(t)} \rho\mathbf{v}(\mathbf{v}-\mathbf{w})\cdot d\mathbf{S} - \int_{S(t)} p d\mathbf{S} + \int_{V(t)} \mathbf{j}\times\mathbf{B}\, dV \tag{3}$$

$$\frac{dE}{dt} = \frac{d}{dt}\int_{V(t)} \varepsilon\rho dV = -\int_{S(t)} \rho\varepsilon(\mathbf{v}-\mathbf{w})\cdot d\mathbf{S} + \int_{S(t)} (\frac{c_p kT}{e}\mathbf{j})\cdot d\mathbf{S} - \int_{V(t)} p\nabla\cdot\mathbf{v}\, dV \tag{4}$$

$$\frac{d\Phi}{dt} = \frac{d}{dt}\int_{\Sigma(t)} \mathbf{B}\cdot d\mathbf{S} = \int_{C(t)} [(\mathbf{v}-\mathbf{w})\times\mathbf{B}]\cdot d\mathbf{l} \tag{5}$$

where **j** is the electrical current density, $\varepsilon = p/\rho(\gamma\text{-}1)$ the internal energy density (c_p is the specific heat at constant pressure) and *e* the charge of the electron. In the last equation *Σ* is an arbitrary open surface bounded by a closed path *C*, which moves according to **w**(**x**,*t*).

The integration domain is represented with a structured irregular mesh, with fixed connectivity made of hexahedral cells. Each cell is surrounded by 6 faces and 8 vertices. Fluid variables are assigned to staggered locations in the mesh. Pressure, internal energy, density, species concentration, cell volume and mass are all assigned to cell centres. Coordinates (*x,y,z*) and velocities (u_x, u_y, u_z) are assigned to cell vertices.

For example, using **w** = 0 the evolution of the cell mass $M \approx \langle\rho\rangle V$, $\langle\rho\rangle$ being a mean value of the density over de cell, and V the volume, is written as

$$\frac{dM}{dt} = -\sum_i \rho_i \mathrm{v}_i \cdot \mathrm{S}_i \tag{6}$$

where $\mathbf{S}_i$, $\mathbf{v}_i$, ρ_i, are the surface, velocity and mass density of the *i*-th face of the cell. Analogously, the momentum of a cell, $\mathbf{P} \approx \langle\rho\rangle\langle\mathbf{v}\rangle V$, evolves according to

$$\frac{d\mathbf{P}}{dt} = -\sum_i \rho_i \mathbf{v}_i (\mathbf{S}_i\cdot\mathbf{v}_i) - \sum_i p_i \mathbf{S}_i + \frac{1}{4\pi}(\sum_i \mathbf{S}_i\cdot\mathbf{B}_i)\times \langle\mathbf{B}\rangle \tag{7}$$

where <**B**> is the mean value of the magnetic field on a cell.

Using the same ideas all integral equations are discretized. In order to set the values on the faces, appropriate mean values over adjacent cells are used.

Global conservation of the above magnitudes is checked during calculation. Also, condition $div(\mathbf{B})=0$ is controlled over the whole integration domain.

The same method may be applied for more complex modelling of plasmas that includes ion viscosity, thermal conduction (electrons and ions), magnetic diffusion, chemical or thermonuclear reactions, Bremsstrahlung radiation, and realistic equation of state (EOS) from the ideal gas to the degenerate electron gas. The integration in time is sequential. This means that each process is integrated with a different uncoupled method during a time step. The overall truncation error is $O(\Delta x,\Delta y,\Delta z,\Delta t)$. Solution does not dependent on the integration order of the terms.

Calculation proceeds as follows:

- Hydrodynamics is integrated using an improved Predictor-Corrector method as described in the next paragraph.
- Diffusion processes (of species concentration and energy) are integrated with ADI methods (Alternating Direction Implicit) for the 2D/3D case, and with full implicit methods for the 1D case.
- Source terms from chemical or nuclear reactions are explicitly integrated in time.
- Mesh vertices are moved arbitrary over the fluid as it is described below.

2.1. Hydrodynamics

By "*hydrodynamic terms*" we refer only to the Euler equations included into the full set of conservation equations. So in this part of the calculation fluid variables are updated in time assuming that the other processes (diffusion, reaction, convection) are temporally frozen during one time step.

The calculations necessary to advance a solution one step in time Δt are separated in two distinct phases. Phase One consists of an explicit Lagrangian calculation, except mesh vertices are not moved. Phase Two consists of an iteration that adjusts the pressure gradient forces to the advanced time level. This phase eliminates the usual Courant–like numerical stability condition that limits sound waves to travel no further than one cell per time step. In an explicit method pressure forces can be transmitted only one cell each time step, that is, cells exert pressure forces only on neighbouring cells. When the time step is chosen so large that sound waves should travel more than one cell, the one cell limitation is inaccurate and a catastrophic instability develops. To overcome this instability time-advanced pressure gradients may be used. Unfortunately the time–advanced pressures depend on the accelerations and velocities computed from those pressures, so an iterative calculation of the set of equations is required. Physically, iteration offers a means by which pressure signals can traverse across more than one cell in a unit time step.

2.1.1. Phase One, Part One. The velocity components of vertex are explicitly advanced in time using pressure gradients computed from the currently available pressure and mesh coordinates.

2.1.2. Phase Two. The target of this phase is to obtain new velocities that have been accelerated from time-advanced pressure gradients. So this is an implicit problem and iteration must be used. The desired pressure for cell i, must be solution of:

$$p_i^L - f\left(\rho_i^L, \varepsilon_i^L\right) = 0 \tag{8}$$

Density and internal energy at the time level $n+1$ can be obtained from the values at the time level n with the following iteration

$$\rho_i^L = \rho_i^n \frac{V_i}{V_i^*} \tag{9}$$

$$\varepsilon_i^L = \varepsilon_i^n + \frac{p_i^n}{\rho_i^n}\left(1 - \frac{V_i^*}{V_i}\right) \tag{10}$$

where V^* is the volume that the cell would have if its vertices were moved using the velocities calculated during Part One.

Finally, a solution for p_i^L can be obtained by applying Newton-Raphson iteration to (8) by performing the following steps:

(a) Compute V^* using the most updated values for the velocity (**v**).

(b) Compute new guesses for ρ_i^L and ε_i^L with (9) and (10).

(c) Compute a pressure correction δp_i according to

$$\delta p_i = -\frac{p_i^L - f\left(\rho_i^L, \varepsilon_i^L\right)}{S_i} \tag{11}$$

where S_i is a relaxation factor computed once for each cell at the beginning of phase two.

(d) Adjust the current guess for cell pressure

$$p_i^L = p_i^L + \delta p_i \tag{12}$$

(e) Adjust the velocities at the corners of cell i to reflect this pressure change:

The mesh is repeatedly swept and calculations (a)–(e) are performed once for each cell, until no cell exhibits a pressure change violating the inequality:

$$\left|\delta p_i^L / p_{\max}\right| < \varepsilon \tag{13}$$

where ε is a suitable chosen small number and p_{max} is the maximum pressure in the mesh.

Phase Two concludes when condition (13) is reached. It can be shown that the number of iterations to reach convergence (N) is approximately:

$$N \approx \left(\frac{u\delta t}{\delta x}\right)\frac{1}{\sqrt{\varepsilon}} \tag{14}$$

where δx is a typical cell dimension. Thus, Phase Two converges in a finite number of iterations, regardless of the actual material sound speed. It is this feature of the ICE method that makes it superior to an ordinary explicit method whose time step must be continually reduced as the sound speed is increased.

2.1.3. Phase One, Part Two. The final values of **v** and p obtained from the Phase Two must be used to update vertex coordinates and cell volume and density. In addition cell energy must be accordingly adjusted.

2.2. Mesh rezoning

Lagrangian cell methods are not adequate for describing flows undergoing large distortions, because cells may undergo severe deformations.

Nuclear burning or combustion structures may present steep gradient of temperature, species concentration and heat release. The adaptive method consists of shifting mesh vertices over the fluid in order to keep a reasonable mesh structure and increase the spatial resolution where the physical solution demands.

The rezone velocity **U** with which vertices are moved over the fluid is calculated in order to increase spatial resolution where steep temperature gradients and high chemical/nuclear heat release are produced. To do so, a distribution function for vertices concentration is calculated over a 1D integration path. The new coordinates for each vertex are calculated in such a way that the Lagrangian distance remains uniformly distributed. Another possibility is to define **U** in order to restore an Eulerian method. This calculation of **U** is 1D, where the coordinate represents the path length following one computational direction. The procedure is repeated over several paths where corrections are needed.

Vertex *i* moves from *i* to *i'* according to

$$\mathbf{x}_i' = \mathbf{x}_i + \mathbf{U}\Delta t \tag{15}$$

where **U** is the rezone velocity. During this process there is an exchange of mass, species, energy and momentum among neighbourhood cells.

Then the following steps must be applied

- Calculate the new coordinates using (15), then calculate the volume exchange among neighbourhood cells
- Associated with the volume exchange there will also be a mass, species concentration and total energy exchange. The mass, concentration or energy per unit mass assigned to this volume are computed using a donor cell flux. Finally, fluid conserved properties are updated.
- Once updated the conserved cell and vertex properties, cell volume, density and species concentration are calculated from the new vertex coordinates.

3. Parametric study of a cylindrical pinch

Self-constricted plasma configurations occur in a number of situations, including geophysics (lightning) and astrophysics (current channels at galactic scales). Tonks [2] introduced the term "pinch" to describe these configurations. It was soon discovered however, that the equilibrium pinch suffered from large number of magneto-hydrodynamic instabilities, including sausage and kink instabilities.

Plasma Focus and Z-pinch are machines based on this principle. They are an interesting source of pulsed radiation (including thermonuclear reactions) with a broad utility in many technological applications.

The scope of this study was, on the one hand, to give an approximated description of the scaling law of the thermonuclear reactions in a Plasma Focus (PF) discharge, and, on the other hand, to explore the optimal conditions for generation and propagation of thermonuclear reactions in a cylindrical fibre (Z-pinch).

3.1. Scaling of a Plasma Focus device

The PF dynamics is usually divided into four stages (or phases): breakdown, run-down, convergence and pinch (or focus). A parametric study requires that the code should be run thousand times, In order to speed up the calculation only the last stage is fully modelled, while the previous phases set the initial values for the pinch phase.

Besides the fact that the detailed physics of a PF is rather complicated, the dynamics of the run-down and convergence phases can be reasonably described using simplified models. The "snow-plough" model (SP) assumes that the CS is pushed forward by the magnetic field against a neutral gas [3], the profile is derived from momentum equilibrium and the inner structure is neglected. The SP model assumes that no diffusion of the magnetic field is present, and that the inertia terms are negligible (see, for example [4]). The snowplough model was successfully confronted against both experimental and numerical results (2D plasma fluid codes were available more than 30 years ago [5-6]). In addition, the simplified model used here assumes that a "good" breakdown is produced (that the initial discharge is concentrated in a small volume close to the insulator surface).

The SP model can be used until the travelling profile arrives to the axis (see figure 1). At this point (pinch phase) the SP model is no longer valid, thus the full 2D code is used.

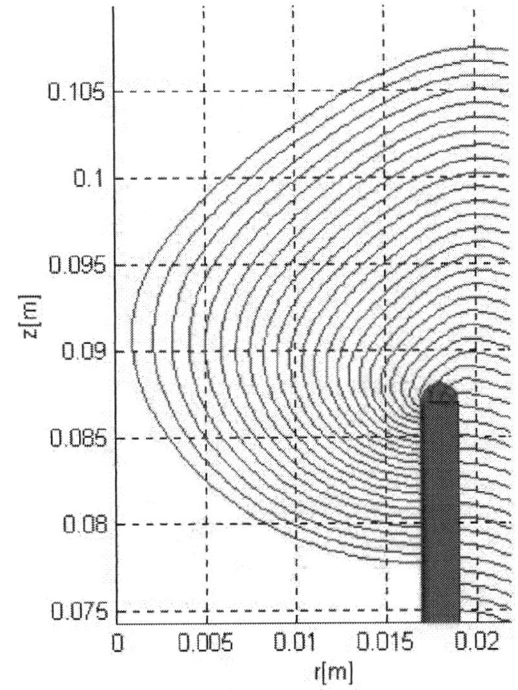

Figure 1. The CS at successive time steps before arriving to the axis ($r = 0$). The inner electrode is also shown. The final step of the snow-plough model is used as the initial condition for the 2D code.

3.2. Pinch stage

During the convergence phase, both the magnetic piston and the shock wave move towards the axis. The pinch phase starts when the shock reflects off the axis. Since the magnetic piston continues its converge, a further compression and heating of the plasma occurs until the reflected shock hits the piston. This stage is usually called "first compression".

Due to the fact that the piston is not a straight cylinder, the maximum compression occurs at different time for different axial positions, leading to the development of an $m = 0$ instability producing several necking (called "second compression" or necking phase).

Nuclear reaction and X-ray emission are produced in the pinch phase. The mechanism of nuclear reaction of a Plasma Focus device remains still unclear. At least three different mechanisms can be devised: thermonuclear production along the whole column as a result of conversion of kinetic energy into internal energy during the first compression, thermonuclear production due to the development of instabilities and further heating (i.e. anomalous resistivity), and beam-target mechanism that can also have different contributions (for example, trapped ions in the pinch or ion beam against the neutral gas). From an experimental point of view some measurements were carried to clarify this point. For example, the anisotropy should be able to differentiate between steady thermonuclear contribution from beam-target or fast moving thermonuclear production. Besides anisotropy there are many other characteristic that are particular for each production: maximum of the neutron production as a function of the filling pressure, scaling with the current, ratio between DD and DT outputs, influence of contaminants, etc. Probably the above mentioned three mechanisms may share one or more of these properties, but quite probably they can not share all the same properties.

3.3. Pinch simulation

In order to get reasonable initial conditions for the compression stage, a simplified snow-plough model was used for reproducing the coaxial and convergence stage. In figure 1, the CS at successive time steps during the convergence stage is shown. The corresponding current at the final time of the convergence is taken as the initial current for the compression.

For the pinch stage a 2D, time dependent, two-fluid (ions and electrons), two-temperature, model was used. This set of equations was coupled to the circuit equation that represents the main discharge of the device, through

$$\frac{d[(L_o + L_b)I]}{dt} = \frac{Q}{C} - RI - \int_b d\mathbf{l} \cdot \mathbf{E}' \tag{16}$$

where Q is the charge of the capacitor bank, C the capacity, Lo and R the inductance the resistance of the (concentrated) circuit, respectively, L_b the inductance calculated at the plasma boundary (b refers to the plasma boundary) and I the current.

In order to characterise the neutron production of the first compression thousands of runs changing the initial parameters must be performed.

We have used data from 5 different Plasma Focus, namely, PFII (BA), Nessi (NE), Speed (SP), Frascati (FR) and Brasimone (BR). The main parameters are (see table I): anode length l and radius a, cathode radius b, bank capacity C, external inductance L_0, charging voltage V_0, initial filling pressure (p_0) and the gas type (only pure D_2 were used in the present calculations). These machines were selected because they cover an ample range of energy and velocity, and results from them are readily available in the literature. The Brasimone Plasma Focus is the solely one that worked in a repetitive fashion.

Table 1. List of parameters (a: inner radius, b: outer radius, l: anode length, C: bank capacity, L_o: external inductance, V: charging voltage).

Case	a(cm)	b(cm)	l(cm)	C(μF)	L_0(nH)	V(kV)
BA	1.8	3.6	12.	10.5	31.	20÷40
NE	3.3	6.25	21.	225.	20.	15÷25
SP	2.5	7	6.5	1.56	40.	120÷220
FR	8.	12.	56.	1250.	16.	20÷40
BR	1.65	4.5	13.5	30.5	100	21

The calculations proceed as follows. For each case and for each possible value of V_0 the initial filling pressure is browsed until the maximum neutron yield is obtained.

In figure 2 the calculated neutron output (Y_n) is plotted as a function of the pinch current. The best fit is given by

$$Y_n = 6 \times 10^9 \, I_p^4 \tag{17}$$

that corresponds fairly well to measured values. Also the neutron output is plotted as a function of bank energy in figure 3. Note that at higher energies Plasma Focus devices are less efficient since the ratio between the output energy (proportional to the thermonuclear yield) and the input energy (electrical energy) diminishes with bank energy.

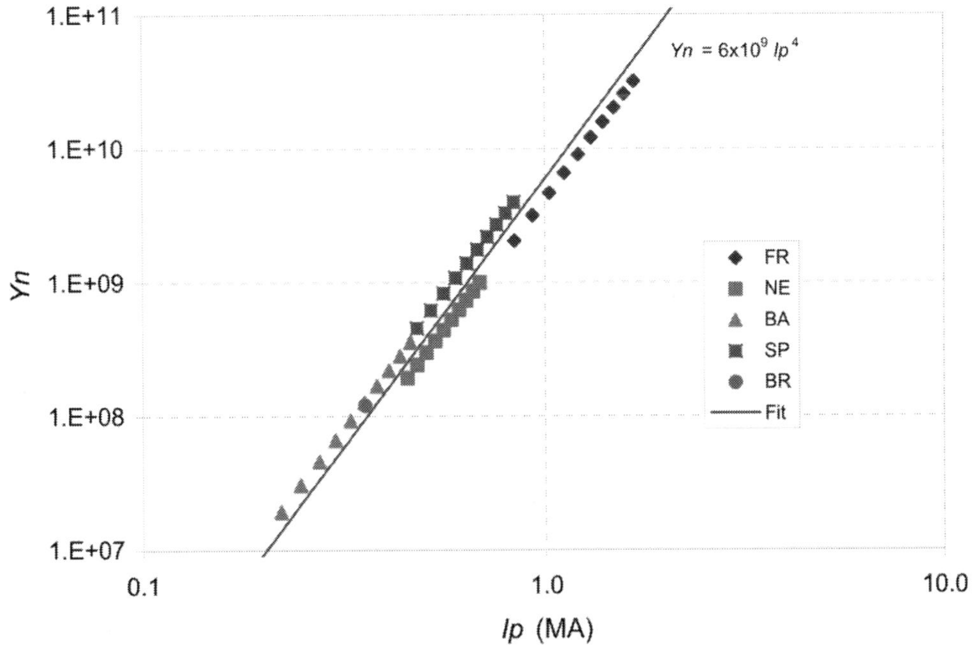

Figure 2. Neutron production (Y_n) vs. pinch current (I_p).

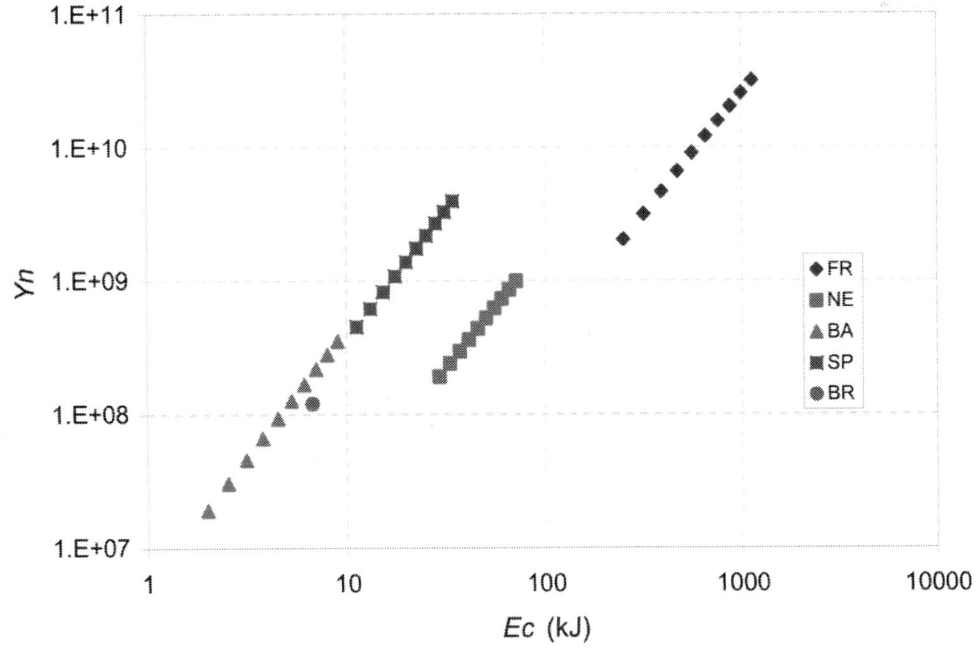

Figure 3. Neutron production (Y_n) vs. bank energy (E_c).

In order to clarify this behaviour, in figure 4 the ratio between ion and electron temperature at the end of the first compression phase are plotted against the energy. During convergence and compression the ions are heated by a shock wave, while electrons are heated by collisions with atoms (joule heating is less important before the development of instabilities). From figure 4 it is apparent that at high energies the equipartition time is relatively smaller than the obtained at low energies, thus the ions are expected to be relatively cooler. This means that high energy machines are less efficient in neutron production than smaller machines. This fact seems to be linked to the necessary relation, set by construction, between the parameters that prevents to optimize all stages at the same time.

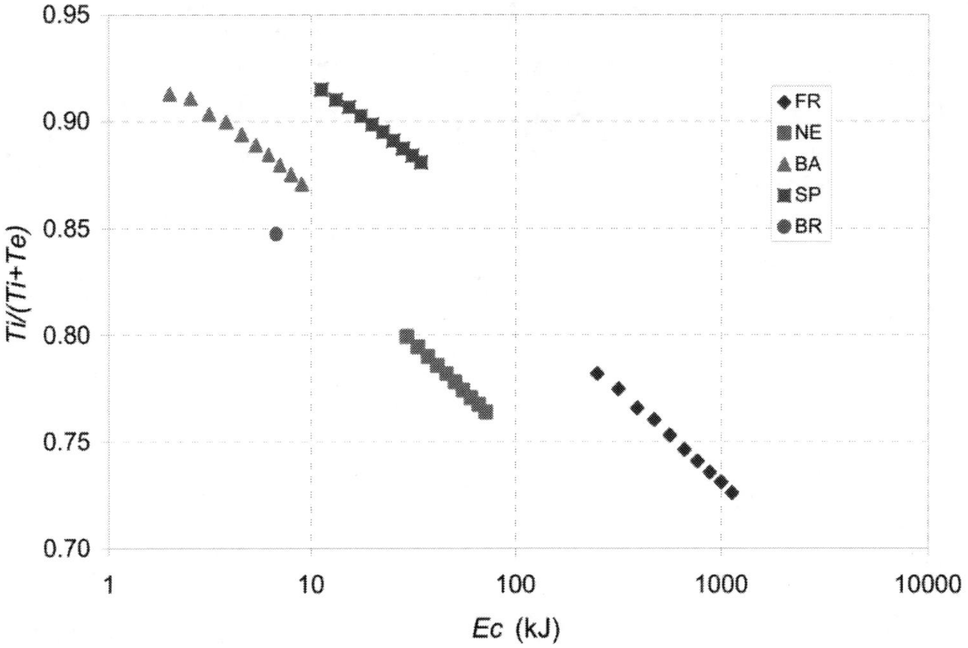

Figure 4. Temperature ratio (T_i/T_i+T_e) vs. bank energy (E_c)

3.4. Pinch analysis
The numerical 2D simulation was a useful tool for the parametric study of a Plasma Focus. The numerical simulation helped to understand the important role of the equipartition time on the neutron production. It is apparent from these results that larger machines are less efficient than smaller ones.

The cooling effect of the equipartition time has not to be confused to the drastic drop in neutron production that happens at higher energies. A possible explanation, based on the results from the 2D code, is that assuming that "good" and "bad" breakdown regions are separated by the line $p_o \propto V_o$ [7] then always will be "crossed" by the pressure of maximum production, $p_o \propto V_o^2$, at higher energies. This behaviour is confirmed by the present parametric study, which means that the standard PF configuration cannot be simultaneously optimized for breakdown and nuclear production.

4. Generation and propagation of thermonuclear reactions.
The idea of a simple nuclear reactor based on a Z-pinch has been discarded because the heating of a whole plasma column to high temperatures will require unrealistic energy input for the plasma heating [8-11]. Another approach [12] to the production of a positive energy yield consists of a small pinch zone heated to high temperature only. Experiments on Z-pinches always concern low and high temperatures and high-density plasma regions that arise spontaneously in Z-pinch necks. A burn wave might be initiated in the Z-pinch column if in one of this small plasma region a Lawson-like condition

is fulfilled. The nuclear energy, which is produced because of a burn wave spread, is controllable and is sufficient for the compensation of energy losses for the pinch production.

Perhaps the most promising application of liner implosions is the volume or spark ignition in cylindrical plasma. During the last decade fast implosions of annular current sheaths (fast Z pinches) have made important progress in reaching high energy densities. New possibilities for a broad spectrum of experiments, from X-ray generation to controlled thermonuclear fusion are devised. At present Z pinches are the most intense laboratory X-ray sources; powers in excess of 200 TW have been obtained [13,14].

At the same time experiments have shown [15,16] that a cylindrical liner can successfully confine a magnetised, hot plasma column [17]. The technical parameters necessary for Inertial Confinement Fusion (ICF) applications (in particular, volume and spark ignition in plasma compressed by an imploding liner) have been investigated through numerical simulations. These calculations provide an insight into the structure of the liner and the plasma and indicate the range of parameters required for ignition and detonation in such a liner-pinch system.

Simple 2D simulations of a D-T nuclear detonation inside a cylindrical liner have been performed by Avrorin *et al* [18].

In order to find the ignition criteria and the correct structure it is necessary to perform two-dimensional numerical simulations using a realistic plasma fluid model and corresponding transport coefficients.

The system of equations is based on a two-fluid model of plasma with transport and viscosity coefficients used in [19] and the fusion cross sections from [20].

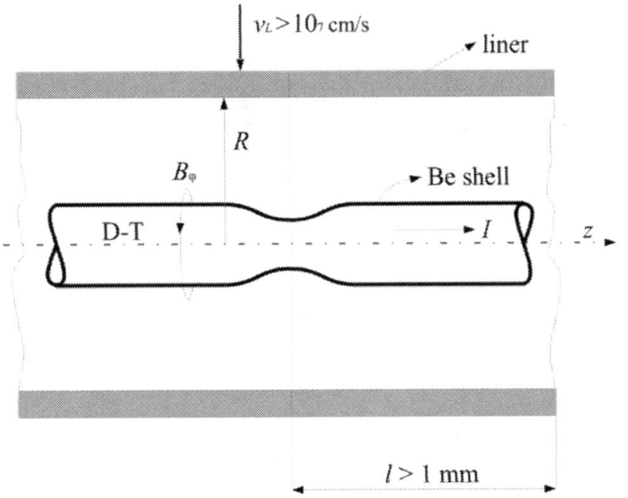

Figure 5. Implosion of cylindrical liner on a Z-pinch.

A promising approach is based on a spark created in a neck of a dense Z-pinch. The neck is formed as a result of an $m = 0$ instability starting from an initial small $m = 0$ axial perturbation. The advantage of this approach is that the spark can be programmed to occur at the moment of maximum pinch current at the correct place and time by choosing suitable initial configuration. The disadvantage is that one must provide an initial magnetic flux and the associated current I_0 and that the current rise time t_r must be short, in most cases $t_r < 10$ ns. Considering the short t_r's required, it seems that the direct use of present day megampere generators [13,14] is inadequate and the only possible means for getting $dI/dt = 10^{15}$ A/s is the compression of a B_φ-flux of a Z-pinch by a fast imploding liner (figure 5).

The spark created in a neck of a dense Z-pinch can ignite a fusion detonation in the adjacent D-T plasma channel. An $m = 0$ instability of a Z-pinch carrying a current of the order of 10 MA, with a rise time inferior to 10 ns can generate a spark capable of igniting a fusion detonation in the adjacent D-T plasma channel. Such μZ-pinch may be produced by a fast implosion of a cylindrical liner, while a conical channel properly chosen can amplify the spark energy. In order to derive some general rules for the parameters of the spark, the transition, the cylinder of advanced fuel and the liner different numerical models were used.

In the following example, the initial pinch parameters are: radius $a = 0.2$ cm, temperature $T = 30$ eV, linear density $N = 5.2\times10^{19}$ part/cm.

Different liner mass and velocity were used. Also, we have used several initial velocities for the liner ranging from slow liners ($v = 2\times10^6$ cm/s) to fast liners ($v = 3\times10^7$ cm/s). The initial current was $I = 1$ MA.

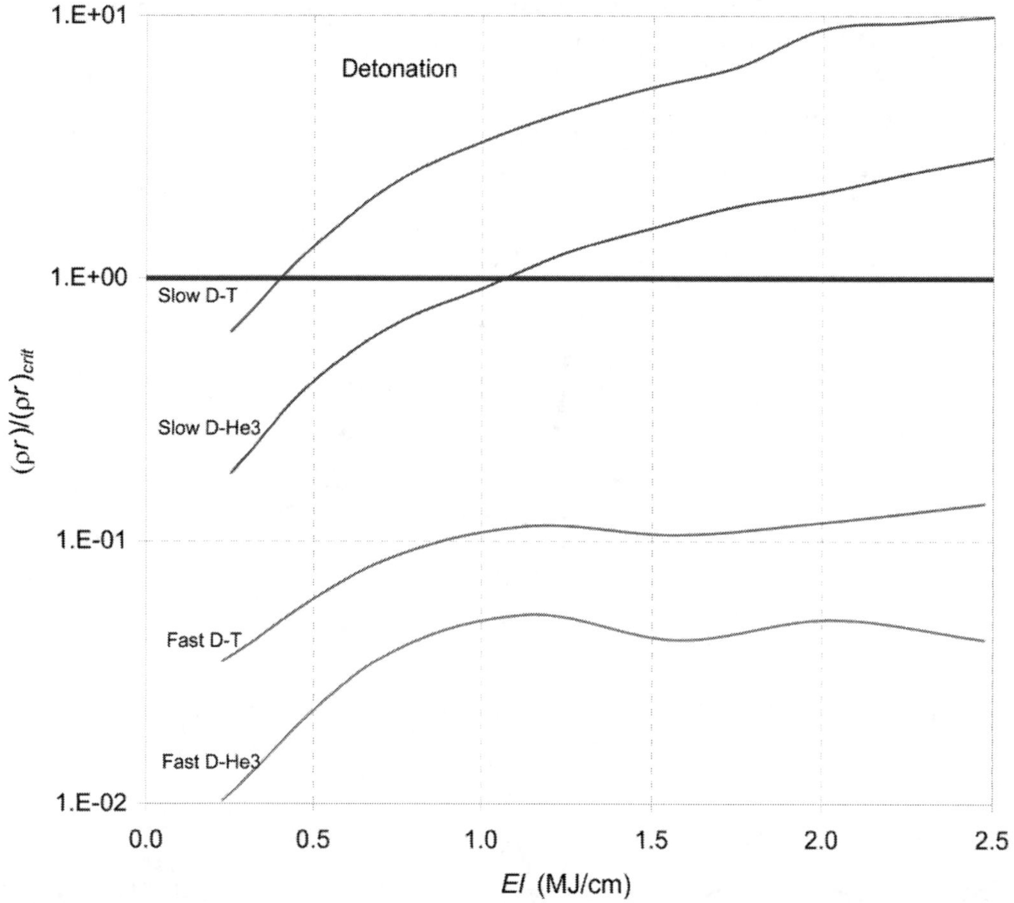

Figure 6. Detonation criteria for slow and fast liner for D-T and D-He3 plasma. Fast liners do not provide enough inertia to confine the plasma up to the detonation conditions.

Note that at the same kinetic energy value of the liner, the slow one provides a larger tamping effect than the faster one, while the faster one produces a larger temperature rise in the pinch than the slow one.

In D-T the condition for detonation propagation is fulfilled at 0.5 MJ/cm liner kinetic energy level (see figure 6). D-He3 plasma requires liner energies above 1.1 MJ/cm in order to reach detonation conditions.

Although the detonation propagation condition is achieved at about 0.5 MJ/cm of liner energy (figure 6), the trigger energy necessary for starting the burn is higher than the required in the previous case. For D-D ignition is obtained at the level of 1.3 MJ/cm, while for D-He3 the ignition criterion seems to be impossible to reach within reasonable liner kinetic energy. Note that, the fast liner produces a larger temperature in the plasma (see figure 7). This has the effect of increasing the thermonuclear output that inhibits a large compression. For this reason both the detonation and ignition are hardly to reach using fast liners.

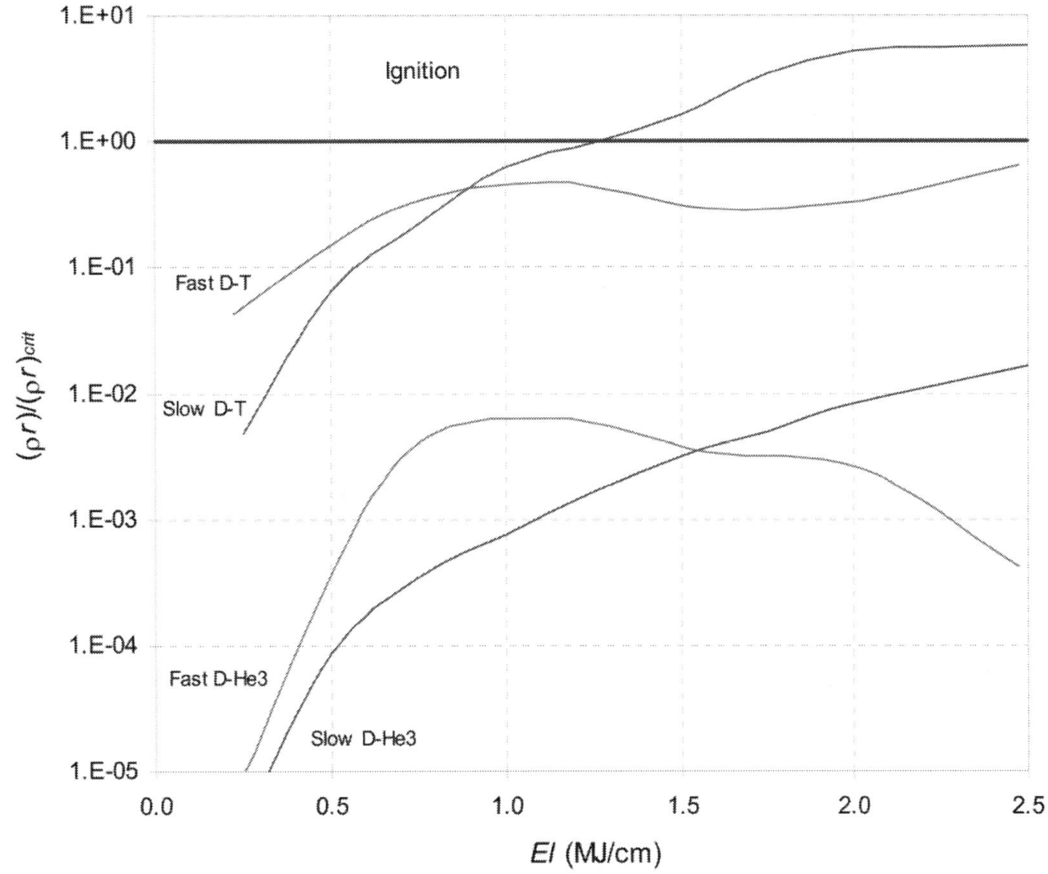

Figure 7. Ignition condition for slow and fast liner for D-D and D-He3 plasma.

Note that for detonation propagation the plasma does not have to be heated to the ignition temperature, and therefore the resulting energy required for detonation is much smaller than that for volume ignition. Provided the spark energy can be neglected with respect to the total energy, the spark ignition scheme is energetically much more advantageous than the volume ignition.

Slow liners can be useful for compressing the plasma up to the detonation propagation condition in both D-T and D-He3. The liner energy required is about 1 MJ/cm. At the same time, the development of the $m = 0$ instability, creates a region where the ignition condition may be reached. At that point a thermonuclear detonation is triggered; it can then propagate along the pinch if the adequate conditions were already achieved. The numerical code can help in finding the best regime to this purpose.

5. Development of the KH instability

Another physical problem studied by means of the numerical code is related to the turbulent mixing layers of plasma at the near equatorial flanks of the terrestrial magnetopause. The numerical 3D code was used to simulate the large amplitude perturbations and waves at the dusk side low latitude boundary layer of the magnetopause generated by an interplanetary tangential discontinuity.

We focused on the motion and structure of the magnetopause/boundary layer observed in response to a joint tangential discontinuity/vortex sheet (TD/VS) observed by the ACE spacecraft on December 7, 2000. Sharp polarity reversals in the east-west components of the field and flow occurred at the discontinuity. These rotations were followed by a period of strongly north-ward IMF. These two factors elicited a two-stage response at the magnetopause, as observed by Cluster spacecraft situated in the boundary layer at the dusk side terminator.

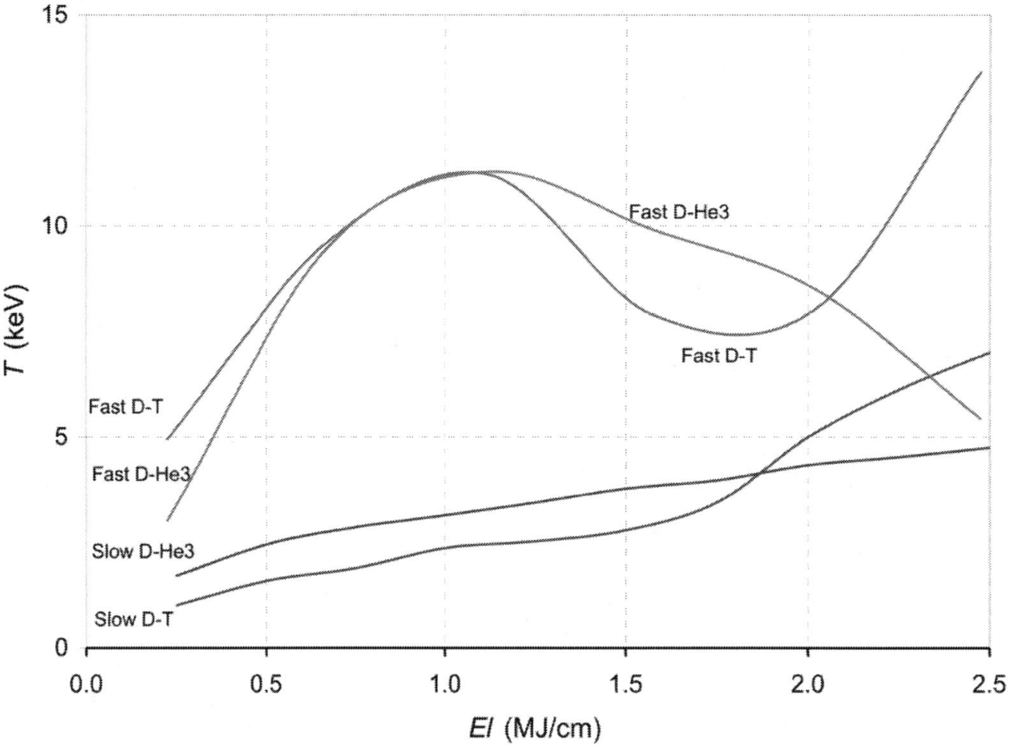

Figure 8. Temperature at the centre of the pinch for slow and fast liner for D-D and D-He3 plasma.

These studies were aimed to clarify some particular physical properties of the instability of velocity shear flows and mixing layers, such as the 3D vorticity amplification mechanism and non conservation of vorticity flux.

The most common of these is the Kelvin-Helmholtz instability feeding on the flow shears which exist across the magnetosheath/magnetosphere interface (see figure 9). Ever since Dungey [21] pointed out the possibility of the magnetopause going Kelvin-Helmholtz unstable, many studies have been devoted to this instability, the surface waves it causes, their coupling to the geomagnetic field giving rise to field line resonances, and how this instability might mediate the entry of magnetosheath plasma into the magnetosphere [22-31].

Attention has been drawn to two other interplanetary parameters whose changes may occasion considerable deformation and motion of the magnetopause. One is a variable orientation of the interplanetary magnetic field (IMF). Fairfield et al. [32] showed that, even when the dynamic pressure is constant, the magnetopause and inner-magnetosphere boundaries may be in motion. They attributed these motions to pressure changes generated in the foreshock region by the IMF changes. These pressure changes subsequently convect through the bow shock and impinge on the magnetopause. The authors thus concluded that even in the presence of a solar wind that is absolutely steady in velocity and density but which carries an embedded interplanetary magnetic field of variable orientation; there will be a variation of pressure exerted on the magnetopause.

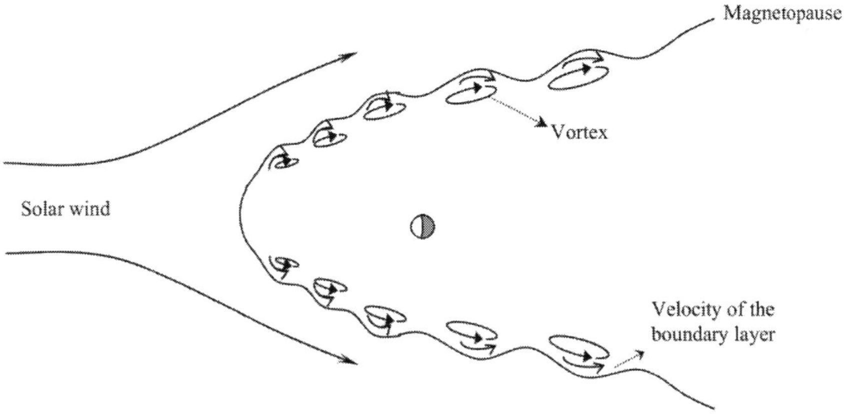

Figure 9. Sketch of vortex formation due to the Kelvin Helmholtz instability in the magnetopause

For the observations at the magnetopause we focused on Cluster 3 data. The Cluster spacecrafts were on an outbound trajectory just tail ward of the dusk terminator and at northern latitudes. (At this early stage of the mission, the spacecraft separation was of order a few hundred km and all spacecraft see the same features.). For some hours prior to our observations, the spacecraft stayed in the vicinity of the magnetopause because the magnetopause boundary was expanding slowly outward due to a slow decrease in dynamic pressure. From 14:00 to 14:30 UT the spacecraft stays within a distance of 0.4 RE from the magnetopause. We may reasonably assume that Cluster 3 is inside the boundary layer according to the model, in agreement with instrument readings.

We carried out a stability analysis of a limited segment of the magnetopause, where the boundary layer may be considered to lie in a local tangent plane. The boundary layer is then modelled as a planar plasma slab with stratified density, flow velocity, and magnetic field. The physical quantities on either side of the magnetosheath-magnetosphere transition are obtained from the Cluster 3 readings at the extremes of the large oscillations recorded after 13:58 UT. Specifically, we selected the second oscillation, where the lowest speed was recorded at maximum T. That is, we assume that at the extremes of the oscillation, Cluster 3 samples alternately the high density-low temperature magnetosheath plasma, and the low density-high temperature magnetosphere plasma. In this way the local instability is examined by parameters measured locally and in temporal proximity.

The full description of the problem and main results are presented in a separated paper in these Proceedings [33]. Anyhow, some sample outputs showing the 3D, complex character of the problem are presented. Figure 10 shows streamlines complexity and formation of a swirling flow in the large eddy at $t = 84$ s. In figure 11 temperature, density, pressure and magnetic lines plots show that the KH waves appear already after a few theoretical e-folding times, due to the great oscillations generated by the tangential discontinuity stress that has reduced the amplification time. In order to get an appropriated resolution, up to 1,000,000 cells were used.

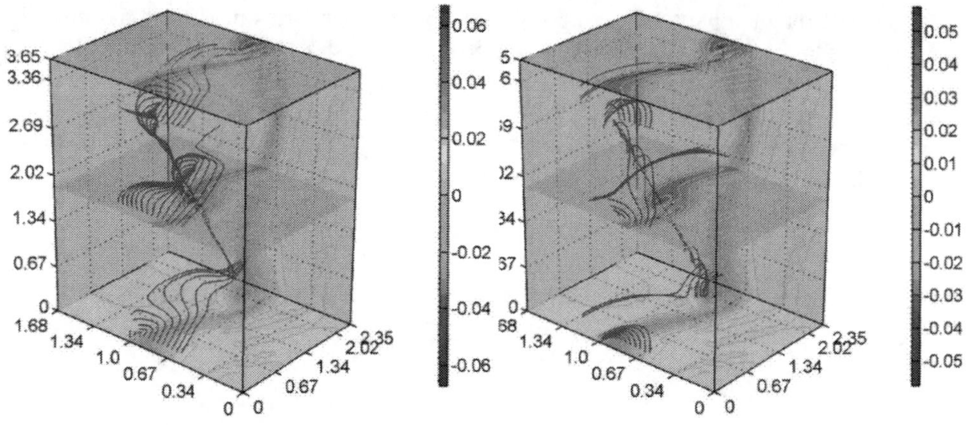

Figure 10. Mixing layer, $t = 84$ s, vorticity contours and streamlines. Comparing streamlines views: left, fixed (Earth) frame, and right, moving frame with average velocity ($U = 78.5$ km/s). In the moving frame magnetosheath and magnetopause plasmas are seen counter streaming.

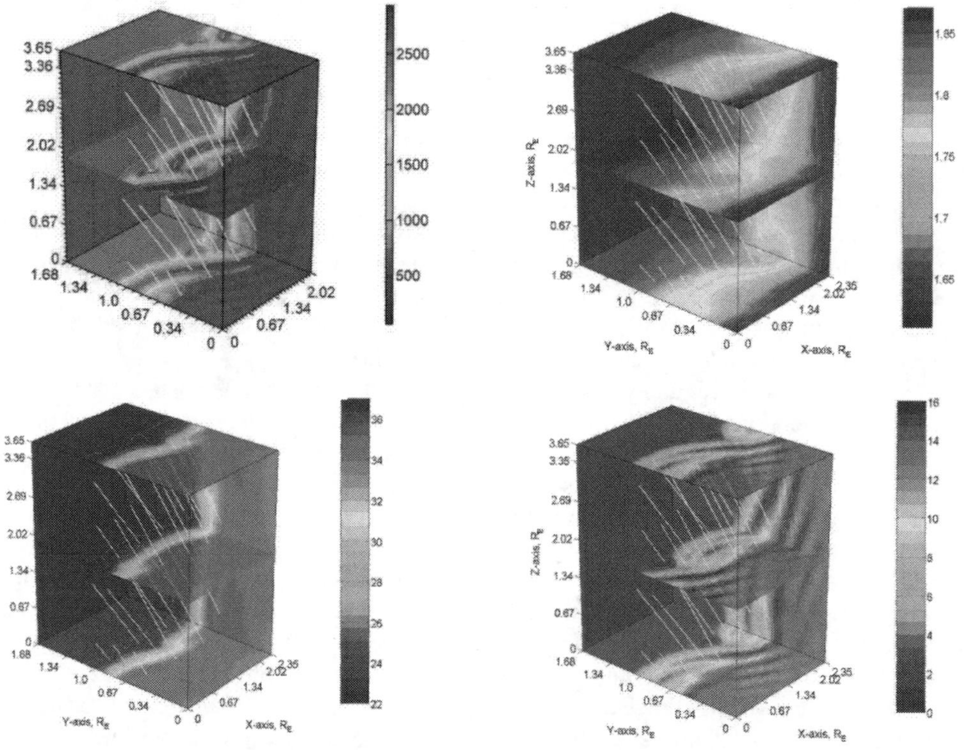

Figure 11. Scalar fields and magnetic lines at $t = 180$ after an e-folding time. Upper left: temperature; upper right, isobars; lower left, $|\mathbf{B}|$; lower right, density.

Some interesting features of the phenomenon are obtained from the numerical results. For example, high temperature rises in a compact vortex core correlated with local density depletion while gas, and magnetic pressure (roughly constant across the structure) are near to surrounding values. Also detected a high density patch in a vorticity structure correlated with a local increase of gas pressure, whereas temperature stays near to surrounding values.

Concomitant with 3-D vortex stretching, kinetic helicity density rises at the vorticity cores. Finally, non conservation of vorticity becomes manifest at about (or even before) one rollover time, where in addition to vortices with positive rotation (the same sign of the original vorticity sheet) other coherent structures with strong negative vorticity also arise.

6. Conclusions

The numerical simulation was a powerful tool for the parametric analysis of pinch compression and development of Kelvin-Helmholtz instability.

The sequential implementation resulted of practical use, because each physical process was treated with a different numerical method.

The improved hydrodynamics method worked well in an ample range of Mach number from subsonic (10^{-3}) to supersonic.

The time step was restricted by efficiency considerations rather than stability conditions. When time step becomes larger, the calculation progress faster because fluid variables, transport properties, etc., are updated with less frequency. But the implicit stage (phase two) of the hydrodynamic advance requires more iterations to perform. An optimum time step has been used for which the minimum computational effort is obtained.

7. Acknowledgments

Work partially supported by grant UBA X206, University of Buenos Aires.

8. References

[1] Bilbao L 1990 *J Comp Physics* **91** 361.
[2] Tonks L 1937 *Trans Electrochem Soc* **72** 167.
[3] Butler T D, Henins I, Jahoda F C, Marshall J, Morse R L 1969 *Phys Fluids* **12** 1904.
[4] Gratton F, Vargas M 1976 *Proc. Energy Storage Compression and Switching* (Bostick, Nardi, Zucker Eds Plenum Press).
[5] Potter D E 1971 *Phys Fluids* **14** 9.
[6] Maxon S, Eddleman J 1978 *Phys Fluids* **21** 10.
[7] Gourlan C, Kroegler H, Maisonnier Ch, Rager J.P, Robouch B, Bertalot L, Gentilini A, Arcipiani B, Pedretti E, Steinmetz K 1978 *Proc 2nd Int. Conf on Energy Storage Compression and Switching* (Plenum Press, New York) 221.
[8] Artsimovich L A 1961 Controlled Nuclear Reactions *Fizmatizdat* Moscow.
[9] Bakshaev Yu L, Blinov P I, Chernenko A S *et al* 2000 *Czech J Phys* **Suppl 50** (3) 121-126.
[10] Batunin A V, Bulatoy A N, Vikharev V D 1990 *Fizika Plasmy* (*Sov J of Plasma Physics*) **16** 1027-1035.
[11] Gol'berg S M, Liberrnann M A, Velikovich A L 1990 *Plasma Phys Contr Fusion* **32** 319-326.
[12] Ryutoy D D, Derzon M S, Matzen M K 2000 *Reviews of Modern Physics* **72** 167-223.
[13] Branitski A V *et al* 1996 *Proc Beams 96* Prague 140.
[14] Quintenz J P and Sandia Team 1998 *Proc Beams 98.*
[15] Maisonnier Ch *et al* 1965 *Proc Plasma Phys And Contr Nucl Fus Reserch* (IAEA Culham) **Vol 2** 345.
[16] Bortolotti A *et al* 1993 *Proc 3rd Int Conf On DZP* (London) **AIP 299** 372.
[17] Degnan J H *et al* 2001 *IEEE Transactions on Plasma Science* **29** 93.
[18] Avrorin E N, Bunatyan A A, Gadzhiev A D *et al* 1984 *Fiz Plasmy* **10** 514 [*Sov J Plasma Phys* **10** 298].
[19] Bilbao L, Linhart J G 1996 *Plasma Phys Reports* **22** 457.
[20] Bosch H S, Hale G M 1992 *Nuclear Fusion* **32** 611.

[21] Dungey J W 1955 "Electrodynamics of the outer atmospheres" Report 69, Ions Res Lab. Pa State Univ, University Park.

[22] Southwood D J 1979 *Proceedings of Magnetospheric Boundary Layers Conference* ESA SP-148 Eur Space Agency Paris 359

[23] Miura A 1984 *J Geophys Res* **89** 801

[24] Chen S H, Kivelson M G, Gosling J T, Walker R J, Lazarus A J 1993 *J Geophys Res* **98** 5727.

[25] Kivelson M G, Chen S H 1995 in *Physics of the Magnetopause* (edited by P Song, B U O Sonnerup and M F Thomsen) *Geophys Monogr Ser* **90** AGU Washington DC 257.

[26] Seon J L, Frank A, Lazarus A J, Lepping R P 1995 *J Geophys Res* **100** (11) 907.

[27] Fairfield D H, Otto A, Mukai T, Kokubun S, Lepping R P, Steinberg J T, Lazarus A J, Yamamoto T 2000 *J Geophys Res* **105** (22),159.

[28] Otto A, Fairfield D H 2000 *J Geophys Res* **105** (A9) 21175-21190.

[29] Farrugia C J, Gratton F T, Contin J, Cocheci J C C, Arnoldy R L, Ogilvie K W, Lepping R P, Zastenker G N, Nozdrachev M N, Fedorov A, Sauvaud J A, Steinberg J T, Rostoker G 2000 *J Geophys Res* **105** (A4) 7639-7667.

[30] Gratton F T, Bender L, Farrugia C J, Gnavi G 2004 *J Geophys Res* **109** (A04211) 1-13.

[31] Farrugia C J, Gratton F T, Torbert R 2001 *Space Sci Rev* **95** (1/2) 443-456

[32] Fairfield D H, Baumjohann W, Paschmann G, Luehr H, Sibeck D G 1990 *J Geophys Res* **95** 3773-3786.

[33] Gratton F T, Bilbao L, Farrugia C J, Gnavi G 2008 "Large eddy simulations in MHD: the rise of counter-rotating vortices at the magnetopause" this *JOP Conf. Series*.

On the influence of the nozzle length on the arc properties in a cutting torch

L Prevosto[1,†], **H Kelly**[2,‡], **M Risso**[1] **and D Infante**[1]

[1] Grupo de Descargas Eléctricas, Departamento Ingeniería Electromecánica, Universidad Tecnológica Nacional, Regional Venado Tuerto, Las Heras 644, Venado Tuerto (2600), Santa Fe, Argentina.
[2] Instituto de Física del Plasma (CONICET), Departamento de Física, Facultad de Ciencias Exactas y Naturales (UBA) Ciudad Universitaria Pab. I, (1428) Buenos Aires, Argentina

E-mail: prevosto@waycom.com.ar

Abstract. In this work, an experimental study on the influence of the nozzle geometry on the physical properties of a cutting arc is reported. Ion current signals collected by an electrostatic probe sweeping across a 30 A oxygen cutting arc at 3.5 mm from the nozzle exit were registered for different nozzle lengths. The temperature and density radial profiles of the arc plasma were found in each case by an inversion procedure of these signals. A comparison between the obtained results shows that the shorter nozzle (R_N = 0.50 mm, L_N = 4.5 mm operated at 0.7 MPa and 35 Nl/min) produces a thinner and hotter arc than the larger nozzle (R_N = 0.50 mm, L_N = 9.0 mm operated at 1.1 MPa and 20 Nl/min). This behavior is attributed to the marked difference of gas flow rate due to the clogging effect. A smaller gas mass flow reduces the convective cooling at the arc border and decreases the power dissipation of the arc column, resulting in small axis temperatures.

1. Introduction

Transferred arc plasma torches are widely used in industrial cutting process of metallic materials because of their ability to cut practically all metal and the high productivity than can be achieved with this technology. Plasma cutting is a process of metal cutting at atmospheric pressure by an arc plasma jet, where a transferred arc is generated between a cathode and a work-piece (the metal to be cut) acting as the anode. In order to obtain a high-quality cut, the arc must be as collimated as possible. To this end, a new generation of cutting torches, the so called "high energy density torch" or "high-definition torch" has been developed. These torches are characterized by an arc current intensity in the range of 30 ÷ 100 A, flat cathodes, oxygen as the plasma gas, very small nozzle diameters (≈ 1 mm) and by the generation of an under-expanded supersonic arc jet with a shock wave close to the nozzle exit [1].

Among the various possible ways of stabilizing an arc by a superimposed convective flow in a tube, gas vortex stabilization has been successfully used in cutting torches. In these torches the arc is

† Fellow of YPF
‡ Member of the CONICET

confined to the center of the nozzle while an intense vortex of a gas is maintained at the arc periphery. Centrifugal forces drive the cold gas toward the walls, which is thus thermally well protected. In addition to the circumferential component of the vortex flow, there is also a superimposed axial velocity component that continuously supplies cold gas. The intense convective cooling produced at the arc border enhances the power dissipation in the arc column, which, in turn, results in high axis temperatures [2]. Due to the clogging effect of the heated gas in the nozzle [3], the gas mass flow rate is typically small and the level of stabilization of the arc may be insufficient to avoid double-arcing problems (where the arc is split into two: one connecting the cathode and the nozzle, the other connecting the nozzle and work piece). This phenomenon is one of the main drawbacks that put a limit to increasing capabilities of plasma arc cutting process [4]. An optimal design of a cutting torch nozzle must satisfy contradictory conditions. In order to provide a fast, highly constricted and hot plasma jet, the nozzle should be long and narrow. On the other hand, such long and narrow nozzle reduces the gas mass flow, with the consequent de-stabilization and asymmetry of the arc. Differences in cutting arc properties related to changes in the nozzle radius were measured using spectroscopic techniques in [5].

In this work, an experimental study on the influence of the nozzle length on the physical properties of a cutting arc is reported. Ion current signals collected by an electrostatic probe sweeping across a 30 A oxygen cutting arc at 3.5 mm from the nozzle exit for two different nozzle lengths was registered. The temperature and density radial profiles of the arc plasma were obtained in each case by an inversion procedure of these signals. A comparison between the results obtained for both nozzle lengths is presented and discussed.

2. Experiment

The cutting torch used in this study consisted of a cathode centered above an orifice in a converging-straight copper nozzle with liquid cooling. The cathode was made of copper (7 mm in diameter) with a hafnium tip (1.5 mm in diameter) inserted at the cathode center. A flow of oxygen gas cooled the cathode and was also employed as the plasma gas. The gas passed through a swirl ring to provide arc stability. Two converging-straight bore nozzle designs were used and compared. In both cases the bore radius of the nozzle cylindrical part was $R_N = 0.5$ mm whereas its length presented a marked difference $L_N = 4.5$ and 9 mm.

To avoid plasma contamination by metal vapors from the anode, a rotating steel disk with 200 mm in diameter and 15 mm thickness was used as the anode [3]. In this study, the disk upper surface was located at 5 mm from the nozzle exit. The arc was transferred to the edge of the disk, and the rotating frequency of the disk was equal to 29.5 Hz. At this velocity, a well-stabilized arc column was obtained, and no noticeable damage on the lateral surface of the anode disc was found. Thus, practically no metal vapors from the anode were present in the arc. A scheme of the torch indicating several geometric dimensions of the nozzle is presented in Fig. 1.

During arc operation the total arc voltage (V_{CA}) and the nozzle floating voltage (V_{NA}) were registered with respect to the grounded anode, either by using a high-impedance (10 MΩ) voltage meter or by registering them on a digitizing oscilloscope. By performing a small orifice (1 mm in diameter) on the cathode lateral surface the pressure in the plenum chamber (p_{ch}) was measured by connecting a pressure meter at the upper head of the cathode. The gas flow rate injected in the torch was also registered. In this experiment, the arc current was fixed to a value of 30 A.

The electrostatic probe system employed for studying the arc was similar to that previously reported in [6] and consisted of a thin tungsten wire with a radius $R_p = 0.1$ mm, sweeping trough an arc plane perpendicular to the current flow direction at a constant velocity v_p = 18 m s^{-1}. Since an arc cutting torch creates an under-expanded sonic flow at the nozzle exit, we studied the arc plane at z = 3.5 mm (where z is the axial position measured from the nozzle exit). At this axial position the arc pressure p can be considered to be near to the atmospheric value [1]. A scheme of the probe circuit is also shown in Fig. 1. To obtain ion current measurements, the probe was biased with a resistance (R_1 = 140 Ω) connected between the probe and the cathode. The ion signal was registered by measuring the probe voltage (V_p) on R_1 using a resistive voltage divider (R_2 = 1 kΩ and R_3 = 12 Ω) with an input resistance value much larger than the R_1 value but with a measuring resistance (R_3) small enough to avoid distortions of the signal due to the discharge of the 500 pF capacitor (corresponding to the measuring coaxial cable) on R_3. This biasing method takes advantage of the voltage distribution occurring during the arc discharge, and was successfully employed in a previous work [6]. To perform probe floating voltage measurements (with respect to the grounded anode), V_f, the values of R_2 and R_3 were changed to R_2 = 56 kΩ and R_3 = 3.3 kΩ, while R_1 was disconnected and the voltage on R_3 was registered (see Fig. 1). The discharge time of the 500 pF capacitor on the last R_3 value was short enough to avoid signal distortions in floating conditions. The employed oscilloscope was a two-channel Tektronix TDS 1002 B with a sampling rate of 500 MS/s, analogical bandwidth of 60 MHz, and an input impedance of 1 MΩ in parallel with a stray capacitance of 20 pF.

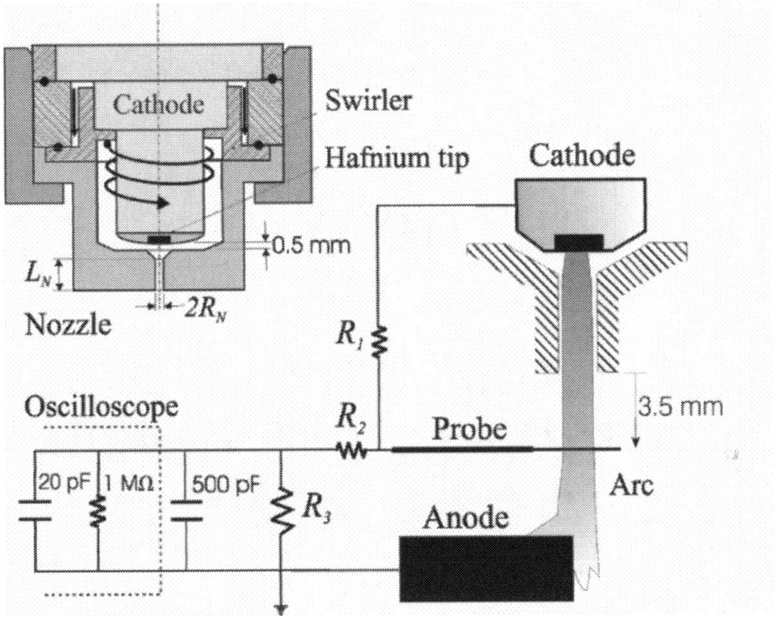

Figure 1. Scheme of the arc torch indicating several geometric dimensions. The probe biasing circuit is also shown.

3. Experimental Results

In Fig. 2 two typical probe ion current waveforms corresponding to nozzles of L_N = 4.5 mm operated at 0.7 MPa and L_N = 9.0 mm operated at 1.1 MPa are presented. The waveforms correspond to different arc runs, but for comparative purposes they were plotted on a single time scale. The registered gas flow rate trough the shorter nozzle was 35 Nl/min whereas only 20 Nl/min was measured for the larger nozzle in spite of its high operating pressure. As can be seen, both signals show an almost square shape but with differences in amplitude and time duration. For the nozzle with L_N = 4.5 mm a signal duration of ~ 50 μs (corresponding to an arc diameter of ~ 0.9 mm) and an ion current amplitude of ~ 0.8 A was registered. For the nozzle with L_N = 9.0 mm a signal duration of ~ 70 μs (corresponding to an arc diameter of ~ 1.3 mm) and an ion current amplitude of ~ 1.3 A was registered. The signal duration refers to the temporal width at the top of the flat-top profiles.

The total arc voltage, the nozzle voltage and the probe floating potential signals (measured at z = 3.5 mm from the nozzle exit in each case) were also measured (see Table 1). Note that the increase in the nozzle length leads to a substantially arc voltage rise (of about 55 V) corresponding to an arc power rise of about 1.65 kW. Statistical fluctuations in the current amplitude of the registered signals were ~ 5 %. The origin of such fluctuations is likely due to arc voltage fluctuations (ripple) that in turn produce corresponding changes in the plasma arc structure.

Table 1. Voltage measurements in the studied nozzles

	L_N = 4.5 mm	L_N = 9.0 mm
V_{CA} (V)	-160 ± 8	-215 ± 11
V_{NA} (V)	-80 ± 4	-95 ± 5
V_f (V)	-22 ± 1	-12 ± 0.6

No noticeable double-arcing damages were registered with the larger nozzle, in spite of the relatively small operating gas mass flow for this case.

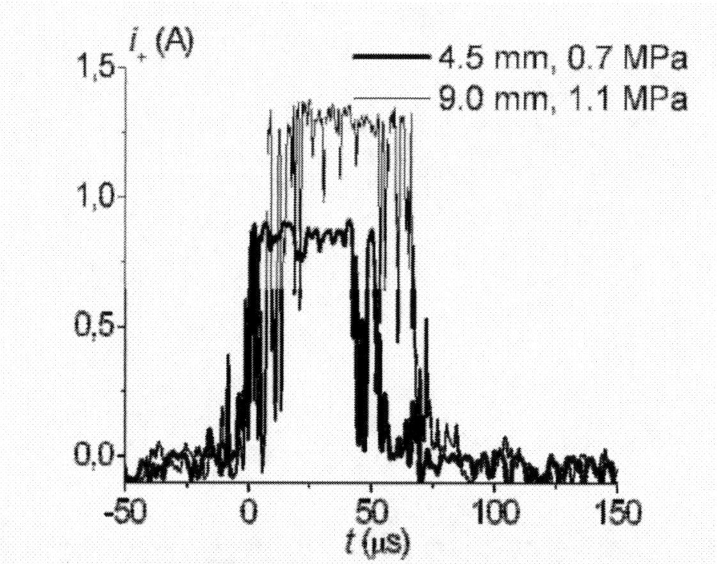

Figure 2. Typical probe ion signals obtained at 3.5 mm from the nozzle exit for the two nozzle lengths.

4. Interpretation of the results

The plasma is assumed in local thermodynamics equilibrium (LTE). Although the effects of non-LTE have been studied in cutting torches [5, 7] a theory based on LTE can in general give satisfactory temperature predictions when compared with experiments [1].

Assuming circular symmetry for the arc section, the arc core can be divided into several concentric elemental annulus (see Fig. 3), so that the ion probe current can be expressed as the sum of contributions from many regions with different plasma quantity values, but at the same probe potential $V_p(x)$; where $x \equiv v_p\, t$ is the coordinate of the probe axis at a generic time t. As the electron thermal velocity $\bar{v}_e \equiv \sqrt{8kT/\pi m}$ (k is the Boltzmann's constant, T is the plasma temperature and m is the electron mass) is very large compared to the fluid-plasma velocity, the usual expression for the electron current collected by a probe in high-pressure quiescent plasmas applies. This current (per-unit probe length) for a cylindrical probe is [8]

$$i'_{-} = 2\pi R_p \frac{1}{\gamma}\frac{e n_e \bar{v}_e}{4}\exp\left(-\frac{e\left|V_p(x)-V_s\right|}{kT}\right) \tag{1}$$

where e is the electron charge modulus, n_e the electron density, V_s the plasma potential and the non-dimensional correction factor γ is given by [8] $\gamma \approx 1+\frac{\lambda_D}{2\lambda_{ea}}\frac{kT}{e\left(V_f - V_s\right)}$, where λ_D is the electron Debye length and λ_{ea} the electron mean free path for collisions with neutrals. The factor γ is very close to unity for the considered conditions ($\lambda_D << \lambda_{ea}$ and $kT/e < |V_p - V_s|$) and so $\gamma = 1$ was taken. The ion current (per-unit length) for a cylindrical probe in high-pressure high-velocity plasma jet, is given by [9]

$$i'_+ = 0.46\ e\, n_i v_B \lambda_{ia} \frac{e|V_p(x) - V_s|}{kT} \tag{2}$$

where n_i is the ion density, $v_B \equiv \sqrt{kT/M}$ the Bohm velocity (M is the ion mass) and λ_{ia} the ion mean free path for collisions between ions and neutrals, defined as $\lambda_{ia} \equiv 1/n_n\sigma_0$; where n_n is the neutral density and σ_0 is the elastic cross section, typically $\approx 5\times10^{-19}\ m^2$ [10].

The plasma potential value V_s can be obtained equating (1) and (2) for $V_p = V_f$,

$$0.18\sqrt{\frac{m}{M}}\frac{\lambda_{ia}}{R_p}\frac{e|V_f - V_s|}{kT} = \exp\left(-\frac{e|V_f - V_s|}{kT}\right). \tag{3}$$

From (3) and for the temperature range of interest $(10 \div 16)kK$ it results $V_s - V_f \approx 8\,V$.

Taking into account (2), and using simple geometrical considerations, the total ion current to the probe in terms of an integral along the radial direction is given by,

$$i_+(x \equiv v_p t) = 2\times0.46\ e \int_{r=|x|}^{R_A} \frac{|V_p(x) - V_s|}{k/e} \frac{n_i\, v_B\, \lambda_{ia}}{T} \frac{r}{\sqrt{r^2 - x^2}} dr\ . \tag{4}$$

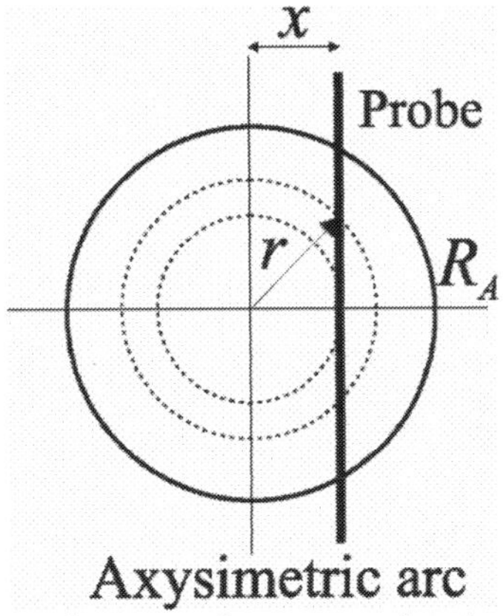

Figure 3. Scheme of the probe sweeping across the plasma at a generic time.

By using the Abel inversion technique, a radial profile of the plasma quantities can be derived from (4) with some additional assumptions. Note that the neutral density and the plasma density and temperature are involved in the integral appearing in (4). These magnitudes can be related through the state equation (assuming that the arc pressure is uniform along the radius) and through the Saha equation, thus closing the system. It is well known that any error in the input data for an Abel inversion technique can be amplified in the computation, and care must be taken to reduce such errors to a minimum. To this end, the input data were fitted with a high-order polynomial before performing the Abel inversion. The radial temperature and plasma charge density profiles were then calculated in accord with the Saha equilibrium and the ideal gas state equation for p = 0.13 MPa. The overpressure value of 0.03 MPa takes into account stagnation effects at the upstream edge of the probe surface for a plasma flow velocity of $\approx 2\times10^3$ m s^{-1} (of the order of the ion sound velocity) and a plasma density corresponding to a pressure of 0.1 MPa and a temperature of 14 kK (as it will be shown later, this is a typical value obtained for T).

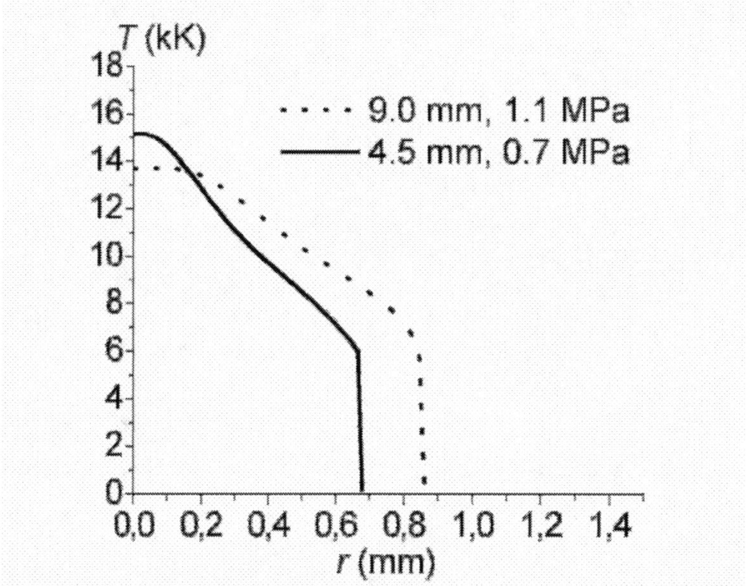

Figure 4. Radial profile of the plasma temperature obtained at 3.5 mm from the nozzle exit for the two nozzle lengths.

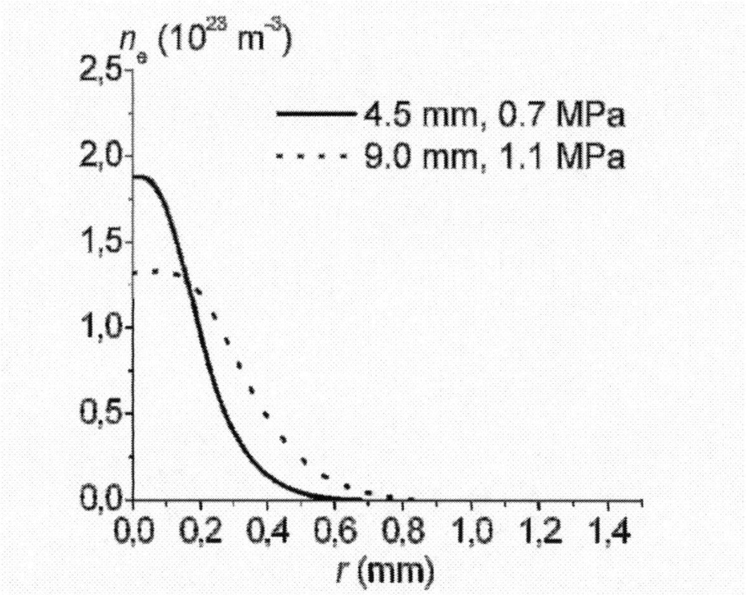

Figure 5. Radial profile of the plasma density obtained at 3.5 mm from the nozzle exit for the two nozzle lengths.

In Figs. 4 and 5 the obtained profiles of the plasma temperature and density are respectively shown for both nozzle geometries. As can be seen, the temperature profiles show a monotonic decrease from the arc center, and with a peak value (T_0) somewhat higher for the shorter nozzle ($T_0 \approx 15$ kK, whereas

$T_0 \approx$ 14 kK for the larger nozzle). Also, the shorter nozzle produce a thinner arc (the profile decays abruptly at $\approx$ 0.7 mm, whereas the corresponding to the larger nozzle at $\approx$ 0.85 mm). It is worth noting that the abrupt decay at the end of the profiles cannot be taken with confidence as it is an artifact of the Abel inversion technique [11]. The plasma density profiles shown in Fig. 5 are obtained from the temperature profiles (Fig. 4) by applying the Saha equation. As expected, the larger density values are concentrated in the high temperature (T > 10 kK) region, and correspondingly, the peak value for the shorter nozzle is higher than that of the larger one. This behavior is attributed to the marked difference of gas flow rate due to the clogging effect. A low gas mass flow reduces the convective cooling at the arc border and decreases the radial heat flux, producing a decrease in the temperature drop from the arc axis. Due to the very strong dependence of the electrical conductivity on T (at 5 ÷ 10 kK), such decrease in the temperature drop results in a thicker arc with smaller axis temperature.

5. Final remarks

In this work, an experimental study on the influence of the nozzle geometry on the physical properties of a cutting arc is reported. Ion current signals collected by an electrostatic probe sweeping across a 30 A oxygen cutting arc at 3.5 mm from the nozzle exit for different nozzle lengths were registered. The temperature and density radial profiles of the arc plasma were found in each case by an inversion procedure of these signals. A comparison between the obtained results shows that the shorter nozzle (R_N = 0.50 mm, L_N = 4.5 mm operated at 0.7 MPa) produces a thinner and hotter arc than the larger nozzle (R_N = 0.50 mm, L_N = 9.0 mm operated at 1.1 MPa). This behavior is attributed to the marked difference of gas flow rate due to the clogging effect. A low gas mass flow reduces the convective cooling at the arc border and decreases the radial heat flux, producing a decrease in the temperature drop from the arc axis. Due to the very strong dependence of the electrical conductivity on T at (at 5 ÷ 10 kK), such decrease in the temperature drop results in a thicker arc with smaller axis temperature.

Acknowledgements

This work was supported by grants from the Universidad de Buenos Aires (PID X111) CONICET (PIP 5378) and Universidad Tecnológica Nacional (PID UTN 25/Z012).

References

[1] Freton P, Gonzalez J J, Gleizes A, Camy Peyret F, Caillibotte G and Delzenne M 2002 *J. Phys. D: Appl. Phys.* **35** 115.
[2] Boulos M, Fauchais P and Pfender E 1994 *Thermal Plasmas, Fundamentals and Applications* Vol. 1 (Plenum Press).
[3] Ramakrishnan S, Gershenzon M, Polivka F, Kearny T N and Rogozinsky M W 1997 *IEEE Trans. Plasma Sci.* **25** 937.
[4] Nemchinsky V A and Severance W S 2006 *J. Phys. D: Appl. Phys.* **39** R423.
[5] Peters J, Heberlein J and Lindsay J 2007 *J. Phys. D: Appl. Phys.* **40** 3960.
[6] Prevosto L, Kelly H and Mancinelli B 2008 *IEEE Trans. Plasma Sci.* **36** 263.
[7] Ghorui S, Heberlein J and Pfender E 2007 *J. Phys. D: Appl. Phys.* **40** 1966.
[8] Raizer Y P 1991 *Gas Discharge Physics* (Berlin, Germany: Springer).
[9] Prevosto L, Kelly H and Minotti F O 2008 *IEEE Trans. Plasma Sci.* **36** 271.
[10] Anderson H L 1989 *A physicist's Desk Reference* Second edition. (New York: AIP).
[11] Gick A E F, Quigley M B C, and Richards P H 1973 *J. Phys. D: Appl. Phys.* **6** 1941.

Journal of Physics: Conference Series **166** (2009) 012022 doi:10.1088/1742-6596/166/1/012022

Supersonic mixing layers: stability of magnetospheric flanks models

G Gnavi[1,4], F T Gratton[1,2], C J Farrugia[3] and L E Bilbao[1]

[1] Instituto de Física del Plasma, Consejo Nacional de Investigaciones Científicas y Técnicas y Facultad de Ciencias Exactas y Naturales, Universidad de Buenos Aires, Ciudad Universitaria, 1428 Buenos Aires, Argentina
[2] Departamento de Física, Facultad de Ciencias Fisicomatemáticas e Ingeniería, Pontificia Universidad Católica Argentina, Buenos Aires, Argentina
[3] Space Science Center, University of New Hampshire, Durham, NH, 03824, USA

[4]E-mail: graciela@tinfip.lfp.uba.ar

Abstract. Compressibility has a strong influence on the stability of velocity shear layers when the difference of velocity ΔV across the flow becomes supersonic. The flanks of the Earth's magnetopause are normally supersonic $M_s > 1$, and super-Alfvénic $M_A > 1$, depending on the distance from the dayside terminator (M_s and M_A are the sonic and Alfvén Mach numbers of the magnetosheath plasma, respectively). The stability of MHD supersonic flows depends, also on several other features, such as the finite thickness Δ of the boundary layer, the relative orientation of velocity and magnetic fields, the density jump across the boundary and the magnetic shear angle. We analyze the MHD stability of some representative flank sites modeled after data from spacecraft crossings of the magnetopause under different interplanetary conditions, complementing these cases with extrapolations of likely conditions upstream, and downstream of the crossing site. Under northward interplanetary magnetic field conditions, there are solar wind regimes such that the near, but already supersonic, flank of the magnetopause may be locally stable. Stability is possible, e.g., when M_s becomes larger than ~1.2-1.4 while M_A remains smaller than 1.2, and there is magnetic shear between the geomagnetic and the interplanetary magnetic field. Solar winds favouring local stability of the boundary layer are cold, not-too-dense plasmas, with strong magnetic fields, so that M_A is smaller, while M_s is larger, than normal values of the magnetosheath flow. A gap between dayside and tail amplifying regions of Kelvin-Helmholtz disturbances over the magnetopause may exist when the above conditions are realized.

1. Introduction: aims and content.

The interplanetary plasma flow around the terrestrial magnetopause (MP) flanks is supersonic $M_s > 1$ and super-Alfvénic $M_A > 1$ ($M_s = U_1/c_s$: sonic Mach number; $M_A = U_1/V_A$: Alfvénic Mach number; U_1: adjacent magnetosheath speed; c_s: sound velocity; V_A: Alfvén velocity). Both Mach numbers increase toward the magnetopause (MP) tail. Possible exceptions are regions at the near flanks close to the terminator, depending on interplanetary conditions.

Plasma compressibility exerts a considerable influence on the stability of a velocity shear flow when ΔV, the velocity difference across the gradient layer, becomes supersonic. Landau [1] in his study of a uniform density gas flow showed that the Kelvin-Helmholtz (KH) modes of a tangential discontinuity (TD: model in which there is an abrupt velocity jump across a surface or a negligibly thin layer) with $\boldsymbol{k} \,//\, \Delta \boldsymbol{V}$ ($\boldsymbol{k}$: wave vector of the modes) are stable when $M_s > \sqrt{2}$ ($M_s = \Delta V / c_s$) while, of course, an incompressible treatment would indicate instability. In a TD model, however, other modes for which the projection $\Delta V_\kappa = \boldsymbol{\kappa} . \Delta \boldsymbol{V}$ (where $\boldsymbol{\kappa} = \mathbf{k}/|\mathbf{k}|$) is subsonic remain unstable. For long wavelength waves the magnetopause can be approximated locally by a TD, but the KH stability of the MP is a more complex issue because it depends on many other factors. The non-uniform density of the transition layer, the existence of magnetic shear across the boundary, the ratio λ/Δ of the mode wavelength ($\lambda = 2\pi/|\boldsymbol{k}|$) to the scale-length of the velocity gradient, Δ, the possible presence of a multiple-scale boundary layer, the distance of the locale from the subsolar point, the physical state of the interplanetary plasma are all factors contributing to the complexity.

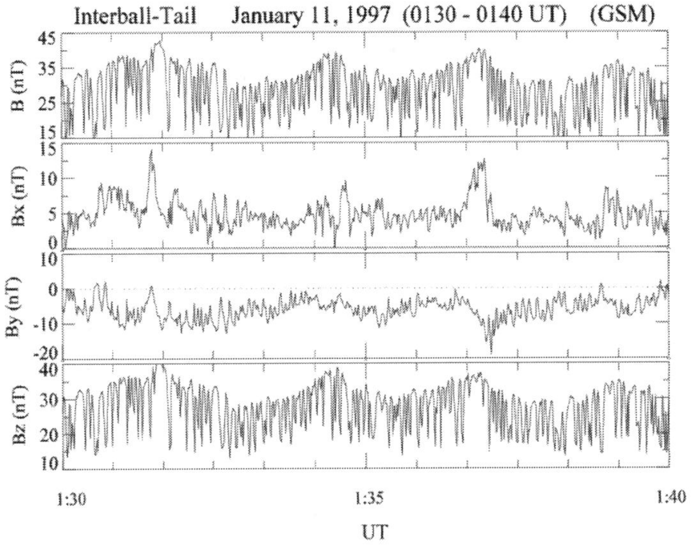

Figure 1. Waves at the near-tail (~13 R_E). Magnetic field data (intensity and *GSM* components) recorded by Interball-Tail on January 11, 1997. Magnetosonic waves in the magnetosheath of $f \sim 0.15$ Hz Doppler shifted with respect to similar waves on the frontside. KH waves on the magnetopause, $\lambda \sim 13\text{-}14\ R_E$ and $f \sim 3.6$ mHz. These waves appear as an envelope modulation of the magnetosonics.

A considerable body of experimental, theoretical, and numerical simulation studies has been published on the KH instability of the MP. For particular cases, analyses of plasma processes beyond ideal MHD, and even an advanced kinetic investigation, have been done to achieve deeper physical insight. We refer the reader to review papers and the literature cited therein, see [2] - [7]. In spite of the valuable and extensive literature the understanding of KH and its consequences for the magnetosphere is not yet complete. In particular the variety of conditions for the KH stability of the MP has not yet been fully explored, and for that purpose ideal MHD is a reliable model.

An important feature of the structure of the MP, both at the flanks as well as on the dayside, is the existence of the low latitude boundary layer (LLBL), a region where plasma quantities take values intermediate between those of the magnetosheath and those of the magnetosphere proper. The shear of the magnetic lines together with a density change often occurs abruptly at the MP, and then in the rest of the LLBL the direction of $\boldsymbol{B}$ stays close to that of the magnetosphere. The velocity, instead, decreases over a longer distance across the boundary layer, without changing its antisunward direction, except occasionally when a weaker counter-flow appears in the magnetosphere.

We may conjecture that the instability mechanism can feed better at the LLBL when the scale-length of the velocity gradient is comparable with the width of this region, because the magnetic tensions inside the LLBL can be switched-off by modes with $B_\kappa = \boldsymbol{B} \,.\, \boldsymbol{\kappa} \approx 0$, known as *flute* modes.

Moreover, at the LLBL the plasma flow is ordinarily in a subsonic regime because the velocity has decreased, while the sound speed has increased, with respect to magnetosheath values. The latter effect is due to the large temperature increment as we move into the magnetospheric plasma. Conversely, the stabilizing effects of magnetic tensions and compressibility are expected to be more effective at the MP interface. These are, however, local properties, while the KH instability depends on a global perturbation that extends across the whole transition layer and beyond. The penetration distance of the modes, i.e., the distance over which the amplitude decreases by *1/e* on either side of the gradient layer, is normally comparable with Δ, so that an overall balance of several effects is what finally decides the stability of the configuration. To find the outcome a full theoretical calculation with continuous field profiles model is necessary.

This paper examines the MHD stability of boundary layer models that describe flow conditions at the near-equatorial magnetopause flanks during periods of northward pointing interplanetary magnetic fields IMF). The cases analyzed are based on actual data returned by spacecraft crossing the MP, when the IMF had a dominant north component. This interplanetary condition is the most favorable for the excitation of KH perturbations, while at the same time no reconnection of magnetic field lines at the front side is possible. Vortices generated by the KH instability at the boundary layer become important agents of momentum, energy, and possibly mass, transport from the solar wind to the magnetosphere. On the other hand, when the IMF points south, reconnection is the dominant coupling process that competes in matters of transport with the velocity shear instability.

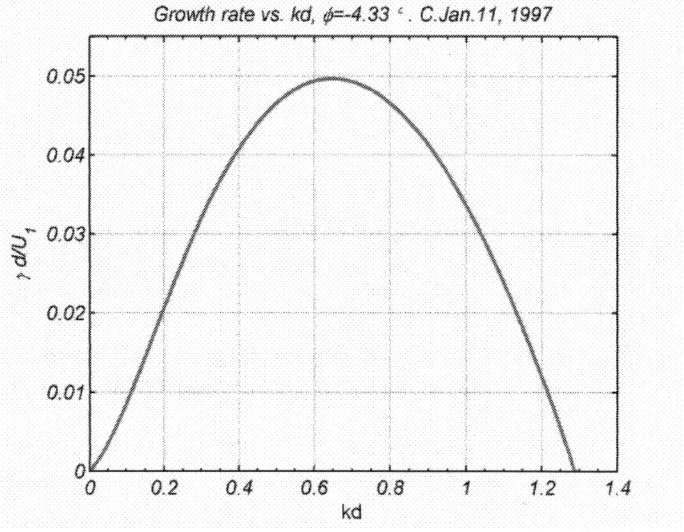

Figure 2. January 11, 1997 event. Growth rate $\gamma d/U_1$ as a function of the wavenumber *kd*, normalized quantities. Modes with **k** to *V* angle for maximum growth rate: $\varphi = -4.33°$.

The paper is, therefore, a study of the Kelvin - Helmholtz instability at the supersonic near-equatorial flanks of the magnetopause for northward IMF and requires a compressible MHD treatment. We focus on the boundary layer configuration during two events observed by spacecraft crossings under northward IMF conditions, and on a hypothetical scenario suggested by flank data. The layout of the paper is as follows. The second section gives the equations of the theory, describes features of the LLBL models employed, and comments upon the stabilizing effects. Section 3 deals with a stability analysis for the January 11, 1997 event, with data from an Interball-tail crossing, extending the study of reference [8] to include compressibility. We also explore possible sunward and tailward scenarios based on arrangement of fields similar to those observed by Interball-tail, but varying the values of M_s and M_A with changing distance from noon. A MP scenario suggested by the competition between compressibility and magnetic shear is the theme of Section 4, where we argue that gaps of local stability in the near flanks are possible when certain physical conditions occur. Section 5 considers the stability of the MP for the November 20, 2001 event recorded by CLUSTER during a long journey in the LLBL. This case is stable from the point of view of a TD approximation (long wavelength limit) but it is in fact unstable when examined by a continuous, finite thickness model. A more realistic two-scale treatment of the same event is also shown, and by comparison with the one-scale model the study reveals the sensitivity of the orientation of $\boldsymbol{k}$ for unstable modes with particular features of the boundary layer structure. Section 7 contains our final remarks.

Among the main points presented we show that local stability of the low latitude boundary layer (LLBL) is possible when the sonic Mach number $M_s=U_l/c_s$ is close to but greater than 1.2 - 1.4 while the Alfvénic Mach number $M_A=U_l/V_A$ is a bit smaller, near to but less than 1.2 (U_l: magnetosheath flow velocity, c_s: sound speed, and V_A: Alfvén velocity, all magnetosheath values). The effect is due to the stabilizing effect of compressibility on the outer supersonic side, combined with magnetic tensions in the inner subsonic part of the LLBL. We conclude, then, that when there is a significant magnetic shear between the geomagnetic and the interplanetary magnetic field, gaps of local stability of the supersonic boundary layer in the near flanks are possible.

2. KH equation, boundary layer models, and main points

2.1 The Kelvin-Helmholtz boundary value problem in MHD.
In this work the theory of the KH instability relies on the set of compressible, ideal (non resistive, and inviscid) MHD equations,

$$\rho\frac{\partial v}{\partial t}+\rho(v.\nabla)v=-\nabla\left(p+\frac{\left|B^2\right|}{8\pi}\right)+\frac{1}{4\pi}(B.\nabla)B, \tag{1}$$

$$\frac{\partial B}{\partial t}=\nabla\times(v\times B), \tag{2}$$

$$\nabla.B=0, \tag{3}$$

$$\frac{\partial\rho}{\partial t}+\nabla(\rho v)=0, \tag{4}$$

$$\left(\frac{\partial}{\partial t}+v.\nabla\right)\left(p\rho^{\gamma_a}\right)=0, \tag{5}$$

where γ_a is the adiabatic ratio. These equations are then linearized with respect to a state that represents a steady state, laminar, plasma flow, or a state defined by the average value of fields and flow if the background is already supporting fluctuations.

For applications to the magnetopause the most common configuration is that of a parallel flow, in which the velocity does not change direction across the MP, so that it is convenient to set the x axis of a local coordinate system along the direction of the velocity field. The magnetic field, instead, in general changes direction from the magnetosheath to the magnetosphere so that the MP is a current sheet. We study the stability of planar stratified flows, where physical quantities are constant in (x,z) planes and vary only along the normal direction y. For the velocity and magnetic field, we set $\boldsymbol{V}=(V_x(y),0,0)$, $\boldsymbol{B}=(B_x(y),0,B_z(y))$, respectively. Similarly, we write $\rho(y)=m_p\, n(y)$ for the mass density, and $T(y)$ for the temperature (n indicates the number of particles density, while ρ is the mass density; m_p is the proton mass).

For ideal MHD perturbations, the amplitude of KH Fourier modes is governed by the following second order differential equation,

$$\frac{d}{dy}\left[H\left(1-\frac{1}{M}\right)\frac{d\zeta}{dy}\right]-k^2H\zeta=0, \tag{6}$$

first derived in [9]. The modes are of the form

$$\Xi=\zeta(y)\exp(-i\omega t+ik_x x+ik_z z), \tag{7}$$

where Ξ is the y component of the Lagrangian displacement of any plasma element from the unperturbed position, and ζ is the corresponding amplitude. The (complex) angular frequency of the modes is denoted by $\omega=\omega_r+i\gamma$. The real part represents the oscillation frequency of the perturbation, and the imaginary part gives the temporal rate of growth. The wave vector of the mode is represented by $\boldsymbol{k}=(k_x,0,k_z)$, and $k=|\boldsymbol{k}|$ denotes the absolute value, so that $k^2=k_x^2+k_z^2$. A complex phase velocity (ω/k) is denoted by c. We also introduce φ, the angle between $\boldsymbol{k}$ and $\boldsymbol{V}$. The coefficients H and M in the ζ equation are defined as follows,

$$H(y)=\rho\left[(c-V_\kappa(y))^2-V_{A\kappa}^{\;2}(y)\right], \tag{8}$$

$$M(y)=1-\frac{\left(V_A^{\;2}+c_s^{\;2}\right)}{(c-V_\kappa)^2}+\frac{V_{A\kappa}^{\;2}c_s^{\;2}}{(c-V_\kappa)^4}, \tag{9}$$

where V_A is the Alfvén speed, c_s is the speed of sound, V_κ, B_κ are the projections on the **k** direction of $\boldsymbol{V}$ and $\boldsymbol{B}$, respectively, $V_\kappa=\boldsymbol{\kappa}\cdot\boldsymbol{V}$, $B_\kappa=\boldsymbol{\kappa}\cdot\boldsymbol{B}$, and $V_{A\kappa}=B_\kappa/4\pi\rho$, with $\boldsymbol{\kappa}=\boldsymbol{k}/|\boldsymbol{k}|$) and all these quantities are functions of y. The analysis follows the classical viewpoint (also called temporal) in which the (real) wavenumber of the mode is given as a Fourier component of the initial perturbation, and the response of the system determines the (complex) value of c. The intricacy of the boundary value problem for the ζ equation arises from the fact that c does not appear as an eigenvalue, but rather as a characteristic value entangled in a non-linear fashion in the coefficients H and M. Moreover, when the direction $\boldsymbol{k}$ changes, the functions $V_\kappa(y)$, $V_{A\kappa}(y)$ also change, and the analysis requires the solution of different differential equations (an example of these functions is shown in figure 8, section 5).

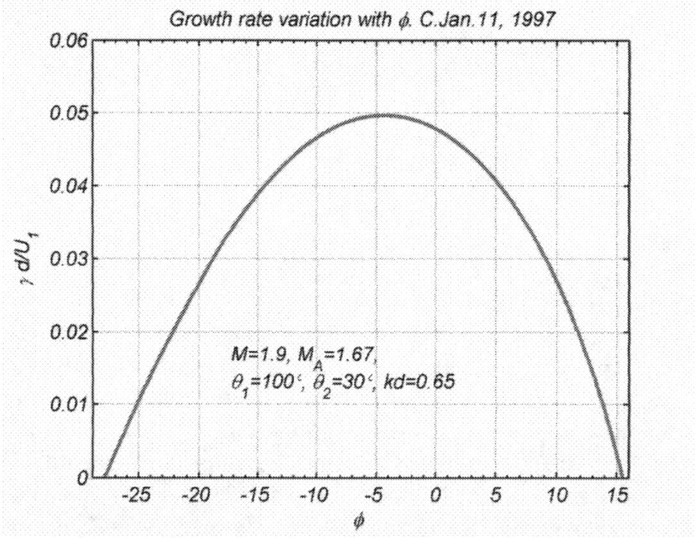

Figure 3. January 11, 1997 event. Normalized growth rate $\gamma d/U_1$ as a function of φ, the $\boldsymbol{k}$ to $\boldsymbol{V}$ angle. The modes are for $kd=0.65$, the wavenumber for fastest growth when $\varphi=-4.33°$.

We solve numerically the boundary value problem that determines the characteristic value c with a conventional shooting method (see, for instance, [10]). The compact form of the stability equation facilitates the implementation of shooting methods. Finite difference methods, instead, are suited to deal with a set of linear first order differential equations as in [11]. The former method needs a good seed value for c. In the latter it is not always easy to discriminate spurious c roots. A combination of both ways, finite differences to find possible seeds and shooting to control the result and refine c, is perhaps the best approach.

2.2 Boundary layer models with hyperbolic functions

One- and two-scale models of the MP interface are built with hyperbolic tangents like tanh(y/d) for the equilibrium fields of the present study. The length d, half the width Δ of the velocity gradient layer, normalizes the y coordinate. In the two-scale model there is also a second length d', half the width δ of the current sheet, which in most cases is smaller than d. The list of all the physical parameters of the boundary layer model is as follows. The values on both sides of the transition (with index 1 for the magnetosheath, and 2 for the magnetosphere) of the vector fields intensities $V = |\boldsymbol{V}|$ and $B = \boldsymbol{B}|$, of the scalar fields $\rho = n\ m_p$ and T, the angle θ formed by $\boldsymbol{B}$ with the x axis, and two basic lengths, the width $\Delta = 2d$ of the velocity gradient layer, and $\delta = 2d'$ the width of the current sheet, which define the ratio of the two scales $r_s = \delta/\Delta$. With these elements, using $Y = y/d$ as a normalized variable, the basic equations of field profiles for the LLBL model are the following,

$$V_x = \frac{1}{2}(U_1 + U_2) + \frac{1}{2}(U_1 - U_2)\tanh(Y), \qquad V_\kappa = V_x \cos(\varphi), \tag{10}$$

$$\rho = \frac{1}{2}(\rho_1 + \rho_2) + \frac{1}{2}(\rho_1 + \rho_2)\tanh\left(\frac{Y}{r_s}\right), \tag{11}$$

$$B = \frac{1}{2}(B_1 + B_2) + \frac{1}{2}(B_1 - B_2)\tanh\left(\frac{Y}{r_s}\right), \tag{12}$$

$$\theta = \frac{1}{2}(\theta_1 + \theta_2) + \frac{1}{2}(\theta_1 - \theta_2)\tanh\left(\frac{Y}{r_s}\right), \tag{13}$$

$$B_x = B\cos(\theta), \quad B_z = B\sin(\theta), \tag{14}$$

$$B_\kappa = B_x\cos(\varphi) + B_z\sin(\varphi) = B\cos(\theta - \varphi). \tag{15}$$

A frequent case has $\boldsymbol{V}_2 = 0$, corresponding to zero velocity on the magnetospheric side. Magnetic field and density may vary on a distance shorter than the velocity gradient scale-length. The temperature function $T(y)$ is a consequence of the previously given profiles and the pressure balance equation across the boundary layer,

$$p + \frac{B^2}{8\pi} = const., \tag{16}$$

which lead to

$$T(y) = \frac{m_p}{2k_B\rho(y)}\left[p_1 + \frac{B_1^2}{8\pi} - \frac{B(y)^2}{8\pi}\right]. \tag{17}$$

This function defines the change of the sound speed across the layer, which cannot be assigned independently from the previous hyperbolic tangent profiles. The ratio $r_s=\delta/\Delta$ is 1 for the one-scale model, or is estimated from data of spacecraft crossings in the two-scale model. We may keep Δ as a free quantity in the computations, and estimate *a posteriori* a value from experimental data (see sections 3, and 5). An example of these functions for a model is given in figure 8, section 5.

2.3 Remarks on stabilizing effects

At large $|y|$ ($|y|/\Delta \gg 1$) all quantities attain their constant values on either side of the boundary layer, so that the solutions are, asymptotically,

$$\zeta \to A_1 \exp\left(-ky\sqrt{\frac{M_1}{M_1-1}}\right), \qquad y>0 \tag{18}$$

$$\zeta \to A_2 \exp\left(+ky\sqrt{\frac{M_2}{M_2-1}}\right), \qquad y<0, \tag{19}$$

(M_1, M_2, are the constant values taken by coefficient M on each side). The normal modes are surface perturbations and must approach zero on both sides of the MP, that is, $\zeta \to 0$ when $y \to \pm\infty$, so that the branch of the square root must be defined as

$$\Re\left(\sqrt{\frac{M_a}{M_a-1}}\right) > 0, \qquad a=1,2. \tag{20}$$

with the exception of possible critical values of c for which either $M_a \approx 0$, or $M_a - 1 \approx 0$ ($a=1,2$). Usually

$$\Re\left(\sqrt{\frac{M_a}{M_a-1}}\right) \sim \mathrm{O}(1), \tag{21}$$

so that the penetration distance of the modes is normally of order $\lambda/2\pi$ on both sides. We may note that in the case of an unstable mode, $\Im(c)>0$, M_a are complex valued, so that the square root has also an imaginary part, and therefore when $|y|/\Delta \gg 1$ the mode oscillates spatially along y, while its amplitude tends to zero.

When $c_s \to \infty$, the coefficient $M \to \infty$, and the KH equation reduces to

$$\frac{d}{dY}\left[H\frac{d\zeta}{dY}\right] - (kd)^2 H\zeta = 0, \tag{22}$$

the incompressible limit (written with $Y=y/d$). The asymptotic behavior of the modes on both sides is $\exp(\pm ky)$, exactly. When the transition layer is very thin, $kd \to 0$, we can derive the dispersion relation of a thin model (also called tangential discontinuity, TD) by integrating the equation across the discontinuity (assumed to be at $Y=0$). We obtain

$$H_1 + H_2 = 0, \tag{23}$$

where H_1 and H_2, are the values taken by H on each side, from which follows a well-known formula for the KH instability of the (incompressible) *thin model*. The extension to the (much more complex) compressible TD in MHD is studied in [12]. The term $\rho(y)V_{A\kappa}^2(y)$ in H represents the magnetic tensions $B_\kappa^2/4\pi$ acting on the KH mode. In fact the roots of $H=0$ give transverse Alfvén waves propagating with respect to the plasma flow

$$c - V_\kappa = \pm V_\kappa(y) \tag{24}$$

Thus, $V_{A\kappa}^2(y)$ acts on the KH instability as a stabilizing effect of magnetic origin. It cannot be switched-off completely except when the magnetic field is unidirectional, a case in which modes with $B_\kappa=0$, known in the literature as *flute modes*, are possible. In general, the directions of the magnetic field lines are different on either side, i.e., a rotation of the magnetic field orientation across the boundary layer occurs, a feature called magnetic shear, which prevents the existence of flute modes and gives always a stabilizing contribution.

Note that the roots of $M=0$ give the phase velocity of MHD fast and slow modes propagating with respect to the local plasma velocity

$$(c-V_\kappa)^2 = \frac{1}{2}\left(V_A^{\,2}+c_s^{\,2}\right) \pm \frac{1}{2}\sqrt{\left(V_A^{\,2}+c_s^{\,2}\right)^2 - 4V_{A\kappa}^{\,2}c_s^{\,2}} \qquad (25)$$

Thus, a coupling of the KH mode with these magnetoacoustic waves is expected to have the coefficient $M \ll 1$, and so a penetration distance much larger than $\lambda/2\pi$, the usual order of magnitude for the KH instability. And since the amplitude grows on both sides by transferring positive energy density of a perturbation on one side to a negative energy density perturbation on the other side, when magnetoacoustic effects are significant the same energy amount is spread on a wider plasma volume. Therefore the growth rate is reduced. The argument gives a qualitative explanation of the stabilizing effect of compressibility.

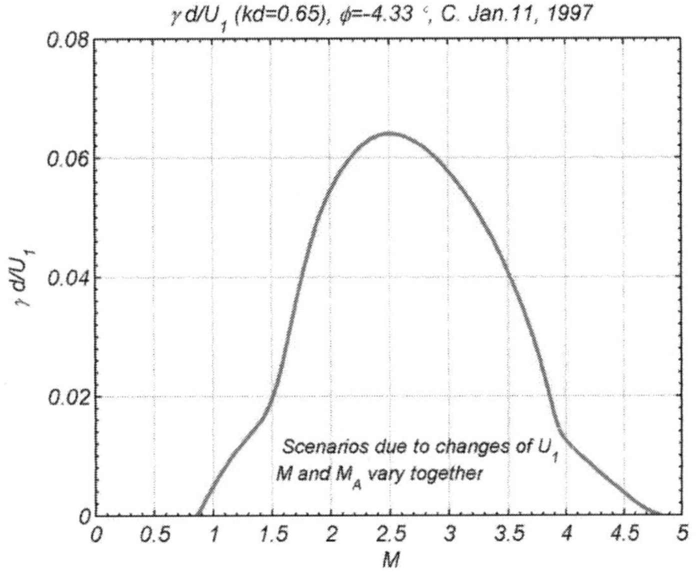

Figure 4. Normalized growth rate $\gamma d/U_1$ for flank scenarios by varying the solar wind speed. The geometry of the fields, the ratio of inner/outer values, and the ***k***-mode remain fixed while M_s and M_A change, keeping a constant ratio.

3. The unstable MP observed by Interball-tail on January 11, 1997

The event occurred during Earth passage of the trailing part of a large interplanetary magnetic cloud. From 1:00-2:00 UT a plasma plug of high density passed the Earth. There was a large increase (~ 30 times normal value) of the dynamic pressure, P_{dyn}, of the solar wind, while the interplanetary magnetic field (IMF) turned to north. In the following hours the pressure subsided, and the clock angle (the angle between the IMF and the geomagnetic north) decreased in stepwise fashion, as the IMF became strongly northward. Interball-tail was left out in the magnetosheath by the huge compression of the boundary. It explored the outer region adjacent to the MP during some time, and then reentered the magnetosphere as the size returned to normality, slowly crossing the low latitude boundary layer (LLBL) at a near dusk flank position, distant ~ 13 R_E in the anti-sunward direction (R_E : Earth radius).

Figure 1 shows a sample of magnetic field fluctuations recorded by Interball-tail during an interval in the magnetosheath, close to the MP boundary. The high frequency waves are identified as mirror modes convected by the magnetosheath flow from dayside regions, while the low frequency amplitude modulation is produced by a surface wave of KH origin traveling on the adjacent MP. Thus, mirror waves in the magnetosheath are riding on envelope KH surface perturbations. A description of the complex event, and a discussion of the data fluctuations observed by the spacecraft over the MP and in the LLBL, which were interpreted as KH waves, is given in [8].

Here we report a stability study with the one-scale model for a period of the event (about 3:00 UT). The calculation extends the earlier analysis by considering compressibility. The basic parameters derived from Interball-tail data are summarized below

Table 1. Input parameters for the period ~ 03:00 U,

Magnetosheath (1)		Magnetosphere (2)	
n_1	10 cm^{-3}	n_2	1 cm^{-3}
B_1	30 nT	B_2	20 nT
U_1	300 km/s	U_2	~ 0 km/s
T_1	0.130 keV	T_2	~ 1.9 keV
θ_{B1V}	100°	θ_{B2V}	30°

The local values of the main Mach numbers based on magnetosheath quantities, are $M_s = 1.90$, $M_A = 1.67$, so that a KH calculation requires a compressible treatment (not given in [8]. The growth rate $\gamma = \Im(\omega)$ is computed from the imaginary part of c. Figure 2 shows the growth rate $g = \gamma d/U_1$ versus the wave number kd ($k = |\boldsymbol{k}|$), both given as normalized quantities. The result is computed for modes with a particular orientation of $\boldsymbol{k}$, which subtends the angle $\varphi = -4.33°$ to $\boldsymbol{V}$. This direction of $\boldsymbol{k}$ gives the maximum growth rate of the instability.

The last statement is illustrated in figure 3 that shows the dependence of g on φ, while $kd = 0.65$ remains fixed at the value associated with the mode of maximum growth rate, $\gamma d/U_1 = 0.05$ of figure 2. We note that modes that deviate by $D\,\varphi \approx \pm 15°$ from the optimal $\boldsymbol{k}$ direction are stable.

We may evaluate approximately the wavelength of the mode of fastest growth as $\lambda \approx 4.8\ R_E$, assuming $\Delta \approx 1\ R_E$, for the LLBL width, an estimate often quoted for the flanks (see also section 5). The corresponding *e-folding time* for the instability is then $\tau_e \approx 3.5$ minutes.

3.1 Possible upstream and downstream scenarios

The KH instability is convective and is carried downstream by the flow, while the growth continues if the supporting background is amplifying. Thus, the perturbations observed by a spacecraft are not generated locally but come from places further upstream. It is therefore of interest to infer, from the evidence obtained at the Interball-tail position, the possible conditions for KH growth at MP sites closer to the dayside, or to conjecture about the amplification properties of the MP further tailward, where the perturbations are finally transported.

Figure 4 shows the result of an exploration of sunward and tailward scenarios with the same basic configuration of magnetic field, density, temperature, and relative fields angles, varying only the solar wind speed U_1. The configuration of the fields and the $\boldsymbol{k}$-mode, remain fixed but M_s and M_A are changed, keeping a constant ratio. We note that the stabilization trend at upstream stations is due to the increasing strength of the magnetic tensions with decreasing M_A, while the compressibility effect becomes weaker. Therefore, upstream stabilization can be attributed to magnetic shear. Conversely, downstream compressibility increases with M_s, while magnetic tensions become weaker as M_A increases. Thus, downstream stabilization is mainly due to compressibility.

These statements must be taken as indications of trends, not as realistic extrapolations of the Interball-tail data: the scenarios are only possible configurations at the flanks. Moving away from the spacecraft locale, the fields change because of the 3-D shape of the magnetopause, and the spatial variation of the interplanetary parameters. In particular, at the distant tail the geomagnetic field is much weaker than at the near flanks. Thus, the magnetic tensions inside the boundary layer are small and *flute* modes with respect to the IMF may switch-off almost completely magnetic effects. As we reach the far tail the MP flank becomes KH unstable again.

Another calculation was done with the same basic configuration of magnetic field, density, temperature, and relative fields angles, but with a reduced magnetosheath speed U_1, such that $M_s = 1.00$, $M_A = 0.88$. This is a possible MP scenario upstream, at a moderate anti-sunward distance from the Interball-tail site in the same event. The KH instability is then carried downstream and continues to grow. At the upstream locale the growth rate *g* is smaller as expected, however it is worth noting that the fastest growing mode occurs at a longer λ.

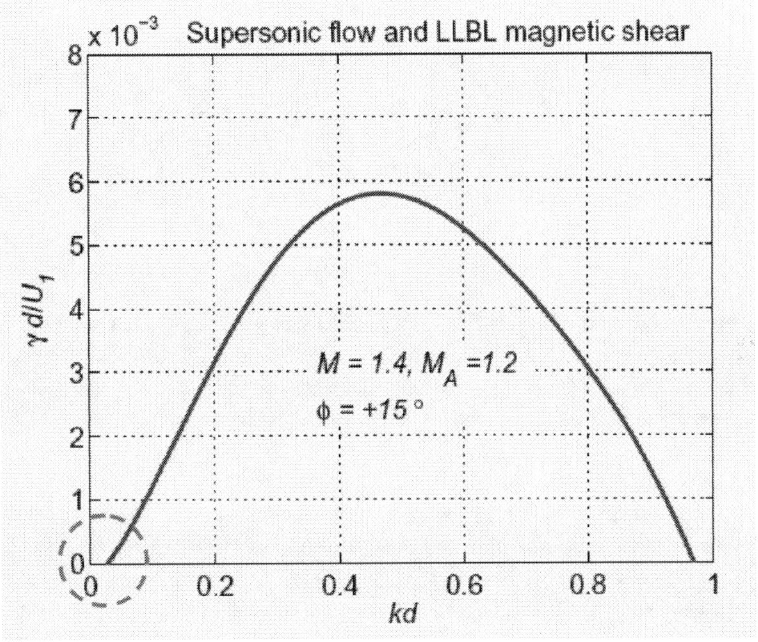

Figure 5. A scenario with supersonic flow and magnetic shear in the LLBL. Normalized growth rate $\gamma d/U_1$ *vs.* *kd* for the ***k***-angle of fastest growth, $\varphi = 15°$.

4. Local stability gaps of the near flanks

The previous case and companion extrapolations suggest a hypothetical LLBL scenario with supersonic flow and large magnetic shear. The geomagnetic field is a bit stronger than the outer field. The flow is supersonic $M_s = 1.4$, and slightly super Alfvénic $M_A = 1.2$. The magnetic shear across the LLBL is $45°$. The maximum growth rate (after optimization over *k* and the ***k***-angle φ) is found to be very small: e-folding time for $\Delta = 6000$ km, $U_1 = 300$ km/s, $\tau_e = 0.5$ hour, which means that the growth is negligible. Upstream, where $M_s = 1.2$ and $M_A = 1$ (U_1 smaller by 20%), no amplification at all is found: the boundary layer is locally stable.

The normalized growth rate $g = \gamma d/U_1$ as a function of the normalized wave number *kd* is shown in figure 5. The result is shown for the mode with optimal direction of the ***k***-vector, $\varphi = +15°$. For very long λ's (the range near $kd = 0$) the growth rate is zero at a small but finite *kd* value. A thin model (tangential discontinuity) for $M_s = 1.4$ and $M_A = 1.2$ fails to predict this instability. Therefore a "*pitfall*" of the TD model (see [13]) occurs also for supersonic flows. This is worth keeping in mind when the popular TD model is used in the flanks for downstream estimates. There, γ could achieve significant values for continuous profiles due to the increase of U_1, while TD might predict stability.

5. A long journey in the boundary layer

Due to an unusual combination of circumstances on November 20, 2001 the CLUSTER flotilla remained about 10 hours in the LLBL of the near magnetospheric dusk flank (at a GSM X position ~ -$4\ R_E$). The IMF was northward during this interval, with long lapses of low (≤ 30°) clock angle. The spacecrafts recorded series of intermittent bursts (3 - 4) of nearly periodic, large amplitude, signals during most of the time of the long LLBL exploration. Very neat wavy features are seen in the data records of the magnetic field, and plasma density. A sample of CLUSTER data for this event is shown in figure 6.

For a stability analysis, the input parameters estimated from CLUSTER data for a period ~ 14:00-17:00 UT, are the following

Table 2. Input parameters for the period ~ 14:00-17:00 UT

Magnetosheath (1)		Magnetosphere (2)	
n_1	10 cm^{-3}	n_2	1 cm^{-3}
B_1	20 nT	B_2	20 nT
U_1	300 km/s	U_2	~ 0 km/s
T_1	0.3 keV	T_2	~ 2 keV
θ_{B1V}	80°	θ_{B2V}	30°

Some values are similar to the Interball-tail case of January 11, 1997, but we note that here $B_1 = B_2$, so that the current sheath only turns the field but does not change its intensity, and the angles of the field configuration are different. The significant changes in the stability analysis between the two events show how sensitive is the KH process to the input values.

5.1 One-scale and two-scale models

In this case $M_s = 1.36$ and $M_A=1.97$, and using a one-scale model the fastest growing mode is found for φ = -28°. Figure 7 shows the normalized growth rate g as a function of kd for a one-scale and a two-scale model. A feature worth noting is that the limit of very long wavelengths $\lambda \gg \Delta$ is stable, while the calculation shows instability at the CLUSTER site for $\lambda \sim \Delta$. This means that a stability analysis performed with a TD model for the MP would arrive at a wrong conclusion, i.e., a stability verdict while in fact the site is unstable (on this issue see [13]).

The average period observed by CLUSTER was $T = 120\text{-}150$ s. Using the one-scale model results, we may estimate both, $\lambda = 2\pi/k$, and Δ, from the experimental value for $\omega_r = 2\pi/T$ assuming that the observed perturbations are due to the KH mode of largest growth ($kd \approx 0.7$). These should pass by CLUSTER with a phase velocity $\Re(c)=v_{ph} \approx 2/3U_1$ (magnetosheet speed $U_1=300$ km/s). We find $\lambda = 3.71$ - $4.64\ R_E$, and estimate the LLBL thickness as $\Delta = 0.8 - 1\ R_E$, depending on the chosen value for T. This estimate is in agreement with values quoted for the near flank in other works.

The data, however, show evidence that the MP structure was stratified with a scale-length for $\boldsymbol{B}$, n, and a different one for $\boldsymbol{V}$. Another calculation was done, therefore, with a two-scale model. An example of field profiles for this case is shown in figure 8. Although it was not easy to ascertain, an estimate of 0.2 for the ratio $r_s=\delta/\Delta$ was employed. A striking result is that the growth rates of the KH modes with a $\boldsymbol{k}$ direction $\varphi = -28°$ is now negligible. This is the $\boldsymbol{k}$ direction of maximum growth rate for the one-scale model. The instability was found instead, by changing the orientation of the $\boldsymbol{k}$-vector until it was nearly perpendicular to the magnetospheric field, i.e., $\varphi = -60°$.

The maximum growth rate is for $kd = 0.5$ (a 1.4×λ increase) and is reduced by a factor ≈ 3.25 with

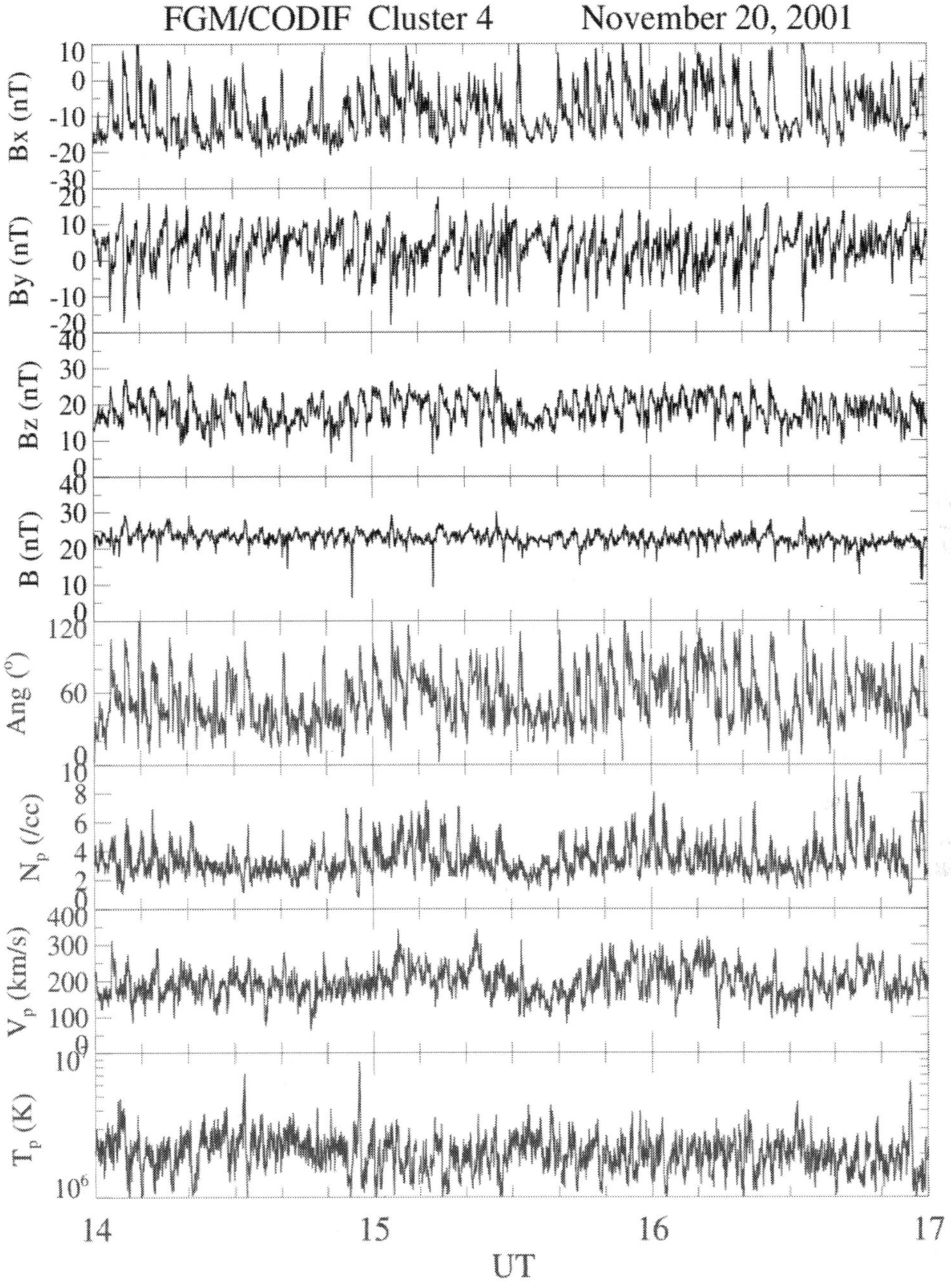

Figure 6. Magnetic field and plasma parameters measured by the FGM and CODIF instruments on Cluster 4 from 14 to17 UT, November 20, 2001. From top to bottom the panels show the components of the magnetic field and the total field, the angle between ***B*** and ***V***, the number density, bulk speed and temperature of the protons.

respect to the one-scale model value, see figure 7. The direction of ***k*** most favorable for instability in the two-scale model is such that the magnetic field in the LLBL and the magnetosphere is fluted by the perturbation, so that the magnetic tensions are turned off on that side.

Comparing results of one- and two-scale models, the important changes found show how responsive is the stability of the MP to fine structure properties of the LLBL. It is worth commenting that the sensitivity of the ***k*** orientation on the structure of the boundary layer has important consequences for the subsequent nonlinear evolution of the KH instability, as shown in [14].

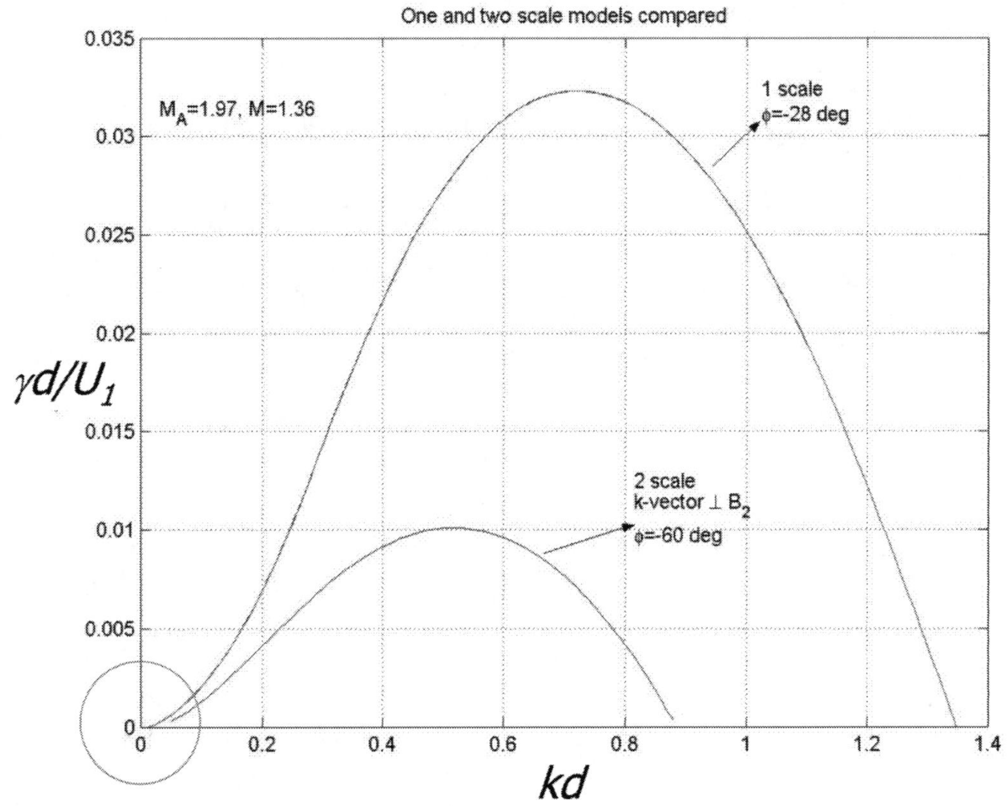

Figure 7. November 20, 2001 event. Results for one-scale and two-scale mixing layer models. Growth rate $\gamma d/U_1$ as a function of wavenumber *kd*, normalized quantities. For the one-scale model we show the modes with most unstable angle $\varphi = -28°$. For the two-scale model the best angle φ for the instability occurs at $-60°$. The growth rate is much less than that of the one-scale model. The circle emphasizes that long wavelength modes are stable (see text).

6. Summary

The KH instability is a major wave source and thus a major contributor to viscous coupling of the magnetosphere with the solar wind. Particularly in periods of IMF $B_z > 0$ of long duration corresponding to the $B_z > 0$ phase of a magnetic cloud passage.

The KH instability that develops in special strips of the front side MP during northward IMF periods generate perturbations that are convected tailwards [15]. The oscillations may enter a neutral stability gap of the MP before they reach the far flank, which is again unstable due to the weakening of magnetic tensions in the LLBL.

A magnetopause flank locale may be locally KH-stable but yet wavy as it receives KH-waves generated in the KH-active strips further upstream. Under conditions shown here the source of the LLBL turbulence is not the upstream neighborhood within a few KH amplification lengths, but rather the far away subsonic regions on the dayside: the unstable MP strips of low magnetic shear. Solar winds with northward IMF favorable to stable gaps at the supersonic flanks have cold, not too dense, plasmas, and strong magnetic fields, so that M_A is larger, while M_s is smaller, than average values.

These conditions may be realized during the passage of the rear of an interplanetary a coronal mass ejection by Earth.

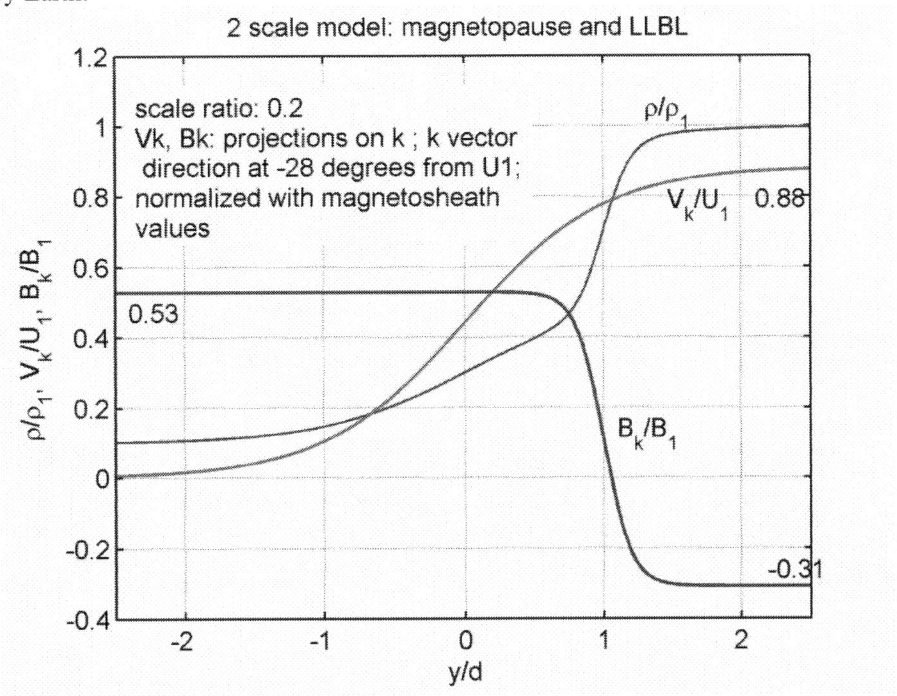

Figure 8. Functions that represent the basic fields for a two-scale model of the boundary layer (see text at sections 2.2 and 6). Normalized density, and normalized projections of the velocity and magnetic fields on the ***k*** vector. The direction of the latter is given by $\varphi = -28°$, the angle of the most unstable modes of the one-scale model.

Due to the stabilizing actions of magnetic shear and compressibility, the examples of flank models considered show that the instability occurs for a set of ***k***-modes with a quite small angular spread around the direction of fastest growth. KH wave fronts and vortices arise in the supersonic boundary layer with a well defined orientation that depends on the field profiles of the transition layer.

The important changes of the instability properties revealed in the comparison between one and two-scale models show how sensitive is the stability of the supersonic MHD boundary layer to fine structure properties.

The study is intended as a contribution to a still not available extensive survey of the stability of representative boundary layers of supersonic flows at the MP flanks.

Acknowledgments
L. Bilbao, G. Gnavi, and F.T. Gratton are grateful to CONICET for research grant PIP 5291 and to UBA for research grant UBACYT X90. C. J. Farrugia is supported by USA NASA grants NNX08AD11G and NNG05GG25G and by NASA CLUSTER grant to UNH.

References

[1] Landau L D 1944 *Dokl. Akad. Nauk. SSSR* **44** 151
[2] Belmont G and Chanteur G 1989 *Phys. Scripta* **124** 124.
[3] Fitzenreiter R J and Ogilvie K W 1995 *Physics of the Magnetopause (Geophys. Monog. Ser.* vol 90*)* ed P Song, B U Ö Sonnerup and M F Thomsen (Washington DC: AGU) p 277

[4] Kivelson M G and Chen S-H 1995 *Physics of the Magnetopause (Geophys. Monog. Ser.* vol 90*)* ed P Song, B U Ö Sonnerup and M F Thomsen (Washington DC: AGU) p 257
[5] Miura A 1995 *Physics of the Magnetopause (Geophys. Monog. Ser.* vol 90*)* ed P Song, B U Ö Sonnerup and M F Thomsen (Washington DC: AGU) p 285
[6] Lotko W and Sonnerup B U Ö 1995 *Physics of the Magnetopause (Geophys. Monog. Ser.* vol 90*)* ed P Song, B U Ö Sonnerup and M F Thomsen (Washington DC: AGU) p 371
[7] Farrugia C J, Gratton F T and Torbert R B 2001 *Space Science Rev.* **95** 443
[8] Farrugia C J *et al.* 2000 *J. Geophys. Res.* A **105** 7639-7667
[9] Gratton J, Gratton F T and González A G 1988 *Plasma Phys. Contr. Fusion* **30** 435-456
[10] Keller H B 1992 *Numerical methods for two point boundary-value problems* (New York: Dover)
[11] Miura A and Pritchett P L 1982 *J. Geophys. Res.* A **87** 7431
[12] González A G and Gratton J 1994 *J. Plasma Phys.* **51** 43-60
González A G and Gratton J 1994 *J. Plasma Phys.* **52** 223-244
[13] Gratton F T, Bender L, Farrugia C J and Gnavi G 2004 *J. Geophys. Res.* A **109** doi: 10.102 V2003JA010146
Gratton F T, Gnavi G, Farrugia C J and Bender L 2005 *Brazilian J. Phys.* **34** 1804-1813
[14] Gratton F T, Bilbao L E, Farrugia C J and Gnavi G 2008 *JOP: Conf. Series* (in press)
[15] Farrugia C J , Gratton F T, Bender L, Biernat H K, Erkaev N V, Denisenko V, Torbert R B and Quinn J M 1998 *J. Geophys. Res.* A **103** 6703

Large eddy simulations in MHD: the rise of counter-rotating vortices at the magnetopause

F T Gratton[1,2,4], L E Bilbao[1], C J Farrugia[3] and G Gnavi[1]

[1] Instituto de Física del Plasma (INFIP), Consejo Nacional de Investigaciones Científicas y Técnicas y Universidad de Buenos Aires, Buenos Aires, Argentina
[2] Departamento de Física, Facultad de Ciencias Fisicomatemáticas e Ingeniería, Pontificia Universidad Católica Argentina, Buenos Aires, Argentina
[3] Space Science Center, University of New Hampshire, Durham, NH, 03824, USA

[4] E-mail: fgratton@infip.org

Abstract. A study of the magnetohydrodinamic (MHD) development of coherent structures in compressible, inhomogeneous, mixing layers due to the velocity shear instability is reported. The non-linear evolution of the original vorticity sheet is computed with 3-D large eddy simulations (LES) of temporal mixing layers tailored to represent distinctive conditions at the terrestrial magnetopause. We find that the boundary layer is characterized by the growth of large-scale vortices and becomes a site of mass mixing and enhanced plasma diffusion. In MHD the Lorentz force and its associated baroclinic term, together with the ordinary baroclinic term, and stratified entropy across the mixing layer, conspire to hinder vorticity flux conservation. In our LES the non-conservation of vorticity becomes manifest after ~ one rollover time when in addition to vortices with positive rotation (the same as the original vorticity sheet) other coherent structures with strong negative vorticity also arise, a noteworthy effect examined here. It is found that the vorticity is concentrated in cores of both signs with absolute values $\sim 4\text{-}5\times\omega_i$, (maximum vorticity of the initial shear layer). Concomitant with 3-D vortex stretching, the kinetic helicity also rises at vorticity cores. Furthermore, high temperature occurs in the cores, $\sim 3\times T_i$ (magnetospheric temperature) correlated with local density depletion, $\sim 0.4\times n_i$ (magnetospheric density), while gas and magnetic pressure remain close to surrounding values. The study is intended as a contribution to the understanding of solar wind interaction with the magnetosphere during periods of northward interplanetary magnetic field.

1. Introduction

We report a study of magnetohydrodinamic (MHD) development of vortices in compressible, non-homogeneous boundary layers, due to the velocity shear instability (Kelvin-Helmholtz, KH). Transition layers between flows of different velocity, density, and magnetic fields, are prominent aspects of space plasmas. In a magnetosphere context the magnetopause boundary layer comes to mind separating the stagnant, hot and tenuous magnetosphere plasma threaded by the Earth's magnetic field from the cold and dense magnetosheath plasma flowing past it. We present results of the non-linear dynamics of the KH instability in MHD obtained with large eddy simulations (LES) using a new 3D+t MHD code of an arbitrary Lagrangian - Eulerian and finite volume class (ALE+FV). The code was developed at our Institute (For technical details see [1]). The study assumes periodic boundary conditions, both along, as well as transverse to, the average flow. The MHD simulations show the roll-off of large-scale vortices in the boundary layer, stretched across the main stream and oriented along the magnetic field lines.

The simulation reveals, among other features, the evolution of pairs of intense counter-rotating vortices of similar strength, which develop from a velocity gradient of much less vorticity all in the same direction. Hot and tenuous plasma is entrained in the core of these vortices. We argue that the rise of counter-rotating vortices in the 3-D simulations is due to the joint presence of non-uniform density and magnetic forces. We also confirm that the addition as an initial disturbance of a trigger velocity field pattern (of sizable amplitude) associated with a fast growing mode (according to linear theory) shortens the time it takes for the KH instability to fully develop, thereby hastening the appearance of large-scale coherent structures.

Fluid dynamics is a significant aspect of our problem, particularly when it deals with mixing of matter. In fluid dynamics a mixing layer is the region between two flows of different velocity, often also of different density (see, e.g., [2], [3]). An important example in a magnetospheric context is the boundary layer, which is often found to have a magnetic field orientation typical of the magnetosphere at the place of observation, but with a plasma population that contains both particles of high energies typical of the magnetosphere, as well as low energy magnetosheath plasma, i.e., a mixture of plasmas from both sides of the layer [4].

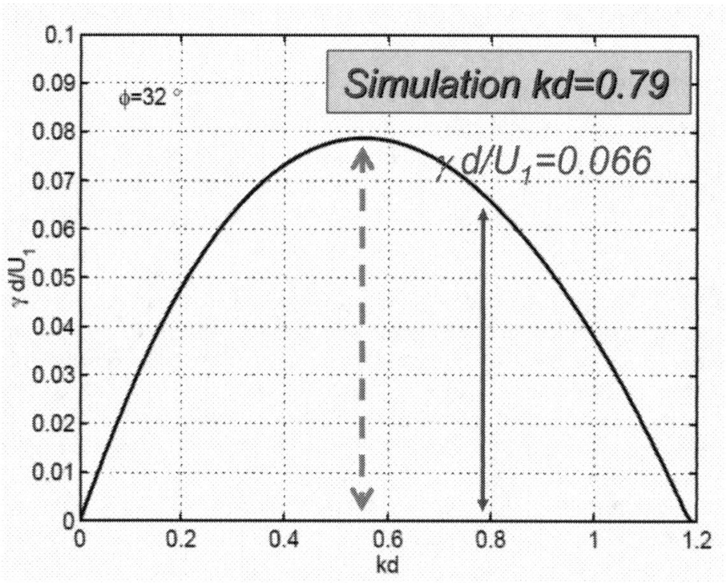

Figure 1. The linear KH instability for a December 7, 2000, boundary layer model. Growth rate versus wavenumber as normalized quantities, for the optimal $\vec{k}$ angle for the instability $\varphi = 32°$. The dashed line indicates the growth rate and wavenumber of the most unstable mode. The solid line indicates the mode used to start the numerical simulation (see text).

In fluid mechanics a mixing layer is frequently turbulent due to the velocity gradient, which renders it prone to instability. The turbulent layer is characterized by the presence of large-scale vortices, and it becomes a site of enhanced mass diffusion (e.g., [2]). Turbulent flow implies one disordered in space and time, but this specification does not necessarily imply a chaotic behavior in all three dimensions. Mixing layers in ordinary fluids are usually a source of quasi 2-D turbulence. Large-scale vortices arise from the rolling-up of the original vortex sheet that separates motions with different speeds. Likewise, quasi 2-D vortical structures also occur in mixing layers in space plasmas, as shown here and in many other studies (e.g., [5], [6], [7]). However, in MHD, vortex generation depends also on the direction and strength of the magnetic field.

In spite of what the term may seem to imply, fluid turbulence is a deterministic phenomenon, i.e., it is governed by deterministic laws and equations, even though the complexity of the non-linear evolution is such that predictability in practice is lost after a finite time. As the Reynolds number R_e increases, direct numerical simulation usually can deal only with the large scale aspects of the flow, and contain errors introduced by numerical schemes, the absence of small scale effects, and imprecise description of initial and boundary conditions. All the errors are amplified by fluid dynamic nonlinearities, and after some time the predicted and the actual turbulent flow can be very different. Therefore, as is widely known, unpredictability and randomness appear even in deterministic systems.

Vortices arising in KH--unstable layers have been extensively studied in ordinary fluid dynamics, both in experiments as well as in computations. A similar body of knowledge on the plasma version of the KH instability has grown also in space physics. Many authors provided observational, theoretical, and computational insights. The progress in our understanding of magnetospheric KH can be followed in surveys, for example, [8], [9], [10], [11]. Still, further research on dynamical properties of MHD mixing layers is desirable in order to attain a deeper understanding comparable to that of classical fluids.

In the magnetospheric context computational MHD of the KH instability began with the pioneering work of Miura [12]. Recent studies using computer simulations (e.g., [13], [14]) strive to understand in what manner precisely does KH lead to mass exchange. In [6] a main thrust was to show that KH rolled up vortices can foster mass transfer, and in the same spirit is [13]. However, the sonic and magnetic Mach numbers, M_s and M_A, respectively, of our numerical experiments correspond to a physical regime different from that of these works. Our simulation models the near flanks where the flow is still subAlfvenic. Their studies are for more distant sites where the plasma is superAlfvenic. While in our study the magnetic field lines show "rigidity", in their works the flow can easily bend the magnetic field lines.

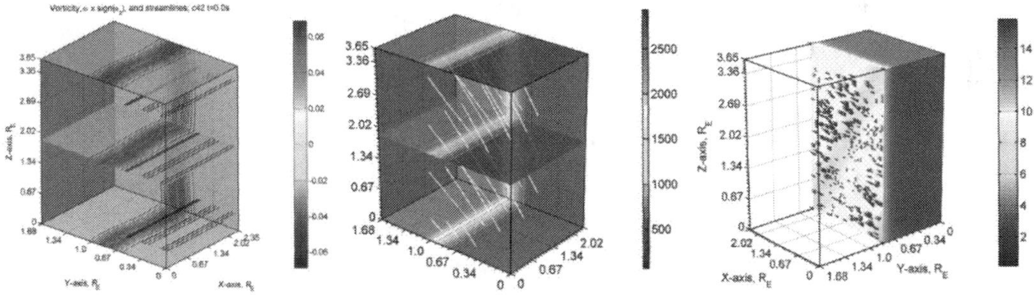

Figure 2. Initial configuration of numerical simulation. Panel *a* (left) gives vorticity contours $\omega = |\vec{\omega}|\mathrm{sign}(\omega_z) = const.$ on three planes $Z = const.$, and on the back of the computational box $X = const.$; sets of streamlines (black solid lines) on $Z = const.$ planes are also represented. Panel *b* (middle) shows (on the same planes as in panel *a*) the initial temperature distribution with $T = const.$ contours. A set of magnetic field lines (yellow solid lines) is also shown here. Panel *c* (right) gives the isodensity surface $n = 9$ p cm^{-3} (with a rotated view). The blue cones indicate the velocity field, which shows an initial perturbation as the vorticity field in panel *a* (see text). The color scales in the panels indicate the intensity of the corresponding scalar quantity. All axes tick marks are in units of Earth radii, R_E.

The computational study of a boundary layer MHD flow presented here is modeled with parameters belonging to the duskside magnetopause, derived from data recorded by Cluster 3 on December 7, 2000. The event is in the class of northward pointing interplanetary magnetic fields (IMF), on which more is said in [16]. Cluster was crossing the low latitude boundary layer (LLBL) of the magnetosphere near the dusk terminator. The phenomenology of that event, which developed after the impact of a tangential discontinuity/vortex sheet (TD/VS) on the magnetosphere, was studied in another paper [15]. We gave there also results of a stability analysis based on input parameters measured in situ. The theoretical study was in agreement with observations interpreted as the growth of surface waves of KH origin. The main physical properties of plasma and fields at the Cluster locale are the subsonic $M_s = 0.57$, and sub-Alfvenic $M_A = 0.70$, flow, and a very small magnetic shear across the boundary layer.

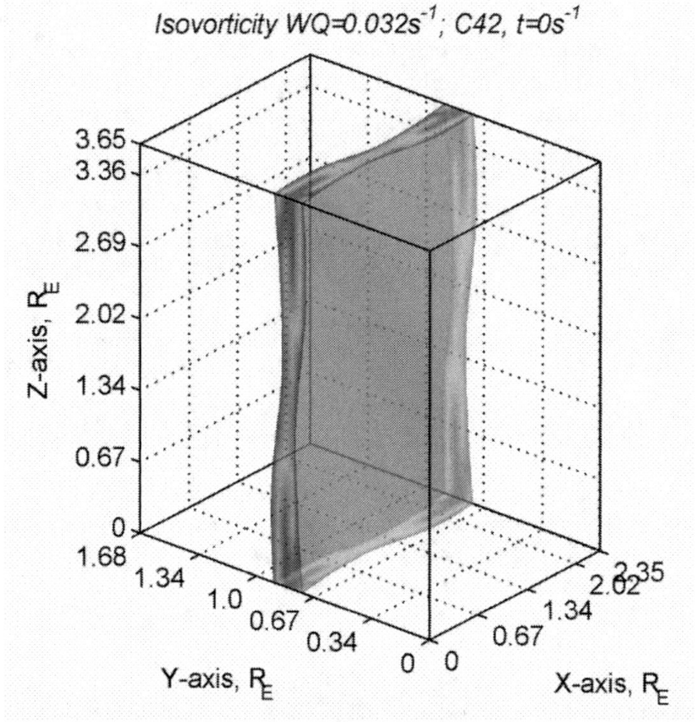

Figure 3. Isovorticity surface with $\omega = |\vec{\omega}|\mathrm{sign}(\omega_z) = 0.032\ \mathrm{s}^{-1}$ at $t = 0.0$ s; the maximum initial $|\boldsymbol{curl}(\vec{v})|$ is $\omega_i = 0.0485\ \mathrm{s}^{-1}$. The figure illustrates the level of initial perturbation.

The LES allows us to achieve a better understanding of the boundary layer physics after the onset of the KH instability in the particular event considered, which was characterized by moderate values of M_s and M_A, both smaller than one. Because of their relevance to energy, momentum, and possibly also mass transfer into the magnetopshere, KH simulations can be particularly instructive. Summing up: besides being of concern for magnetofluidynamics in revealing aspects of vortex formation in boundary layers, our simulations are significant also for magnetospheric physics.

The plan of the paper is as follows. In section 2 we review briefly the bases and the limitations of LES. We comment on some fluid dynamics and MHD aspects of the problem, and explain the reasons for the non-conservation of the vorticity flux in our computational study. In section 3 we report the initial and boundary conditions for the 3D+t numerical simulation. We give the parameters derived from Cluster data for December 7, 2000, on which the boundary layer model is tailored to. We report also on the linear stability results for the same model. Section 4 shows the results of the numerical experiment. The computations bring forward novel features of the process that broadens, and strain the low latitude boundary layer (LLBL), the outer coat through which external plasma enters the magnetosphere. Section 5 contains a summary and conclusions.

2. LES, MHD aspects, and non-conservation of vorticity flux

The precise details of the KH instability conditions in compressible MHD are intricate, but speaking in general one can say that the boundary layer becomes unstable when the magnetic shear angle across the layer is small, and/or when the Alfvenic Mach number $M_A = U/V_A$ (where U denotes the velocity of the external flow and V_A is the Alfvén velocity) is large, both factors limiting the stabilizing *B*-forces. The rolling-up of the initial vorticity stratum by the Kelvin-Helmholtz mechanism generates large vortices, and the vorticity of these structures is further modulated, and increased by a 3-D vortex stretching mechanism.

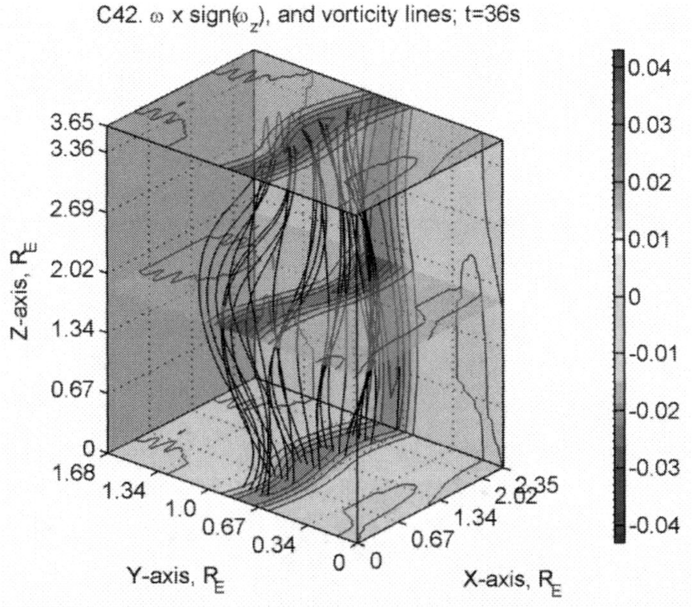

Figure 4. Vorticity contours, $\omega = |\vec{\omega}|\ \mathrm{sign}(\omega_z) = const.$, and shape of vorticity lines at $t = 36$ s ($\approx 2/5$ rollover time τ_r). The pattern of the initial perturbation is still visible albeit with a larger amplitude.

Let l denote a typical scale length of large vortices, while v is a characteristic velocity of the turbulence (for instance, a typical rotational velocity of the eddy), and let ν_{eff} stand for the effective diffusion coefficient for momentum transfer in the plasma. Then we can introduce a large-scale Reynolds number with $R_e = l\mathrm{v}/\nu_{eff}$. When the rollover time $\tau_r = l/\mathrm{v}$ of the large eddies is much smaller than the diffusion time l^2/ν_{eff}, i.e., when $R_e >> 1$, which is normally the case in space plasmas, a range of large-scale dynamics exists that can be studied with the ideal MHD theory, supported during the nonlinear stage by computational fluid dynamics (CFD). It is only at the small scale length that dissipative effects become important. But their influence on large-scale phenomena is reflected after lapses much longer than the turnover time, which is not the case of this study where the simulations last up to $t \sim \tau_r$. The CFD of large structures ignoring dissipative effects is technically known as large eddy simulations (LES) [2].

The computational work is based on the set of compressible, ideal (non-resistive, and inviscid) MHD equations. It is not necessary to report the MHD set here: the equations are shown in the integral form adequate for the numerical ALE+FV code in [1], and are written in differential form in [16] (this *JOP Conf. Series*).

For any vector field $\vec{X}$ it is customary to define the helicity density, and the total helicity (over a volume), of $\vec{X}$ as

$$h_X = \vec{X}.\mathbf{curl}\left(\vec{X}\right), \qquad H_X = \int_{Vol} \vec{X}.\mathbf{curl}\left(\vec{X}\right) d^3x \tag{1}$$

When $\vec{X} = \mathrm{v}$, $\vec{\omega} \equiv \boldsymbol{curl}(\vec{\mathrm{v}})$ is the vorticity, and the kinetic helicity density and the kinetic helicity are introduced as

$$h_V = \vec{V}.\vec{\omega}, \qquad H_V = \int_{Vol} \vec{V}.\vec{\omega}\, d^3x, \tag{2}$$

quantities considered in the analysis of the simulations. The kinetic helicity density h_v is an indicator of the presence of swirling flow. Note that the kinetic helicity H_v is not conserved in MHD, and that in classical fluids it is an invariant only when the density ρ is uniform (see, for instance, [19]).

We may also mention that when $\vec{X} = \vec{A}$, where $\vec{A}$ is the vector potential of the magnetic field $\vec{B} \equiv \boldsymbol{curl}(\vec{A})$, the magnetic (or potential) helicity is introduced. The total magnetic helicity is conserved in ideal MHD, while the total cross helicity, i.e., the integral over a volume of the cross helicity density $\vec{B}\cdot\vec{\mathrm{v}}$, is an MHD invariant only under special conditions. However, we do not examine magnetic or cross helicities in this work.

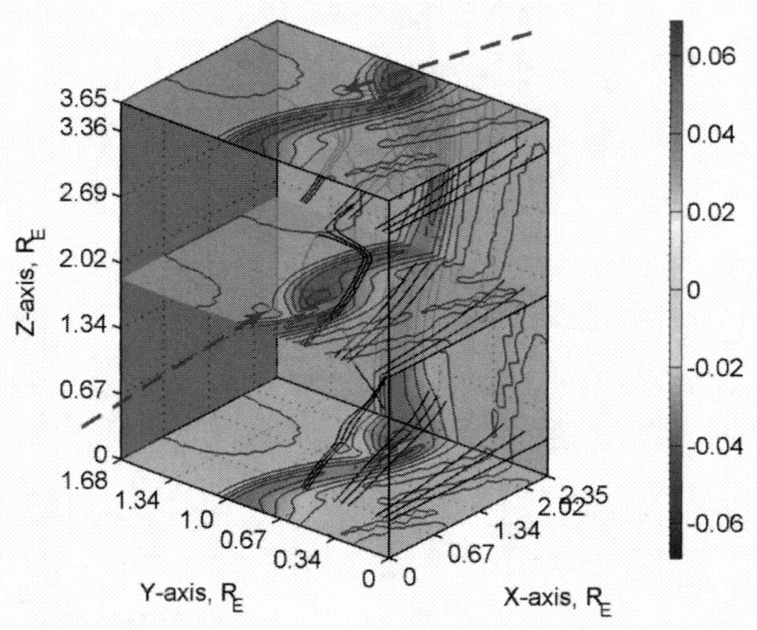

Figure 5. Vorticity contours and some streamlines at t= 89 s, near to ½τ_r. The maximum vorticity value increased, and the vorticity sheet begins to rollover. Dotted arrows point to a spot where a seed of negative vorticity begins to grow.

While the large-scale dynamics of vortices can be studied ignoring dissipation, further vorticity changes take place in a MHD mixing layer because the Lorentz force $\vec{F}_L = \vec{j} \times \vec{B}/c$ is not in general curl-free, and because its associated baroclinic term $-\left(1/\rho^2\right)\nabla\rho \times \vec{F}_L$ is not zero in a perturbed LLBL. Moreover, note that in a stratified LLBL the plasma entropy S across the boundary is non-uniform. Together with ρ, p, T, and $\vec{B}$, S is stratified in the direction transverse to the main flow. And due to the pressure balance condition $p + B^2/8\pi = c_1$, the isoentropic condition $p\rho^{-\gamma} = c_2$ (where c_1, c_2, denote constants) does not hold across the boundary layer, regardless of the fact that the entropy S of all plasma elements is preserved along their trajectories during the ensuing dynamic evolution, in accord with ideal MHD. Therefore, even the baroclinic term $-\left(1/\rho^2\right)\nabla\rho \times \nabla p$ of classical fluid dynamics ([20]) is active in this context. And since it can be shown that

$$\mathbf{curl}\left(-\frac{1}{\rho}\vec{\nabla}p\right)=\vec{\nabla}T\times\vec{\nabla}S, \tag{3}$$

then, although in equilibrium the gradients of temperature and entropy are initially parallel, the RHS of this equation does not stay at zero value during the evolution. For instance, under perturbed conditions the appearance of a streamwise component of ∇T in conjunction with the transverse component of ∇S, generates a spanwise **curl** component.

All the mentioned terms appear in the MHD vorticity evolution equation, and may hinder the conservation of the vorticity flux. For compressible, ideal MHD plasma, the following equation holds

$$\frac{d}{dt}\left(\frac{\vec{\omega}}{\rho}\right)-\left(\frac{\vec{\omega}}{\rho}\right).\vec{\nabla}.\vec{v}=\frac{1}{\rho^{3}}\vec{\nabla}\rho\times\vec{\nabla}p+\frac{1}{\rho}\mathbf{curl}\left(\vec{j}\times\vec{B}/\rho c\right), \tag{4}$$

where the left hand side terms lead to the well known conservation of the flux of $\vec{\omega}/\rho$ when the right hand side is zero. The latter terms are responsible instead for the non-conservation of the vorticity flux during the large eddies' evolution. In our computational study, the non-conservation of vorticity becomes manifest since in addition to the vortices with positive rotation (with the same sign as the original vorticity sheet) other vortices with strong negative vorticity also arise, a striking effect that we report in this paper. Vorticity concentrated in cores of both signs, with absolute value several times larger than the maximum vorticity of the initial shear layer, are found in our LES.

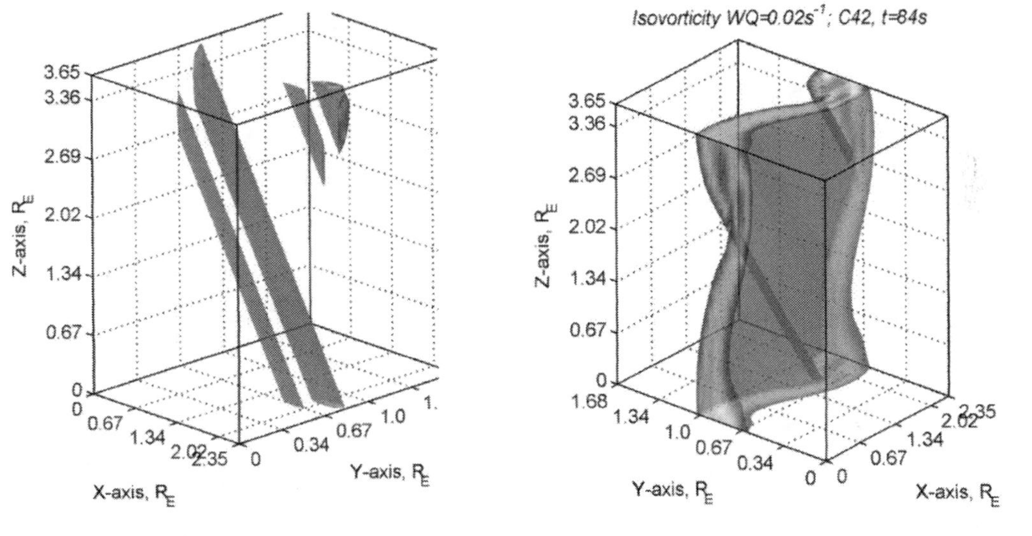

Figure 6. Vortex development at t= 89 s in two panels: left a) isovorticity surface ω = 0.0485 s^{-1} (equal to ω_i, the maximum value of the initial vorticity layer) with a rotated view that helps to reveal a twofold structure; right b) isovorticity surface ω = 0.02 s^{-1} ($\approx$2/5 ω_i), regular view.

In CFD a mixing layer with periodic boundary conditions in the streamwise (or mean flow) direction is called a temporal mixing layer, which is the kind studied here. As mentioned, unstable temporal mixing layers develop large vortex-like structures. In fluid dynamics these are called coherent structures because they last for long periods of time, and can be found far downstream with roughly similar shapes (see, e.g., [20], [2]). As found also in preceding MHD computational studies, and again in this paper, the vortex core is a concentration of vorticity around which convected plasma layers of different density values wind-up, mixing outer with inner matter. The coherent structures have an important influence on the boundary layer properties, like the downstream broadening of the layer width, or the lengthening of the transit time of plasma elements through the local site. As Moffat liked to state "vortices are the sinews of turbulence" [21].

3. Local boundary layer model and stability

We study the evolution of a planar stratified flow in which the physical quantities are initially constant in (x, z) planes, and vary only perpendicular to these planes (y-direction). Initially we assume for the velocity and magnetic fields, $\vec{v} = (v_x(y), 0, 0)$, $\vec{B} = (B_x(y), 0, B_z(y))$, respectively. Similarly, we write $\rho(y) = m_p n(y)$ for the mass density, and $T(y)$ for the temperature (n is the number of particle density, while ρ is the mass density; m_p being the proton mass). The planar flow is intended to model a local site of the magnetopause boundary layer. The initial field profiles, that represent a steady (or average) state, are built with hyperbolic tangents like $\tanh(y/d)$, where d is a normalization length for the y coordinate.

The physical parameters that define the initial and boundary conditions of the problem are the values on either sides of the transition (with subindex 1 for the magnetosheath, and 2 for the magnetosphere) of the vector fields intensities $v \equiv |\vec{v}|$, $B \equiv |\vec{B}|$, the scalar fields, ρ (or n), T, the angle θ of $\vec{B}$ with the x axis (aligned with the main flow), and the width $\Delta = 2d$ of the gradient layer as a basic scale length. We could keep Δ as a free quantity, but in view of the geophysical application we assume $\Delta = 0.5$ R_E for the output of the simulations, and measure distances taking the Earth radius R_E as unit. With these elements, using $Y=y/d$ as a normalized coordinate, and assuming zero velocity on the magnetospheric side of the boundary layer, the initial field profiles of the model are as follows,

$$V_x = \frac{1}{2}U_1\left(1+\tanh(Y)\right), \tag{5}$$

$$\rho = \frac{1}{2}(\rho_1+\rho_2)+\frac{1}{2}(\rho_1-\rho_2)\tanh(Y), \tag{6}$$

$$B = \frac{1}{2}(B_1+B_2)+\frac{1}{2}(B_1-B_2)\tanh(Y), \tag{7}$$

$$\theta = \frac{1}{2}(\theta_1+\theta_2)+\frac{1}{2}(\theta_1-\theta_2)\tanh(Y), \tag{8}$$

$$B_x = B\cos(\theta), \quad B_z = B\sin(\theta), \tag{9}$$

The temperature function $T(y)$ is a consequence of the preceding profiles and the pressure balance equation across the boundary layer, which lead to

$$T(y) = \frac{m_p}{2k_B\rho(y)}\left[p_1+\frac{B_1^2}{8\pi}-\frac{B(y)^2}{8\pi}\right]. \tag{10}$$

Following up the findings of [15], we show results of MHD LES tailored to the December 7, 2000 event. Specifically, we apply boundary conditions derived from data observed by Cluster 3 at a dusk flank of the MP on December 7, 2000, from ~14:00 to 14:29 UT, as in [15]. While the initial profiles develop and change spanwise completely during the LES, the asymptotic values $Y\rightarrow\pm\infty$ of the quantities on either side are fixed as boundary conditions.

The main parameters for the case of December 7, 2000 are listed in Table 1. From these values it follows that at the Cluster position the sonic Mach number ($M = U_1/c_s$) is $M = 0.57$, and the Alfvenic Mach number ($M_A = U_1/V_A$) is $M_A = 0.70$, both evaluated (conventionally) on the magnetosheath side.

Furthermore, the magnetic shear angle $\delta\theta = \theta_2 - \theta_1$ (difference of magnetic field lines direction on either sides) turns out to be very small ≈5°, i.e., they are nearly aligned. This is an important property of the local configuration, quite favorable to the excitation of KH activity, since the stabilizing influence of the magnetic field critically depends on the existence of a substantial shear angle.

Table 1. Parameters for the case of December 7, 2000

14:05 UT	Magnetosheath	14:03 UT	Magnetosphere
n_1 (cm^{-3})	13.6	n_2 (cm^{-3})	2.3
B_1 (nT)	34.6	B_2 (nT)	22.1
U_1 ($km\ s^{-1}$)	139	U_2 ($km\ s^{-1}$)	~0
T_1 (keV)	0.38	T_2 (keV)	~2.6
$\theta_1 = \angle \vec{B}_1\ \vec{v}_1$	121°	$\theta_2 = \angle \vec{B}_2\ \vec{v}_1$	126°

Figure 1 illustrates the growth rate γ of the unstable modes, represented by $g \equiv \gamma d/U_1$, as a function of kd, where k stands for the absolute value of the wave number of the modes. Both quantities are embodied in dimensionless numbers using $d = \Delta/2$ (Δ ~ width of the velocity gradient layer), and U_1 the plasma velocity at the magnetosheath close to the MP. The results correspond to a particular value of the angle $\varphi \equiv \angle \vec{k}\ \vec{v}_1 = 32°$, the most favorable for the growth of the perturbation. The maximum growth rate is given by $g_m \equiv \gamma_m d/U_1 = 0.079$, and corresponds to the mode with $k_m d = 0.55$ [15].

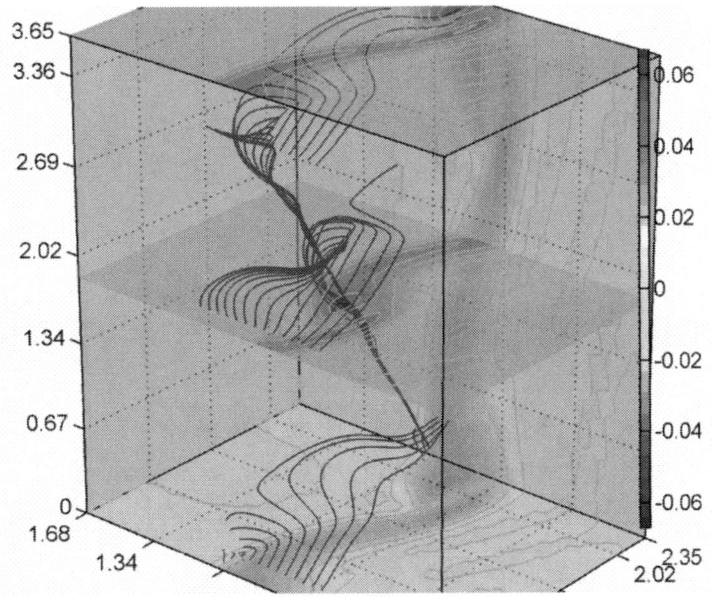

Figure 7. Complexity of the 3-D flow: a close-up of vorticity contours with streamlines shows the formation of a swirling flow in the large eddy.

Since $M_A^2 \approx 0.5$ the magnetic tensions are strong and govern the development of the instability. As a consequence, the wavevector $\vec{k}$ of the most unstable mode is nearly perpendicular to the average direction of the magnetic field (both in magnetosheath and magnetosphere) because magnetic stabilization is then minimal (almost null since $\delta\theta \sim 0$). We found that changing the $\vec{k}$ direction by ~ ±10° from $\varphi = 32°$ (or from $\varphi = 32° + 180°$) the modes are stable. Therefore, perturbations with a specific orientation on the magnetopause, determined by the local direction of the magnetic field, grow faster than any other and become the dominant pattern of the KH excitation with the passage of time.

The e-folding time of the instability was evaluated as $\tau_e \approx 89$ s (or somewhat more, depending on estimates of the boundary layer thickness) [15]. Since the KH activity was observed soon after a sudden change of direction of the IMF that turned due north, in a time lapse comparable to the theoretical exponentiation time, we conjectured that the TD/VS impact on the magnetopause, by adding substantial perturbations to the boundary, acted as a trigger that accelerated the development of the KH instability. By analogy, we introduce a similar feature in the simulation: a significant perturbation in the initial conditions of the LES. This is done adding a *k*-mode with $\varphi = 32°$, and 10% velocity amplitude, to the initial steady state velocity field. The other fields are left unperturbed.

The computational box is periodic in *X*, *Z*, streamwise and transverse to the flow, so that the LES described here correspond to a temporal mixing layer. In the computational box k_x, k_z, can take only restricted values, so that the modes are quantified. Due to the choice of side lengths the fastest growing mode for $\varphi = 32°$ that fits in the numerical box corresponds to $kd = 0.79$, as indicated by the dashed line in figure 1 with $g = 0.066$, somewhat smaller than g_m. The computational box and the initial state are described in the three panels of figure 2: *a*) vorticity contours $\omega = |\vec{\omega}|\mathrm{sign}(\omega_z)$, and streamlines; *b*) temperature contours *T* (eV) and magnetic field lines; *c*) isodensity surface $n = 9$ p/cm^{-3}, and velocity cones. The level of initial perturbation at $t = 0$ s can be appreciated in figure 3 where we show the isovorticity surface with $\omega = 0.032$ s^{-1}. The maximum initial $|\boldsymbol{curl}(\mathbf{V})|$ is $\omega_i = 0.0485$ s^{-1}. The colour scale indicates the value of the quantities. As time goes by, the same kind of plots that represent pictorially the non-linear evolution will be shown and commented.

4. Numerical experiments

The numerical experiments run for about one turnover time. Controls by increasing the number of cells give satisfactory results, significant features and values are reproduced with increased precision.

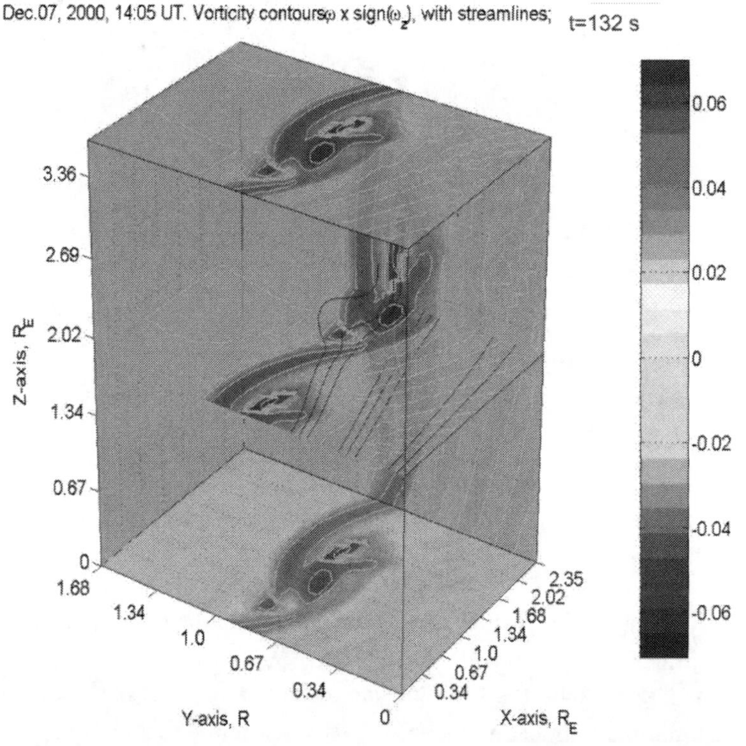

Figure 8. The vorticity contour plot at $t = 132$ s ($\approx 3/4\,\tau_r$) shows the rise of counter-rotating vortices, and the pairing of negative and positive vorticity cores of similar intensity.

Figure 4, at t= 36 s, about 2/5 of τ_r, shows the aspect of vorticity contours, $\omega = const.$, and the shape of a set of vorticity lines. We observe that the pattern of the linear perturbation is still present, although with a larger amplitude.

Significant changes appear in the following plots, drawn at t= 89 s, about ½τ_r. Figure 5 shows vorticity contours and a few streamlines. The maximum vorticity has increased, and the vorticity sheet begins to rollover. The dotted arrows point to a spot where a seed of negative vorticity just starts to grow.

Also at t= 89 s, figure 6 gives details of the vortex development in two panels: *a*) shows the isovorticity surface ω = 0.0485 s^{-1} (equal to ω_i, the maximum value of the initial vorticity layer) displayed with a rotated view that helps to exhibit a twofold structure; *b*) back to the regular view, shows the isovorticity surface ω = 0.02 s^{-1}($\approx 2/5\omega_i$). The changes in the flow and the complexity of the 3-D dynamics are illustrated in a close-up, figure 7, where we can observe in the pattern of the streamlines the formation of a swirling flow in the large eddy.

At t =132 s ($\approx 3/4\,\tau_r$) the rise of negative vorticity is evident in the vorticity contours of figure 8. There is pairing (or coupling) of negative and positive vorticity cores. The evolution of temperature and density can be appreciated in the two panels of figure 9, *a* and *b*, where the progress of mixing of hot and tenuous magnetospheric matter with the cold and dense plasma of the magnetosheath is apparent. Also at t=132 s, figure 10 shows kinetic helicity density contours together with streamlines, which furnish evidence for the complex swirling pattern associated with the vortex core. The rise of

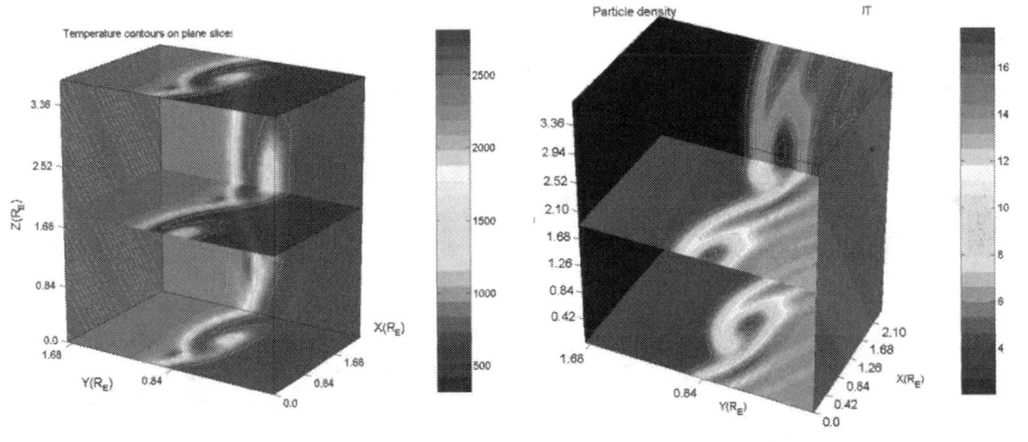

Figure 9. Two panels show at t=132 s from left to right contour plots with the evolution of: *a*) temperature (eV), and *b*) density p cm^{-3} (rotated view). Mixing of hot - tenuous with cold - dense plasma is apparent.

these coherent structures has important consequences for the properties of the mixing layer. There is a broadening of the boundary layer's width, and a substantial lengthening of the transit time of plasma, compared with that of a stable laminar boundary layer.

At about one turnover time, t= 180 s the 3-D stretching mechanism has intensified the initial vorticity level by a 4-5 fold factor. This can be noted in figure 11 *a*) with the shape of the vorticity contours and the change of values in the scale of color. Figure 11 *b*) shows a correlated advance of mixing via the isodensity surface n = 9 p cm^{-3}. It is evident that convected layers of different density wind-up around the large vortices, mixing the outer with the inner matter.

An important characteristic of the physical regime of the simulation can be seen in figure 12 that shows temperature contours, and a set of magnetic field lines. At t= 180 s, as at earlier times, the magnetic field lines are straight and do not bend with the time-varying plasma flow. The slope of the lines shows only moderate oscillations during the simulation. Another noteworthy feature of figure 12 is the magnetic line associated with the positive vorticity center: it shows that the core of the vortex is aligned with the magnetic field.

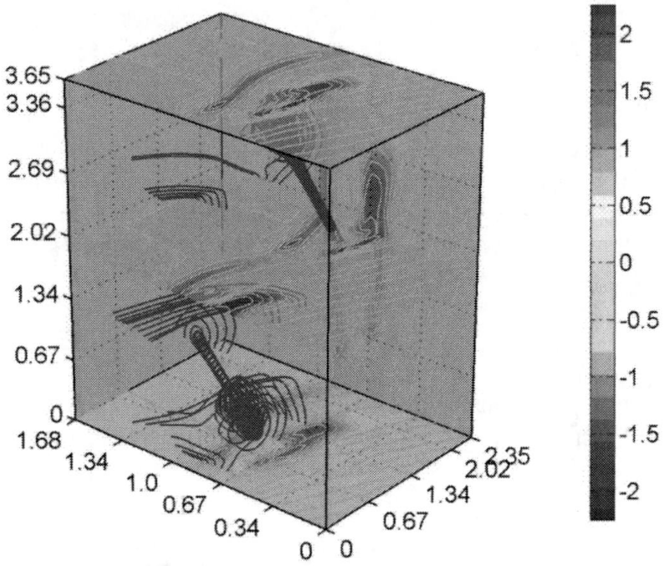

Figure 10. Time t=132 s, shows kinetic helicity density contours (km s^{-2}), and streamlines. Evidence of a complex 3-D swirling pattern at the vortex core.

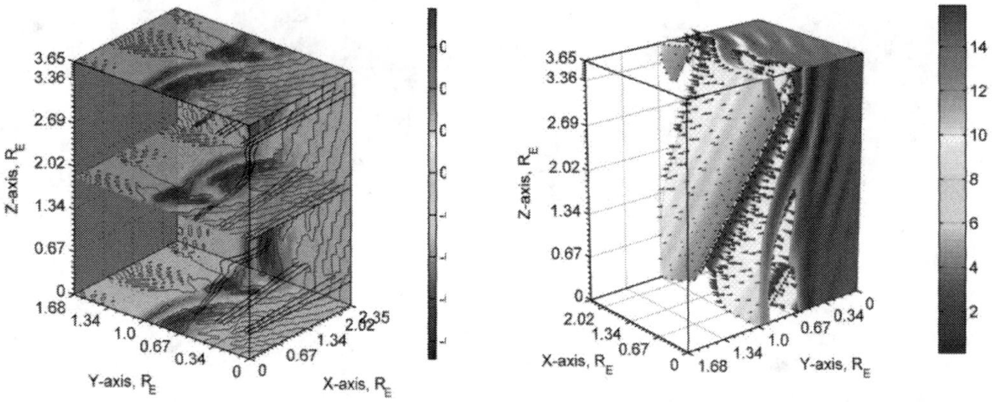

Figure 11. At t= 180 s (≈ one turnover time), left to right, two panels: *a*) with vorticity contours (scale of color in the interval -0.2 to 0.2 s^{-1}), and *b*) progress of mixing via the isodensity surface n= 9 p cm^{-3} (rotated view).

Also for t= 180 s figure 13 shows the left panel *a*) where density contours reveal that the same vortex core is considerably depleted with respect to the magnetopause density (p cm^{-3}). It also shows a high density patch in the vorticity structure, ~ 40% larger than the magnetosheath density. The site is correlated with a local increase of gas pressure, whereas temperature remains near to the surrounding values. In figure 13 the right panel *b*) gives another close-up, where temperature contours (eV) show that the positive vortex core is very hot, with temperatures several times larger than those of the magnetopause plasma. Gas and magnetic pressure are roughly constant across the vortex, and close to the surrounding values.

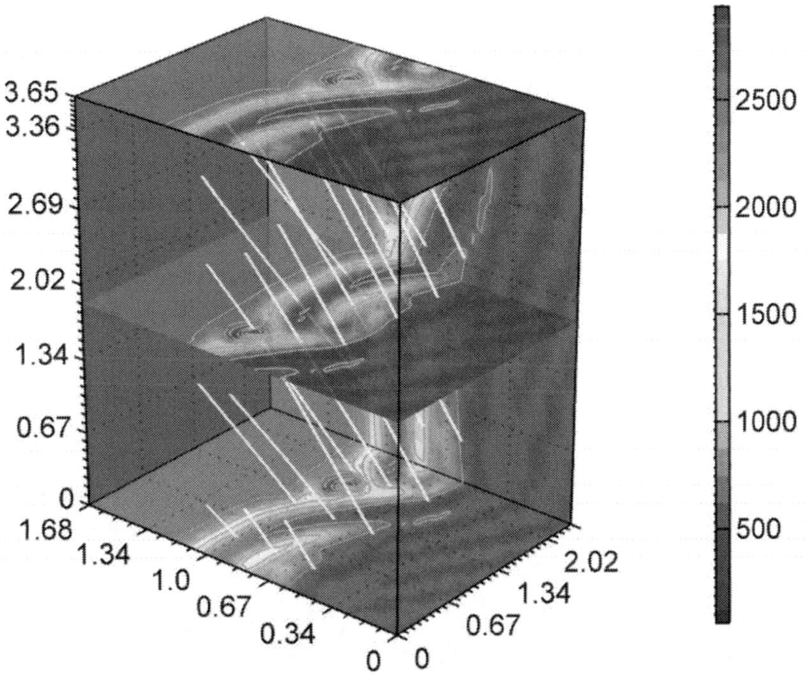

Figure 12. Temperature contours (eV) and magnetic field lines (solid yellow) at t =180 s

Finally we briefly comment on other numerical runs with the same configuration, but starting from random perturbations. The experiments did not show a development of the instability. Neither the $\vec{k}$ directions nor the wavelengths ($2\pi/k$) of the random perturbations that fit into the computational box were in proper conditions to excite the KH process. The perturbations could not comply with the favorable orientation of $\vec{k}$, which is close to 32° (≈ flute modes). After a turnover time, $t \sim 180$ s only stable modes could be observed, with a small fluctuation level.

5. Summary and discussion

MHD simulations have been very fruitful in elucidating features of the KH instability at the magnetopause [12], [8], [17], [6], [13], [18]. Our work is intended as a contribution to this line of research, modeling the boundary layer with data derived from the Dec. 7, 2000, event, during a period of northward IMF [15]. The study complements works done for a different physical regime with $M_A >> 1$. In the limit $M_A \to\infty$ the magnetic field behaves like a passive "contaminant", and yields to flow convection. In [6] the magnetic field lines where found to wind up inside the vortex structure, and small scale current sheets developed with a potential to generate tearing processes.

We reported here a study of the MHD development of vortices in compressible, non-homogeneous, mixing layers due to the velocity gradient instability. The non-linear evolution was computed with 3-D LES of temporal mixing layers tailored to represent distinctive conditions of the terrestrial magnetopause. The boundary layer is characterized by the growth of large-scale vortices, and becomes a site of mass mixing favoring enhanced plasma diffusion. The rise of counter-rotating vortices from the evolution of the original vortex sheet is one of the noteworthy processes revealed by the LES.
The nonlinear numerical simulation of the KH excitation was performed with a new 3D+t MHD code of ALE-FV type. The main results obtained can be summarized as follows.

i) The time for the development of a significant deformation of the boundary layer and, with it, the growth of important non-linear characteristics, is shorter than, or comparable to, the e-folding time of the linear theory when the initial structure starts from a condition already significantly perturbed: "trigger and acceleration of the instability".

ii) Our simulations show the formation of counter-rotating plasma cores with a concentrated vorticity much larger than the vorticity initially stored in the boundary layer. Non conservation of vorticity becomes manifest at about (or even before) ~ one rollover time in our LES, where in addition to vortices with positive rotation (the same sign of the original vorticity sheet) other coherent structures with strong negative vorticity also arise. It is found that vorticity concentrates in cores of both signs with absolute values ~ 4-5×ω_i, the maximum vorticity of the initial shear layer. Concomitant with 3-D vortex stretching, also kinetic helicity density rises at the vorticity cores.

iii) The plasma in the vortex centers is hotter ~ 3 times T_i (magnetospheric temperature), and more tenuous ~ 1/3 n_i (magnetosferic particle density) than that of the adjacent magnetosphere. The longer-term evolution of these singular structures, which would be particularly relevant to cases when the IMF is northward-pointing for a long time, is the subject of future work.

iv) We observed a kind of "rigidity" of magnetic field lines, at least for the M_A =0.7 value examined, which is evident in the strong influence of the magnetic field direction on the formation and shape of the non-linear features.

Although the overall aspect may look 2-D at first sight, on closer inspection many 3-D properties become manifest, so that the study of the process requires in fact a 3-D treatment. This aspect of the non-linear stage takes a different tack with respect to linear theory, which can be reduced to a 2-D picture, as indicated by the stability equation of KH modes (see [16]).

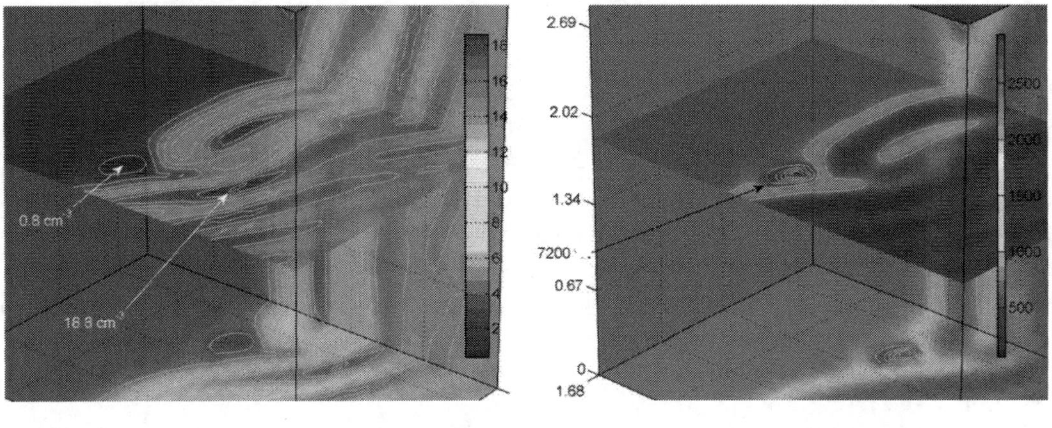

Figure 13. Time *t*= 180 s, two panels, left to right, *a*) close-up of density contours (p cm^{-3}) showing a depleted vortex core and patches of particle concentration; *b*) close-up of temperature contours (eV) indicating a very hot vortex core (positive ω).

v) There is a clear indication, yet to be quantified, of an increment in the time the flowing plasma stays in the boundary layer region deformed by the instability due to the complex pattern of the streamlines.

vi) The broadening of the boundary layer, the plasma mixing in the rolling-off of the vorticity layers, and the increased transit time of the plasma elements, are all factors conducive to an enhancement of ion diffusion through the magnetopause.

All these factors bear on the issue of how effective the development of the KH instability is in fostering mass transfer into the magnetopause. In our numerical experiments the runs have not been extended to lapses much longer than one turnover time in view of the absence of dissipative terms, or subgrid complements, in the present version of the code. The computation of particle trajectories and an estimate of diffusion is a desirable goal, albeit not yet achieved. This, and several other issues need further work, which is partly underway. Our study is intended as a contribution to the knowledge of vortex formation in MHD, and to the understanding of solar wind interaction with the magnetosphere during periods of northward IMF.

Acknowledgments. L E Bilbao, G Gnavi and F T Gratton are grateful to CONICET for the support of research grant PIP 5291, and to UBACYT research grant X90. C.J. Farrugia is supported by USA NASA grants NNX08AD11G, NNG05GG25G, and also by Cluster grant to UNH.

References

[1] Bilbao L E 2006 *Plasma and Fusion Science* (*AIP Conf. Proc.* vol 875) ed J J E Herrera Velázquez
Bilbao L E 2008 this *JOP Conf. Series*
[2] Lesieur M 1990 *Turbulence in Fluids* (Dordrecht : Kluwer Academic Pub)
Lesieur M and Métais O 1996 *Annual Rev. Fluid Mechanics* **28** 45
[3] Huerre P and Rossi M 1998 in *Hydrodynamics and Nonlinear Instabilities* ed C Godrèche and P Manneville (Cambridge: Cambridge Univ. Press)
[4] Paschmann G, Baumjohann W, Scopke N, Phan T D, Lühr H 1993 *J. Geophys. Res.* **98** 13409
[5] Farrugia C J *et al.* 2000 *J. Geophys. Res.* A **105** 7639–67
[6] Otto A and Fairfield D H 2000 *J. Geophys. Res.* A **105** 21175
Fairfield D H, Otto A, Mukai T, Kokubun S, Lepping R P, Steinberg J T, Lazarus A J and Yamamoto T 2000 *J. Geophys. Res.* A **105** 21159
[7] Hasegawa H *et al.* 2004 *Nature* **430** 755
[8] Belmont G and Chanteur G 1989 *Phys. Scripta* **124** 124
[9] Kivelson M G and Chen S-H 1995 *Physics of the Magnetopause (Geophys. Monog. Ser.* vol 90*)* ed P Song, B U Ö Sonnerup and M F Thomsen (Washington DC: AGU) p 257
[10] Miura A 1995 *Physics of the Magnetopause (Geophys. Monog. Ser.* vol 90*)* ed P Song, B U Ö Sonnerup and M F Thomsen (Washington DC: AGU) p 285
[11] Farrugia C J , Gratton F T, Torbert R B 2001 *Space Science Rev.* **95** 443
[12] Miura A 1984 *J. Geophys. Res.* A **89** 801
Miura A 1987 *J. Geophys. Res.* A **92** 3195
[13] Otto A and Nykyri K 2003 *Earth's Low-Latitude Boundary Layer Magnetopause* (*Geophys. Monog. Ser.* vol 133) ed P T Newell and T Onsager (Washington DC: AGU) p 53
[14] Smets R, Delcourt D, Chanteur G and Moore T E 2002 *Annales Geophysicae* **20** 757–769
[15] Farrugia C J *et al.* 2008 *J. Geophys. Res.* A **113** A03208
[16] Gnavi G, Gratton F T, Farrugia C J and Bilbao L E 2008 this *JOP Conf. Series*
[17] Miura A 1992 *J. Geophys. Res.* A **97** 10655
[18] Gratton F T, Farrugia C J, Bilbao L E, Gnavi G and Lund E 2006 *Plasma and Fusion Science* (*AIP Conf. Proc.* vol 875) ed J J E Herrera pp 300–303
[19] Moffat H K 1978 *Magnetic Field Generation in Electrically Conducting Fluids* (Cambridge: Cambridge Univ. Press)
Biskamp D 1993 *Nonlinear Magnetohydrodynamics* (Cambridge: Cambridge Univ. Press)
[20] Tritton D J 1988 *Physical Fluid Dynamics* (Oxford: Clarendon Press)
[21] Moffat H K, Kida S and Okhitani K 1994 *J. Fluid Mechanics* **259** 241–264

Institute of Physics
Dirac House, Temple Back
Bristol BS1 6BE UK